Mineral Chemistry of Granitoids: Constraints on Crystallization Conditions and Petrological Evolution

Mineral Chemistry of Granitoids: Constraints on Crystallization Conditions and Petrological Evolution

Guest Editors

Ignez de Pinho Guimarães
Jefferson Valdemiro de Lima

Basel • Beijing • Wuhan • Barcelona • Belgrade • Novi Sad • Cluj • Manchester

Guest Editors

Ignez de Pinho Guimarães
Graduate Program in
Geosciences
Federal University of
Pernambuco (UFPE)
Recife-PE
Brazil

Jefferson Valdemiro de Lima
Faculty of Geology, Federal
University of Southern and
Southeastern Pará
(UNIFESSPA)
Marabá-PA
Brazil

Editorial Office
MDPI AG
Grosspeteranlage 5
4052 Basel, Switzerland

This is a reprint of the Special Issue, published open access by the journal *Minerals* (ISSN 2075-163X), freely accessible at: https://www.mdpi.com/journal/minerals/special_issues/88J39G8T1D.

For citation purposes, cite each article independently as indicated on the article page online and as indicated below:

Lastname, A.A.; Lastname, B.B. Article Title. *Journal Name* **Year**, *Volume Number*, Page Range.

ISBN 978-3-7258-3287-3 (Hbk)
ISBN 978-3-7258-3288-0 (PDF)
https://doi.org/10.3390/books978-3-7258-3288-0

Cover image courtesy of Jefferson Valdemiro de Lima

© 2025 by the authors. Articles in this book are Open Access and distributed under the Creative Commons Attribution (CC BY) license. The book as a whole is distributed by MDPI under the terms and conditions of the Creative Commons Attribution-NonCommercial-NoDerivs (CC BY-NC-ND) license (https://creativecommons.org/licenses/by-nc-nd/4.0/).

Contents

About the Editors vii

Ignez de Pinho Guimarães and Jefferson Valdemiro de Lima
Editorial for Special Issue "Mineral Chemistry of Granitoids: Constraints on Crystallization Conditions and Petrological Evolution"
Reprinted from: *Minerals* 2025, 15, 65, https://doi.org/10.3390/min15010065 1

Jefferson Valdemiro de Lima, Ignez de Pinho Guimarães, José Victor Antunes de Amorim, Caio Cezar Garnier Brainer, Lucilene dos Santos and Adejardo Francisco da Silva Filho
A Review of the Mineral Chemistry and Crystallization Conditions of Ediacaran–Cambrian A-Type Granites in the Central Subprovince of the Borborema Province, Northeastern Brazil
Reprinted from: *Minerals* 2024, 14, 1022, https://doi.org/10.3390/min14101022 6

Ane K. Engvik, Fernando Corfu, Ilka C. Kleinhanns, Heinrich Taubald and Synnøve Elvevold
Mineralogical and Geochemical Response to Fluid Infiltration into Cambrian Orthopyroxene-Bearing Granitoids and Gneisses, Dronning Maud Land, Antarctica
Reprinted from: *Minerals* 2024, 14, 772, https://doi.org/10.3390/min14080772 34

Michel Cathelineau and Zia Steven Kahou
Discrimination of Muscovitisation Processes Using a Modified Quartz–Feldspar Diagram: Application to Beauvoir Greisens
Reprinted from: *Minerals* 2024, 14, 746, https://doi.org/10.3390/min14080746 57

Bo Liu, Shengkai Jin, Guanghao Tian, Liyang Li, Yueqiang Qin, Zhiyuan Xie, et al.
Mesoproterozoic (ca. 1.3 Ga) A-Type Granites on the Northern Margin of the North China Craton: Response to Break-Up of the Columbia Supercontinent
Reprinted from: *Minerals* 2024, 14, 622, https://doi.org/10.3390/min14060622 71

Mohamed M. Ghoneim, Ahmed E. Abdel Gawad, Hanaa A. El-Dokouny, Maher Dawoud, Elena G. Panova, Mai A. El-Lithy and Abdelhalim S. Mahmoud
Petrogenesis and Geodynamic Evolution of A-Type Granite Bearing Rare Metals Mineralization in Egypt: Insights from Geochemistry and Mineral Chemistry
Reprinted from: *Minerals* 2024, 14, 583, https://doi.org/10.3390/min14060583 90

Junsheng Jiang, Wenshuai Xiang, Peng Hu, Yulin Li, Fafu Wu, Guoping Zeng, et al.
Petrogenesis of the Newly Discovered Neoproterozoic Adakitic Rock in Bure Area, Western Ethiopia Shield: Implication for the Pan-African Tectonic Evolution
Reprinted from: *Minerals* 2024, 14, 408, https://doi.org/10.3390/min14040408 114

Michel Cathelineau, Marie-Christine Boiron, Andreï Lecomte, Ivo Martins, Ícaro Dias da Silva and Antonio Mateus
Lithium-, Phosphorus-, and Fluorine-Rich Intrusions and the Phosphate Sequence at Segura (Portugal): A Comparison with Other Hyper-Differentiated Magmas
Reprinted from: *Minerals* 2024, 14, 287, https://doi.org/10.3390/min14030287 134

Hao Li, Xuguang Li, Jiang Xin and Yongqiang Yang
Zircon U-Pb and Whole-Rock Geochemistry of the Aolunhua Mo-Associated Granitoid Intrusion, Inner Mongolia, NE China
Reprinted from: *Minerals* 2024, 14, 226, https://doi.org/10.3390/min14030226 160

Yunbiao Zhao, Fan Huang, Denghong Wang, Na Wei, Chenhui Zhao and Ze Liu
U-Pb Geochronology, Geochemistry and Geological Significance of the Yongfeng Composite Granitic Pluton in Southern Jiangxi Province
Reprinted from: *Minerals* **2023**, *13*, 1457, https://doi.org/10.3390/min13111457 . . . . **177**

Ahmed A. Abd El-Fatah, Adel A. Surour, Mokhles K. Azer and Ahmed A. Madani
Integration of Whole-Rock Geochemistry and Mineral Chemistry Data for the Petrogenesis of A-Type Ring Complex from Gebel El Bakriyah Area, Egypt
Reprinted from: *Minerals* **2023**, *12*, 1273, https://doi.org/10.3390/min13101273 . . . . **197**

About the Editors

Ignez de Pinho Guimarães

Ignez de Pinho Guimarães holds a bachelor's (1978) and master's degree (1982) from the Federal University of Pernambuco (UFPE) and a Ph.D. from the University of London (1989). She completed postdoctoral research at Kansas University, USA (1996–1997), and the Australian National University (2008), focusing on SHRIMP zircon analyses.

A retired Full Professor at UFPE, she has held roles such as Head of the Department of Geology, Coordinator of the Graduate Program in Geosciences, and President of the Northeast Branch of the Brazilian Geological Society (SBG). Her research focuses on igneous rocks, geochemistry, and geochronology.

At present, she is a permanent professor in UFPE's Graduate Program in Geosciences and has contributed to research projects funded by CNPq, FINEP, and PADCT.

Jefferson Valdemiro de Lima

Jefferson Valdemiro de Lima holds a degree in Geology from the Federal University of Pernambuco (UFPE), in addition to a Master's degree and PhD in Geosciences, with a focus on Geochemistry, Geophysics, and Crustal Evolution, as part of the Graduate Program in Geosciences at UFPE. He previously served as a substitute professor of General Geology at the Department of Geology at UFPE (2018.1–2019.2; 2022.2–2024.1) and is currently an Adjunct Professor at the Institute of Geosciences and Engineering at the Federal University of Southern and Southeastern Pará (UNIFESSPA). His academic and scientific activities cover petrology, geochemistry, and isotopic geology of igneous rocks, with an emphasis on studies of the Borborema Province, Brazilian granitoids, and the relationship between magmatism and shear zones.

Editorial

Editorial for Special Issue "Mineral Chemistry of Granitoids: Constraints on Crystallization Conditions and Petrological Evolution"

Ignez de Pinho Guimarães [1] and Jefferson Valdemiro de Lima [2,*]

[1] Graduate Program in Geosciences, Federal University of Pernambuco, Recife 50740-540, PE, Brazil; ignezdpg@gmail.com
[2] Faculty of Geology, Federal University of Southern and Southeastern Pará, Marabá 68500-000, PA, Brazil
* Correspondence: jefferson.valdemiro@unifesspa.edu.br; Tel.: +55-81-982170432

The origin of granitoids has fascinated geologists since the famous meeting of the Geological Society of France in 1847 [1]. This interest stems from the fact that granitic rocks are the most abundant rock type in the upper continental crust, with their source rocks located in the lower crust and/or upper mantle. Moreover, they are important to geologic evolution, as their diversity has been interpreted to result from tectonic processes and the compositional variability of the crust. Granitic rocks are primarily composed of quartz, alkali feldspar, and plagioclase, which together define their fundamental felsic mineralogical framework. Additionally, these rocks may contain smaller and variable amounts of biotite, amphibole, and/or pyroxenes, which contribute to their mineralogical diversity and reflect variations in their magmatic and tectonic histories. Granitoids can include a large number of accessory phases such as tourmaline, garnet, apatite, zircon, apatite, titanite, epidote, topaz, monazite, and metal ore minerals. Deposits of precious and rare metals, such as Au, Mo, Sn, and many others, are directly or indirectly associated with granites.

The mineral chemistry of the major mineral phases constitutes an important tool for assessing the geochemical affinities and magma sources of the host granitoid and estimating the crystallization conditions of the magma from which they originated [2–8]. The chemical composition of accessory minerals also contributes to the understanding of the crystallization conditions of granitic magmas. The titanite composition is used to estimate pressure and temperature [9]; Fe-Ti oxide compositions are dependent on the redox conditions and magma nature [10] and were used in the process of granite classification [11]; epidote has been used to estimate the crystallization pressure, fO_2, and ascension rate of epidote-bearing magmas [12,13]. It is important to highlight zircon because, in addition to its important geochronologic and isotopic tracer (U-Pb/Lu-Hf), its trace elements composition has been the subject of numerous studies to infer the physicochemical conditions of magma, the components involved in the source, and hydrothermal alterations [14–17], among other factors.

This Special Issue "Mineral Chemistry of Granitoids: Constraints on Crystallization Condition and Petrological Evolution" includes ten contributions on mineral chemistry, whole-rock geochemistry, isotopic chemistry, tectonic setting, and mineral deposits associated with granitic rocks.

The paper by El-Fatah et al. (Contribution 1) presents a detailed study of the Neoproterozoic post-collisional El Bakriyah granitic ring complex, syenogranite core, and alkali feldspar granite rim, intruded into monzogranites, from the Central Eastern Desert in

Egypt. The studied granites are peraluminous, ferroan, and A-type, and they have F, B, Nb, and Ta in considerable concentrations, either in the form of rare metal dissemination or as veins of fluorite and barite. The authors concluded that the granitic magma was generated by high-T dehydration melting of a mixed crust–mantle source, primarily composed of metasediments and amphibolite. The formed magma underwent high fractionation, producing felsic magma emplaced in a within-plate setting. The monzogranitic country rock was emplaced in the transition between the arc and anorogenic settings.

The paper by Zhao et al. (Contribution 2) deals with the study of a highly differentiated I-type composite granitic pluton from Southern Jiangxi Province in China and its relationship with mineralization. Whole-rock geochemistry and the chemistry of biotite and zircon were used to analyze the rock-forming physiochemical and the genetic type of granite. The achieved temperatures (774–777 °C) were lower than those of A-type granites but were close to those reported for highly differentiated I-type granites. The three zircon U-Pb crystallization ages of the two granites were similar: 152 ± 1 Ma, 151 ± 1 Ma, and 149 ± 1 Ma. The authors concluded that the studied granites were formed during Late Jurassic large-scale magmatic activity in an extensional tectonic setting, following the subduction of the Pacific Plate into the Nanling hinterland during the post-orogenic stage. They also found that the high differentiation, high F content, and low oxygen fugacity recorded in the studied granites are conducive to the large-scale mineralization of Sn, Mo, and fluorite.

The paper by Li et al. (Contribution 3) analyzes LA-ICP-MS zircon U-Pb ages and whole-rock geochemistry data from Mo-associated granitoids of the Aolunhua intrusion in Inner Mongolia, NE China, and discusses the Mo-associated granitoids with granite porphyry and the source of the ore-forming rocks of the deposit. The zircon data reveals the crystallization age of the studied granite at 135 ± 1 Ma, correlating it with the widespread Yanshanian intermediate–felsic magmatic activity. Based on the crystallization age, zircon trace element data, and whole-rock geochemistry, the authors concluded that the granite porphyry was formed by the crystallization of crust-derived magmas during a transitional tectonic setting, from compression orogeny to back-arc extension. The Aolunhua ore deposit is within the Cu-Mo metallogenic belt at the northern flank of the Xilamulun River deep fracture. Based on a comparison with other Mo deposits along the banks of the Xilamulun River, the authors proposed that the Tianshan–Linxi constitutes an important Mo metallogenic belt.

Cathelineau et al. (Contribution 4) present a study on two aplite and associated pegmatite dyke swarms from Segura, Portugal. The studied granitoids have high concentrations of fluxing elements (Li, P, F), ranging from 1.5 to 5.0 compared to highly differentiated peraluminous granites, and present a wide variety of Li, Na, Fe-Mn, and Ca-Sr phosphates. Micro-X-ray fluorescence chemical imaging was used to establish the phosphate crystallization sequence. The magmatic differentiation resulted in a P- and Li-rich melt, with the primary crystallization of the amblygonite–montebrasite series and Fe-Mn phosphates. The primary phosphates were replaced by lacroixite during a stage of high Na activity, and external Ca- and Sr-rich hydrothermal fluids replaced the primary Li-Na phosphates with phosphates from the goyazite–crandallite series, followed by apatite formation.

The paper by Jiang et al. (Contribution 5) presents the mineral chemistry of K-feldspar, plagioclase, and biotite, along with Hf zircon isotopic data, Sr-Nd isotopes of the whole rock, and zircon trace element data to discuss the source and physicochemical conditions during the formation of Neoproterozoic Bure adakitic rock in the Western Ethiopian Shield. Based on the results, the authors concluded that the magma was emplaced at a depth of approximately 6.39~10.2 km (1.75~2.81 kbar), under relatively high oxygen fugacity (logfO_2

varying from −18.5 to −4.9), with a crystallization temperature ranging from 659 to 814 °C. They also concluded that the magma was generated by the melting of a Neoproterozoic juvenile crust, coeval with early magmatic stages in the Arabian Nubian Shield.

Ghoneim et al. (Contribution 6) present a systematic study of Neoproterozoic alkali feldspar granite from the Arabian Nubian Shield. This study involved petrography, mineral chemistry, and whole-rock geochemistry. These granites contain significant rare metal mineralization, including thorite, uranothorite, columbite, zircon, monazite, xenotime, pyrite, rutile, and ilmenite. The studied granites are highly fractionated peraluminous, with calc-alkaline affinity and A-type characteristics post-collision, emplaced under an extensional regime of within-plate environments. The authors concluded that the granitic magma evolved through a significant degree of fractionation and that coeval basaltic magmas supplied the necessary heat to melt the crust and provided volatile substances that seeped into the lower crust, resulting in the formation of A-type granite through partial melting of the crust.

Liu et al. (Contribution 7) present a systematic study of A-type Mesoproterozoic porphyritic granites on the northern margin of the North China Craton, including whole-rock geochemistry, zircon Hf isotopes, and zircon U-Pb geochronological data. Two of the studied granites have similar crystallization ages of 1285 ± 3 Ma and 1279 ± 6 Ma. The whole-rock geochemistry classified the studied granites as weakly peraluminous A_2-type granites. The Hf isotopic data suggest that the magmas were derived by partial melting of ancient crustal material. They concluded that the studied granites formed in an intraplate tectonic setting during continental extension and rifting of the north margin of the North China Craton, associated with the late break-up stage of the Columbia supercontinent.

The paper by Cathelineau and Kahou (Contribution 8) used a Modified Quartz–Feldspar Diagram to discriminate muscovitisation processes in the Beauvoir Greisen and compared them to representative series of greisen data from the literature: Cligga Head, Cinovec, Panasqueira, Zhengchong, and Hoggar. Whole-rock geochemical data were obtained by ICP-OES and ICP-MS in unaltered and altered Beauvoir granites from the French Central Massif. Composite and elemental maps using micro-X-ray fluorescence were used to quantitatively determine the relative mineral proportions. They concluded that whole-rock geochemistry provides important information on the main trends of water–rock information, and that substantial muscovite formation, as recorded in the studied area, is explained by a fracture network that channelized fluids in disequilibrium with the mineral assemblage of the granites, particularly albite.

The paper by Engevik et al. (Contribution 9) presents bulk-rock geochemical data, including O-, H-, Sr-, and Nd-isotopic data and zircon and titanite U-Pb geochronological data for Proterozoic and Early Palaeozoic dry gneisses and granitoids in Dronning Maud Land, Antarctica. The granitoid crystallization ages were defined by zircon U-Pb at 520 ± 1.0 Ma, while the U-Pb titanite age of 485 ± 1.4 Ma was interpreted as the alteration age. The gneiss samples were dated by whole-rock Rb-Sr at 517 ± 6 Ma and Sm-Nd at 536 ± 23 Ma. The Sr and Nd isotopic data suggest that the gneiss was derived from a relatively juvenile source but underwent a significant metasomatic effect that introduced radiogenic Sr into the system. The granitoid isotopic data indicate a derivation from Mid-Proterozoic crust, with some additions of mantle components. This paper also highlights the importance of fluids intruding into the studied rocks, causing changes to the rock's appearance, mineralogy, and chemistry.

The paper by Lima et al. (Contribution 10) presents a review of mineral chemistry and crystallization conditions of Ediacaran–Cambrian (580–525 Ma) A_2-type granites from the central sub-province of Borborema Province, Northeastern Brazil. This study reviews published whole-rock and mineral chemistry data from thirteen Ediacaran–Cambrian A-type intrusions and a related dike swarm, presenting new zircon trace element data for five of the intrusions. The studied granitoids are ferroan, predominantly metaluminous, and mostly alkalic-calcic, crystallized under low fO_2 conditions, with temperatures ranging from 990 to 680 °C and pressures of 4 to 7 kbar (crustal depths of 12 to 21 km). The zircon trace elements data suggest post-magmatic hydrothermal processes, which the authors interpreted as being associated with shear zones reactivation.

Conflicts of Interest: The authors declare no conflicts of interest.

List of Contributions

1. El-Fatah, A.A.A.; Surour, A.A.; Azer, M.K.; Madani, A.A. Integration of Whole-Rock Geochemistry and Mineral Chemistry Data for the Petrogenesis of A-Type Ring Complex from Gebel El Bakriyah Area. Egypt. *Miner.* **2023**, *13*, 1273. https://doi.org/10.3390/min13101273.
2. Zhao, Y.; Huang, F.a.n.; Wang, D.; Wei, N.; Zhao, C.; Liu, Z. U-Pb Geochronology, Geochemistry and Geological Significance of the Yongfeng Composite Granitic Pluton in Southern Jiangxi Province. *Minerals* **2023**, *13*, 1457. https://doi.org/10.3390/min13111457.
3. Li, H.; Li, X.; Xin, J.; Yang, Y. Zircon U-Pb and Whole-Rock Geochemistry of the Aolunhua Mo-Associated Granitoid Intrusion, Inner Mongolia, NE China. *Minerals* **2024**, *14*, 226. https://doi.org/10.3390/min14030226.
4. Cathelineau, M.; Boiron, M.-C.; Lecomte, A.; Martins, I.; Silva, I.D.; Mateus, A. Lithium-, Phosphorus-, and Fluorine-Rich Intrusions and the Phosphate Sequence at Segura (Portugal): A Comparison with Other Hyper-Differentiated Magmas. *Minerals* **2024**, *14*, 287. https://doi.org/10.3390/min14030287.
5. Jiang, J.; Xiang, W.; Hu, P.; Li, Y.; Wu, F.; Zeng, G.; Guo, X.; Zhang, Z.; Bai, Y. Petrogenesis of the Newly Discovered Neoproterozoic Adakitic Rock in Bure Area, Western Ethiopia Shield: Implication for the Pan-African Tectonic Evolution. *Minerals* **2024**, *14*, 408. https://doi.org/10.3390/min14040408.
6. Ghoneim, M.M.; Gawad, A.E.A.; El-Dokouny, H.A.; Dawoud, M.; Panova, E.G.; El-Lithy, M.A.; Mahmoud, A.S. Petrogenesis and Geodynamic Evolution of A-Type Granite Bearing Rare Metals Mineralization in Egypt: Insights from Geochemistry and Mineral Chemistry. *Minerals* **2024**, *14*, 583. https://doi.org/10.3390/min14060583.
7. Liu, B.; Jin, S.; Tian, G.; Li, L.; Qin, Y.; Xie, Z.; Ma, M.; Yin, J. Mesoproterozoic (ca. 1.3 Ga) A-Type Granites on the Northern Margin of the North China Craton: Response to Break-Up of the Columbia Supercontinent. *Minerals* **2024**, *14*, 622. https://doi.org/10.3390/min14060622.
8. Cathelineau, M.; Kahou, Z.S. Discrimination of Muscovitisation Processes Using a Modified Quartz–Feldspar Diagram: Application to Beauvoir Greisens. *Minerals* **2024**, *14*, 746. https://doi.org/10.3390/min14080746.
9. Engvik, A.K.; Corfu, F.; Kleinhanns, I.C.; Taubbald, H.; Elvevold, S. Mineralogical and Geochemical Response to Fluid Infiltration into Cambrian Orthopyroxene-Bearing Granitoids and Gneisses, Dronning Maud Land, Antarctica. *Minerals* **2024**, *14*, 772. https://doi.org/10.3390/min14080772.
10. Lima, J.V.; Guimarães, I.P.; Amorim, J.V.A.; Brainer, C.C.G.; Santos, L.; Silva Filho, A.F. A Review of the Mineral Chemistry and Crystallization Conditions of Ediacaran–Cambrian A-Type Granites in the Central Subprovince of the Borborema Province, Northeastern Brazil. *Minerals* **2024**, *14*, 1022. https://doi.org/10.3390/min14101022.

References

1. Pitcher, W.S. *The Nature and Origin of Granite*; Chapman & Hall: London, UK, 1993; p. 321.
2. Anderson, J.L.; Smith, D.R. The Effects of Temperature and FO2 on the Al-in-Hornblende Barometer. *Am. Miner.* **1995**, *80*, 549–559. [CrossRef]
3. Anderson, J.L.; Barth, A.P.; Wooden, J.L.; Mazdab, F. Thermometers and Thermobarometers in Granitic Systems. *Rev. Miner. Geochem.* **2008**, *69*, 121–142. [CrossRef]
4. Harrison, T.M.; Watson, E.B. The Behavior of Apatite during Crustal Anatexis: Equilibrium and Kinetic Considerations. *Geochim. Cosmochim. Acta* **1984**, *48*, 1467–1477. [CrossRef]

5. Hammarstrom, J.M.; Zen, E. Aluminum in Hornblende: An Empirical Igneous Geobarometer. *Am. Miner.* **1986**, *71*, 1297–1313.
6. Blundy, J.D.; Holland, T.J.B. Calcic Amphibole Equilibria and a New Amphibole-Plagioclase Geothermometer. *Contrib. Miner.Petrol.* **1990**, *104*, 208–224. [CrossRef]
7. Schmidt, M.W. Amphibole composition in tonalite as a function of pressure: An experimental calibration of the Al-in-hornblende barometer. *Contrib. Miner. Petrol.* **1992**, *110*, 304–310. [CrossRef]
8. Putirka, K. Amphibole Thermometers and Barometers for Igneous Systems and Some Implications for Eruption Mechanisms of Felsic Magmas at Arc Volcanoes. *Am. Miner.* **2016**, *101*, 841–858. [CrossRef]
9. Enami, M.; Suzuki, K.; Liou, J.G.; Bird, D.K. Al-Fe3+ and F-OH substitution in titanite and constraints on their P-T dependence. *Eur. J. Miner.* **1993**, *5*, 219–231. [CrossRef]
10. Wones, D.R. Significance of the assemblage titanite + magnetite + quartz in granitic rocks. *Am. Miner.* **1989**, *74*, 744–749.
11. Ishihara, S. Magnetite-series and ilmenite-series granitic rocks. *Min. Geol.* **1977**, *27*, 293–305.
12. Schmidt, M.W.; Thompson, A.B. Epidote in calc-alkaline magmas: An experimental study of stability, phase relationships, and the role of epidote im Magmatic evolution. *Am. Miner.* **1996**, *81*, 424–474. [CrossRef]
13. Schmidt, M.W.; Poli, S. Magmatic epitote. *Rev. Mineral. Geochem.* **2004**, *56*, 399–430. [CrossRef]
14. Xie, L.; Wang, R.; Chen, X.; Qiu, J.; Wang, D. Th-Rich Zircon from Peralkaline A-Type Granite: Mineralogical Features and Petrological Implications. *Chin. Sci. Bull.* **2005**, *50*, 809–817.
15. Pérez-Soba, C.; Villaseca, C.; Gonzáles del Tánago, J.; Nasdala, L. The composition of zircon in the peraluminous Hercynian granites of the Spanish central system batholith. *Can. Miner.* **2007**, *45*, 509–527. [CrossRef]
16. Breiter, K.; Lamarão, C.N.; Borges, R.M.K.; Dall'Agnol, R. Chemical Characteristics of Zircon from A-Type Granites and Comparison to Zircon of S-Type Granites. *Lithos* **2014**, *192*, 208–225. [CrossRef]
17. Loucks, R.R.; Henríquez, G.J.; Fiorentini, M.L. Zircon and Whole-Rock Trace Element Indicators of Magmatic Hydration State and Oxidation State Discriminate Copper Ore-Forming from Barren Arc Magmas. *Econ. Geol.* **2024**, *119*, 511–523. [CrossRef]

Disclaimer/Publisher's Note: The statements, opinions and data contained in all publications are solely those of the individual author(s) and contributor(s) and not of MDPI and/or the editor(s). MDPI and/or the editor(s) disclaim responsibility for any injury to people or property resulting from any ideas, methods, instructions or products referred to in the content.

Review

A Review of the Mineral Chemistry and Crystallization Conditions of Ediacaran–Cambrian A-Type Granites in the Central Subprovince of the Borborema Province, Northeastern Brazil

Jefferson Valdemiro de Lima [1,*], Ignez de Pinho Guimarães [2], José Victor Antunes de Amorim [3], Caio Cezar Garnier Brainer [2], Lucilene dos Santos [4] and Adejardo Francisco da Silva Filho [2]

[1] Faculdade de Geologia, Universidade Federal do Sul e Sudeste do Pará, Marabá 68500-000, PA, Brazil
[2] Programa de Pos-Graduacao em Geociencias, Universidade Federal de Pernambuco, Recife 50740-540, PE, Brazil; igneezdpg@gmail.com (I.d.P.G.); caiocgbrainer@gmail.com (C.C.G.B.); afsf56@gmail.com (A.F.d.S.F.)
[3] School of Earth Sciences, The University of Western Australia, Perth, WA 6009, Australia; josevictor.antunesdeamorim@research.uwa.edu.au
[4] Departamento de Geologia, Universidade Federal do Ceara, Fortaleza 60355-636, CE, Brazil; lucilene.santos01@gmail.com
* Correspondence: jefferson1901@gmail.com or jefferson.valdemiro@unifesspa.edu.br; Tel.: +55-81-982170432

Abstract: Ediacaran–Cambrian magmatism in the Central Subprovince (Borborema Province, NE Brazil) generated abundant A-type granites. This study reviews published whole-rock and mineral chemistry data from thirteen Ediacaran–Cambrian A-type intrusions and a related dike swarm. It also presents new mineral chemistry and whole-rock data for one of these intrusions, along with zircon trace element data for five of the intrusions. Geochronological data from the literature indicate the formation of these A-type intrusions during a 55 Myr interval (580–525 Ma), succeeding the post-collisional high-K magmatism in the region at c. 590–580 Ma. The studied plutons intruded Paleoproterozoic basement gneisses or Neoproterozoic supracrustal rocks. They are ferroan, metaluminous to peraluminous and mostly alkalic–calcic. The crystallization parameters show pressure estimates mainly from 4 to 7 kbar, corresponding to crustal depths of 12 to 21 km, and temperatures ranging from 1160 to 650 °C in granitoids containing mafic enclaves, and from 990 to 680 °C in those lacking or containing only rare mafic enclaves. The presence of Fe-rich mineral assemblages including ilmenite indicates that the A-type granites crystallized under low fO_2 conditions. Zircon trace element analyses suggest post-magmatic hydrothermal processes, interpreted to be associated with shear zone reactivation. Whole-rock geochemical characteristics, the chemistry of the Fe-rich mafic mineral assemblages, and zircon trace elements in the studied granitoids share important similarities with A_2-type granites worldwide.

Keywords: A-type granite; post-collisional; Borborema Province

Citation: de Lima, J.V.; Guimarães, I.d.P.; de Amorim, J.V.A.; Brainer, C.C.G.; dos Santos, L.; da Silva Filho, A.F. A Review of the Mineral Chemistry and Crystallization Conditions of Ediacaran–Cambrian A-Type Granites in the Central Subprovince of the Borborema Province, Northeastern Brazil. *Minerals* **2024**, *14*, 1022. https://doi.org/10.3390/min14101022

Academic Editor: Clemente Recio

Received: 2 September 2024
Revised: 6 October 2024
Accepted: 8 October 2024
Published: 11 October 2024

Copyright: © 2024 by the authors. Licensee MDPI, Basel, Switzerland. This article is an open access article distributed under the terms and conditions of the Creative Commons Attribution (CC BY) license (https://creativecommons.org/licenses/by/4.0/).

1. Introduction

Granitic magmatism is essential for continental crust differentiation [1–4]. The origin of granitic magmas is either explained by fractionation of mantle-derived magmas, partial melting of different crustal lithologies induced by mantle heat flow, or the mixing of mantle- and crust-derived melts [5–9]. Heat transfer, as hot magmas ascend through the continental crust, may result in additional partial melting processes and the incorporation of the surrounding rocks, increasing the chemical and petrological complexity of granitic rocks [10–13]. Granites are widespread worldwide and have intruded into the Earth's crust throughout all geologic periods, in a manner associated with several geodynamic processes [14–17]. Despite their relatively simple mineralogy, granites commonly show a wide array of modal and chemical compositions, reflecting the different processes associated with their emplacement and genesis [18–20].

The term A-type granite was introduced to define a specific group of iron-enriched granites with alkaline affinities, characterized by an anhydrous mineral assemblage, formed under low oxygen fugacity (fO_2) conditions, interpreted as intruded in an anorogenic tectonic setting [21]. However, since this definition, studies have shown that A-type plutons have a broader compositional range and may form in many tectonic settings, which has sparked a long-standing debate about their genesis and tectonic significance [22–33].

Overall, the mineral assemblage of A-type granites is composed of iron-rich (e.g., annite, siderophyllite, Fe-hedenbergite, Fe-hastingsite, fayalite) and/or alkali-rich (e.g., aegirine, arfvedsonite, and riebeckite) mafic silicates, associated with perthitic feldspars [18,23,28]. Geochemically, A-type rocks have high total alkalis ($K_2O + Na_2O$); high field strength (HFS) and rare earth element (REE) contents; high Fe# [$FeOt/(FeOt + MgO)$] values; and low MgO, CaO, Eu, and Sr contents [22,23,31].

In contrast with the initially proposed anorogenic origin [21], most studies indicate that A-type granites are emplaced in a variety of environments, both in post-collisional (e.g., orogenic collapse, strike-slip shear zones) and extensional settings (e.g., within-plate, rift, back-arc) [23,32,34]. Although it is generally accepted that A-type granites crystallize at higher temperatures and under more reducing conditions than other types of granite, the sources and crystallization conditions for A-type granite formation are also debatable [22,27,29]. Three main processes have been proposed to produce most A-type compositions, as summarized by Frost and Frost [22]: extreme differentiation from a basaltic source (with incremental degrees of crustal assimilation), partial melting of tonalitic–granitic crust, or a combination of the previous two processes.

In the Borborema Province of NE Brazil, abundant Ediacaran–Cambrian A-type granites have been reported in the Central Subprovince [35–40]. These granites are essential for understanding the multiple episodes of post-collisional deformation during the final stages of the Brasiliano Orogeny and its Pan-African counterparts [35,37,40]. Furthermore, the expressive A-type magmatism in the region provides valuable insights into the processes and conditions associated with A-type granites.

In this study, we review the mineral chemistry and whole-rock data of 13 A-type granitic plutons and a related dike swarm in the Central Subprovince of the Borborema Province, NE Brazil. Most of the data were compiled from scientific articles, but unpublished data, including those obtained from academic works (dissertations and theses), were also included. Additionally, we present new zircon trace element data for five of these granitic intrusions. The aim is to contribute to the discussion and enhance knowledge of the magmatic processes involved in the genesis of A-type granites and provide new insights into the geodynamic evolution of the later stages of the Brasiliano–Pan-African Orogeny in the Borborema Province.

2. The Borborema Province Geological Background

The Borborema Province (BP), located in northeastern Brazil, is limited to the north, the east, and the west by Phanerozoic sedimentary basins, and to the south by the São Francisco Craton. The BP (Figure 1A) is an orogenic belt formerly located in the northwestern–central part of West Gondwana, formed via the convergence and collision of the West Africa, Congo-São Francisco, and Amazonia Cratons during the Cryogenian–Ediacaran (c. 600–630 Ma), in an event known as the Brasiliano–Pan-African Orogeny [41–44].

The BP is divided by the large E-trending Patos and Pernambuco shear zones into three subprovinces (Figure 1B), i.e., North, Central, and South, and each subprovince is subdivided into tectonic domains [43,45]. The studied intrusions are located in the Central Subprovince (Figure 1B).

Overall, the geology of the Borborema Province can be summarized into Paleoproterozoic gneiss–migmatite basement complexes (c. 1.98–2.20 Ga) with Archean nuclei [45–48]; partly overlain by Neoproterozoic metasedimentary–metavolcanic supracrustal sequences [49–53];

intruded by widespread Neoproterozoic predominantly granitic magmatism [54–60]; and cut or bounded by large transcurrent shear zones trending NE or E, with possible continuation into the African continent [61–65].

The first stages of the Neoproterozoic tectonic evolution of the Borborema Province have been the subject of continuous debate (e.g., [42,50]). Some authors propose an accretionary model involving terrane accretion episodes throughout the Neoproterozoic [66–70]. In contrast, Neves [50,71] proposed an intracontinental model with the closure of small basins with limited oceanic crust, in agreement with proposals of a contiguous basement beneath the proposed terranes since the Paleoproterozoic [72–75]. Alternatively, a model consisting of a complete Wilson cycle with continental rifting, followed by subduction of large oceanic realms and a subsequent collision has also been proposed [42,52,56,64]. Diverging interpretations regarding the tectonic setting and evolution of the Tonian tectonothermal event (c. 1.0 Ga, Cariris Velhos Event), and a lack of agreement regarding the interpretation and significance of potential Neoproterozoic suture zones and magmatic arcs, make it difficult to achieve a consensus regarding the tectonic evolution of the BP and are beyond the scope of this review.

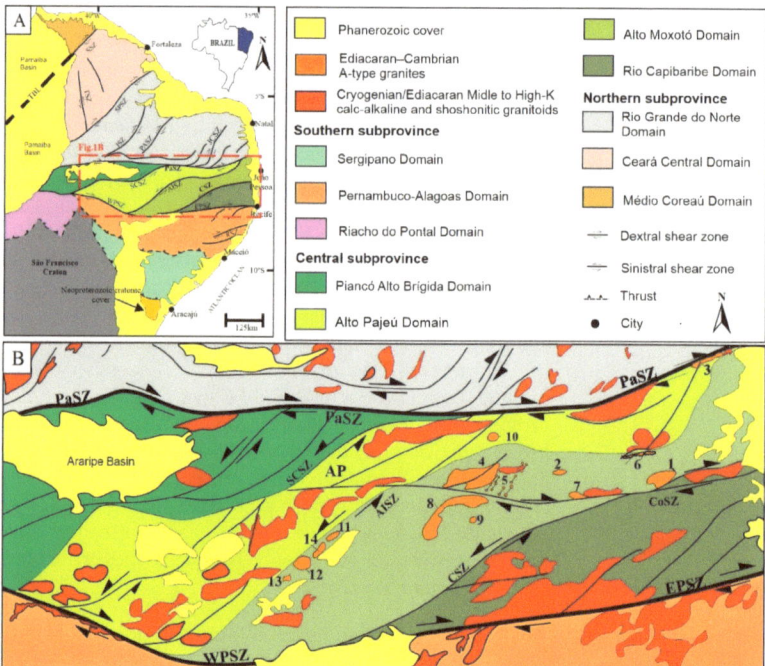

Figure 1. Geological maps illustrating the regional context and the location of the studied plutons [73]. (**A**) Major domains and shear zones of the Borborema Province [73]. Abbreviations—PaSZ: Patos Shear Zone; WPSZ and EPSZ: West and East Pernambuco Shear Zones; SSZ: Sobral Shear Zone; SPSZ: Senador Pompeu Shear Zone; JSZ: Jaguaribe Shear Zone; JCSZ: João Câmara Shear Zone; PASZ: Portoalegre Shear Zone; RSZ: Riachão Shear Zone; TSZ: Tauá Shear Zone; (**B**) Central Subprovince of the Borborema Province with studied plutons highlighted in orange (adapted from Lima et al. [73]). Abbreviations—CoSZ: Coxixola Shear Zone; CSZ: Congo Shear Zone; AISZ: Afogados da Ingazeira Shear Zone; SCSZ: Serra do Caboclo Shear Zone. Studied plutons: 1—Aroeiras Pluton; 2—Bravo Pluton; 3—Pilõezinhos Pluton; 4—Serra Branca Pluton; 5—Serra Branca dike swarms; 6—Queimadas Pluton; 7—Marinho Pluton; 8—Prata Complex; 9—Serra da Engabelada Pluton; 10—Serrote Santo Antonio Pluton; 11—Pereiro Pluton; 12—Serra do Velho Zuza Pluton; 13—Açude do Caroá Pluton; 14—Boqueirão Pluton.

During the post-collisional stages, the evolution of the BP was primarily controlled by strike-slip tectonics, with a network of fault zones that favored the formation of local extensional zones. During this period, the BP was affected by the widespread intrusion of high-K post-collisional, 'A-type' plutons and dike swarms, mineralizing events associated with pegmatites, and the influx of hydrothermal fluids during shear zone reactivation, as well as the development of sag basins with the collapse of orogenic chains and rupture of the continental crust [35,40,76–78]. The different groups of A-type granites that intruded the Central Subprovince at c. 580–525 Ma, mark episodes of tectonic shift and provide insights into the post-collisional evolution of the Borborema Province and are the subject of this review [35,37,73].

3. Geological Context of the Studied A-Type Granites

3.1. Introduction

The studied granitoids are primarily intruded within the Alto Moxotó Domain, often near the boundaries with the Alto Pajeú and Rio Capibaribe Domains, whereas the Serrote Santo Antonio and Pilõezinhos plutons are intruded in the Alto Pajeú domain (Figure 1B). The Alto Moxotó Domain comprises Paleoproterozoic units enclosing some Archean nuclei, overlain by Neoproterozoic supracrustal rocks [47,75,79]. The A-type granitic intrusions are associated with E-trending (Aroeiras, Queimadas, Serra Branca, and Marinho e Pilõezinhos) and NE-trending (Boqueirão, Pereiro, Açude do Caroá, Serra do Velho Zuza, Bravo, and Prata Complex) strike-slip shear zones. The Serra da Engabelada intrusion forms an isolated pluton proximal to the Prata Complex (Figure 1B).

3.2. Aroeiras Complex

The Aroeiras Complex intrudes Paleoproterozoic orthogneisses and migmatites, and Cryogenian–Ediacaran metasedimentary rocks and older granitic plutons (Figure 1B). It comprises a 100 km^2 sigmoidal-shaped igneous complex that was emplaced during the development of an extensional site associated with the synchronous activity of the E-trending dextral Timbauba-Coixixola and NE-trending sinistral Batista shear zones. It comprises felsic sheets dated at 585 ± 6 Ma (U-Pb zircon LA-ICPMS) and small intermediate bodies cut by late felsic dikes dated at 545 ± 4 Ma [35] (U-Pb zircon SHIRIMP).

The Aroeiras Complex comprises porphyritic to equigranular biotite–hornblende monzogranite to biotite syenogranite (Figure 2A), with prismatic allanite and zircon, acicular apatite, and ilmenite mantled by titanite as the main accessory phases. Intermediate lithotypes comprise hornblende–biotite diorite and quartz diorite. Ovoid microgranular mafic enclaves with crenulated borders, double-enclave relations, and hybrid rocks with a rapakivi-like texture are interpreted as evidence for mingling (Figure 2B,C) and mixing processes in the complex [35].

3.3. Bravo Pluton

The Bravo Pluton is a ~40 km^2 intrusion emplaced at 581 ± 2 Ma (U-Pb zircon LA-ICPMS) between the NE-trending sinistral Cabaceiras and E-trending dextral São José dos Cordeiro shear zones [39]. It comprises leucocratic, coarse-grained to porphyritic biotite syenogranite to monzogranite, with K-feldspar phenocrysts surrounded by a medium-grained matrix composed of quartz, feldspars, and mafic minerals (biotite and amphibole), with titanite, apatite, allanite, Fe-Ti oxides, and zircon as the main accessory phases [39]. Oval-shaped microgranular enclaves with a granodiorite to diorite composition and typical features of hybridization are commonly found along the margin of the pluton. Fine-grained monzogranitic rocks and aplite dikes locally cut the coarse-grained lithotypes.

Figure 2. Field aspects of the studied granitoids. (**A**) Porphyritic syenogranite from the principal facies of the Aroeiras Complex; (**B,C**) the mingling of diorite with the syenogranite in the Aroeiras Complex; (**D**) the porphyritic syenogranite of the Pilõezinhos Pluton enclosing the dioritic enclave; (**E**) a general view of the other facies of the Pilõezinhos Pluton, characterized by fine-grained granitic rocks enclosing an intermediate-composition enclave; (**F**) typical leucocratic granitic rock from the Serra Branca Pluton; (**G**) a leucocratic dike from the Serra Branca Suite; (**H**) the mega-dike field aspects of the Queimadas Pluton. (Red line—contact zone with the basement orthogneiss/migmatite; green line—shear zone; (**I**) Diorite as enclaves enclosed by felsic granite of the Marinho Pluton. The felsic granite also encloses pockets of mesocratic granite, interpreted as a hybrid rock, which in turn encloses elongated oriented enclaves of mafic diorite (red circles); (**J**) the dike of rapakivi granite (Marinho Pluton), cut by narrow dikes (up to 20 cm wide) of leucogranite (LD). Red arrows—K-feldspar surrounded by plagioclase of oligoclase composition; (**K**) the typical features of magmatic interaction between mafic and felsic magma of the Prata Complex; (**L**) porphyritic granite of the Pereiro Pluton, cut by veins of leucogranites. K-feldspar (red arrows) occurs as euhedral crystals oriented by magmatic flux processes.

3.4. Pilõezinhos Pluton

The c. 566 ± 3 Ma (U-Pb zircon LA-ICPMS) Pilõezinhos Pluton comprises an ENE-elongated intrusion of c. 100 km^2, located south of the Remígio-Pocinhos shear zone along the boundary between the Northern and Central subprovinces of the Borborema

Province [38]. The Pilõezinhos granitoids consist of equigranular to porphyritic, fine- to coarse-grained syenogranite to monzogranite (Figure 2D,E), intruded into the late Neoproterozoic metasedimentary rocks and Tonian orthogneisses of the Alto Pajeu Domain, in extensional sites created by the synchronous movement of the E-trending dextral Remígio-Pocinhos shear zone (ZCRP) and the NE-trending sinistral Matinhas shear zone, which allowed for the accommodation of the granitic magmas [38].

These granitoids contain quartz as anhedral crystals, recrystallized or subgrain aggregates, microcline phenocrysts, and subhedral and often zoned plagioclase crystals. The main mafic phases are large biotite lamellae and amphiboles. The accessory minerals comprise titanite as the primary crystals or forming coronas around opaque minerals, zoned allanite, and opaque minerals, mainly euhedral ilmenite.

3.5. Serra Branca Suite

The Serra Branca Suite consists of a primary body of ~300 km^2 (Serra Branca Pluton) and a swarm of granitic dikes (Serra Branca-Coixixola dike swarm) intruded into Paleoproterozoic to Archean gneiss migmatites [35,80]. These granitic intrusions are located north of the E-trending dextral Timbauba-Coixixola Shear Zone and west of the NE-trending sinistral Cabaceiras Shear Zone. U-Pb dating of the Serra Branca Pluton gave a 560 ± 5 Ma crystallization age (U-Pb zircon SHIRIMP) [80].

The Serra Branca Pluton comprises small intrusive bodies and sheets of leucocratic, equigranular, medium- to fine-grained biotite syenogranites to monzogranites (Figure 2F), locally containing the xenoliths of surrounding rocks and biotite clots. Quartz, alkali feldspar (perthitic orthoclase and microcline), and plagioclase constitute the mineral framework of these granitoids, along with biotite, the predominant mafic mineral. Apatite, allanite, and zircon are accessory minerals, forming the euhedral to subhedral crystals included within the main mineral phases. Ilmenite is the main Fe-Ti oxide present in these granitoids. The magmatic fabric developed parallel to the NE-trending shear zone foliation is interpreted as evidence that the Serra Branca Pluton is a syn-tectonic intrusion [80]. Magmatic layering with cross-bedding-like features has also been reported in the pluton.

The Serra Branca dikes (Figure 2G) intruded primarily as a NE-trending felsic dike swarm cross-cutting the earlier flat-lying foliation of the basement rocks and the Neoproterozoic metamorphosed supracrustal rocks [35]. Near the Timbauba-Coixixola Shear Zone, the dikes intrude parallel to the steeply dipping mylonitic foliation but show only incipient deformation. The dike swarm comprises porphyritic hornblende–biotite granite to equigranular biotite granite. A Concordia age of 545 ± 3 Ma is interpreted as the crystallization age of these dikes [35].

3.6. Queimadas Pluton

The Queimadas Pluton forms a ~50 km^2 intrusion dated at c. 550 ± 6 Ma (U-Pb Zircon SHRIMP). It intrudes basement gneisses and migmatites parallel to the E-trending foliation associated with the Campina Grande shear zone (Figure 2H), near the boundary between the Alto Pajeú and Alto Moxotó domains [35,36]. A NE-trending dextral shear zone disrupts the pluton into a mega-boudin-like shape.

The pluton comprises biotite ± amphibole porphyritic monzogranites and granodiorites as the main lithotypes, typically enclosing microgranular mafic enclaves (MMEs) and cut by late fine-grained leucogranite dikes [36]. Biotite and amphibole make up less than 10% of the mode. Allanite and apatite form the euhedral crystals included in the main ferromagnesian minerals, and zircon forms prismatic euhedral or round crystals. Biotite- and amphibole-hosted subhedral crystals of ilmenite, and rare monazite are accessory phases. Mafic microgranular enclaves (MMEs) range in composition from porphyritic quartz-monzonitic to quartz-monzodioritic and are mainly located close to the contact with basement rocks. The Queimadas granitoids show S-C dextral foliation, with the C-foliation plan aligned with the E-trending branch of the Campina Grande dextral shear zone. Biotite kinks, sigmoidal plagioclase porphyroclasts, quartz ribbons, mosaic texture, boudins, and necking indicate that the pluton was deformed under brittle–ductile conditions [36].

3.7. Marinho Pluton

The Marinho Pluton is an ENE-trending elongated intrusion of ~18 km^2, composed of syenogranites and monzogranites, intruded into Neoproterozoic metasedimentary rocks and Tonian orthogneisses of granodioritic composition. The Marinho Pluton consists of a granitic intrusion, comprising a small stock and associated dikes, intruded in an extensional site, related to the synchronous activity of the dextral E-trending Coixola shear zone and sinistral NE-trending Carnoio shear zone [81,82].

It comprises two main petrographic facies of a monzogranite to syenogranite composition. The main lithotype, dated at c. 550 ± 3 Ma (U-Pb zircon SHRIMP), is porphyritic and medium-grained and exhibits dioritic enclaves and flow structures [81]. Syn-plutonic dikes and enclaves of a dioritic composition indicate mingling processes. Fine-grained, slightly oriented dikes (~25 m width) of porphyritic rapakivi-like biotite syenogranites dated at c. 527 ± 6 Ma (U-Pb zircon SHRIMP) intrude the main facies of the Marinho Pluton and are cut by fine-grained leucocratic syenogranites (Figure 2I,J) [82].

Petrographically, these granitoids are characterized by perthitic orthoclase and microcline phenocrysts surrounded by a fine-grained matrix comprising subgrain aggregates or ribbon-shaped quartz, feldspars, amphibole, and biotite. Especially in the biotite syenogranite dikes, microcline crystals surrounded by plagioclase rims particularly highlight the rapakivi-like texture. The main mafic phase is biotite partially altered to chlorite. Amphibole locally forms green prisms with bluish-green rims. Titanite, allanite, apatite, and zircon are the main accessory minerals; titanite and allanite often occur as euhedral inclusions in biotite and amphibole, and apatite as acicular crystals.

3.8. Prata Complex

The Prata Complex comprises a boomerang-like pluton separated in two parts by a mafic body, intruded into the Archean–Paleoproterozoic migmatite basement [37]. Hollanda et al. [40] divided the Prata Complex into two distinct intrusions: a northern intrusion with a U-Pb crystallization age of c. 534 ± 3 Ma (U-Pb zircon SHIRIMP) (Sumé Pluton) and a southern intrusion with a crystallization age of c. 533 ± 4 Ma (U-Pb zircon SHIRIMP) (Santa Catarina Pluton).

The Sumé Pluton, about 250 km^2 in size, comprises medium- to coarse-grained hornblende–biotite monzo- to syenogranites, with abundant hybrid rocks resulting from mixing with dioritic magma (Figure 2K). The granitoids contain plagioclase, perthitic microcline locally showing plagioclase mantling, amphiboles, and biotite. Subordinate minerals include allanite rimmed by epidote, titanite, apatite, and zircon. Round to angular, sometimes elongated mafic enclaves are frequent and have a dioritic–quartz dioritic composition.

The Santa Catarina Pluton, covering approximately 170 km^2, primarily consists of coarse-grained porphyritic biotite syenogranite, which is the dominant lithotype and frequently encloses mafic enclaves. This lithotype is characterized by alkali-feldspar phenocrysts exhibiting rapakivi texture. Weak foliation is observed along the southeastern and southern margins of the pluton. In its southern region, a swarm of N-S-trending rhyolite and diabase dikes has intruded into the surrounding gneisses and migmatites, extending into the pluton itself, where the dikes cut the granitoids. Dioritic enclaves in the central–eastern part of the pluton exhibit crenulated contacts, indicating magma mingling and mixing processes between felsic and mafic magmas. These diorites contain rounded and acicular hypersthene crystals, up to 10 mm long and 0.8 mm wide, which are surrounded by augite and hornblende. Additionally, they feature quartz phenocrysts mantled by hornblende and biotite, acicular apatite, small clusters of hornblende and biotite, and poikilitic K-feldspar, all of which indicate magma mixing processes. Allanite is the most abundant accessory mineral in the syenogranites, occurring in modal amounts of up to 5% and reaching lengths of up to 5 mm.

3.9. Serra da Engabelada Pluton

Near the Prata Complex, there is one small granitic pluton (Serra da Engabelada), along with a gabbro stock and swarms of rhyolite to gabbro-norite dikes [37]. The Serra da Engabelada Pluton comprises a rounded intrusion of approximately 50 km^2, intruded into the Paleoproterozoic migmatized orthogneisses, located east of the Prata Complex. It consists mainly of coarse-grained, equigranular biotite syenogranites with rare mafic enclaves.

3.10. Serrote Santo Antonio Pluton

The Serrote Santo Antônio Pluton is a 75 km^2 igneous intrusion, primarily composed of leucocratic, medium- to coarse-grained biotite syenogranite. This pluton is located to the north of the Serra Branca Pluton and is one of the few A-type plutons discussed in this study that is emplaced within the Alto Pajeú Domain, where it intrudes Early Neoproterozoic metamorphosed supracrustal rocks.

3.11. Serra do Pereiro, Serra do Velho Zuza, Açude do Caroá, and Boqueirão Plutons

These plutons were emplaced along the NE-SW sinistral Afogados da Ingazeira Shear Zone, which makes contact between the Alto Pajeú and Alto Moxotó domains of the Central Subprovince in the Borborema Province [83].

The Serra do Velho Zuza (538 ± 23 Ma; U-Pb zircon TIMS), Serra do Pereiro (543 ± 7 Ma; U-Pb zircon TIMS), Açude do Caroá, and Boqueirão plutons show roughly rounded shapes, with the Pereiro and Boqueirão plutons being more elongated, with their major axes parallel to the Afogados da Ingazeira Shear Zone [58]. These plutons are intruded into orthogneisses of the basement and Neoproterozoic supracrustal rocks of the Sertânia and São Caetano complexes [84]. The Serra do Velho Zuza and Boqueirão plutons mainly comprise gray, medium to coarse-grained porphyritic hornblende–biotite monzogranite to syenogranite, while the Serra do Pereiro Pluton comprises hornblende–biotite monzogranite to quartz syenite (Figure 2L). They are composed of K-feldspar phenocrysts in an interstitial matrix composed of quartz, plagioclase, and K-feldspar. Yellow to dark brown biotite and amphibole are the main mafic mineral phases. Epidote, titanite, and opaque minerals are accessory phases.

All plutons contain a small volume of microgranular mafic enclaves. The MMEs are fine-grained and range from diorite to granodiorite. In the Pereiro Pluton, elongated mafic enclaves parallel to the oriented K-feldspar tabular megacrysts define a magmatic fabric.

The Açude do Caroá Pluton stands out from the others by being composed of more mafic rocks, sharing macroscopic and microscopic characteristics with the enclaves found in the other plutons. It comprises mesocratic, fine- to medium-grained biotite quartz-diorite, quartz monzodiorite, and granodiorite, and often contains amphibole- and biotite-rich clots.

4. Whole-Rock Geochemistry

To highlight the chemical characteristics of the studied granitoids, we compiled whole-rock chemical data available in the scientific literature (Aroerias Pluton [35]; Bravo Pluton [39]; Piloezinhos Pluton [38]; Serra Branca Suite [35,80]; Queimadas Pluton [36]; Prata Complex, Serra da Engabelada Pluton, and Serrote Santo Antônio Pluton [37]. In addition, we included unpublished data (Marinho Pluton) and data obtained from a doctoral thesis (Pereiro, Velho Zuza, Açude do Caroá, and Boqueirão plutons [84]). All data are shown in Supplementary Table SI.

The studied granitoids exhibit a wide range of silica contents, with the less evolved rocks (MME and the Açude do Caroá Pluton) showing SiO_2 ranging from 51.3 to 64.8 wt%, while the more acidic granitoids, which are predominant among the studied rocks, have SiO_2 up to 75 wt%. According to the geochemical classification by Frost et al. [19], these granitoids are metaluminous to slightly peraluminous (Figure 3A), with Alumina Saturation Index (ASI) values ranging from 0.78 to 1.17, and essentially belong to the ferroan series (Figure 3B), with Fe# (FeO$_t$/FeO$_t$ + MgO) values ranging from 0.78 to 0.98, which are commonly associated with magmas that evolved under reducing conditions (ilmenite-series granites [85]). Only five samples show discrepant Fe# values (0.59–0.78); these correspond to samples from enclaves within the Pilõezinhos and Serra Branca plutons, as well as the less evolved facies of the Prata Complex. The granitoids show high total alkali (K$_2$O + Na$_2$O) contents (6–10 wt%,) and K$_2$O/Na$_2$O ratios ranging from 0.7 to 2.4 (typically > 1), with compositions ranging from alkali-calcic to alkalic in the MALI (modified alkali–lime index) versus the SiO$_2$ diagram (Figure 3C), with only a few plotting in the calc-alkalic field. This set of geochemical characteristics indicates that the studied plutons are A-type granites. They plot primarily within the field of granites originating in post-collisional settings (post-COLG) in the Rb vs. (Y + Nb) diagram [86] and show (Zr + Nb + Ce + Y) values up to 1700, typical of A-type granites [87] (Figure 4). Additionally, the studied granitoids dominantly plot within the A2-type granite field on the diagram of Eby [34] or straddling the A1-A2 boundary (Figure 5), distinguishing them from pure A1-type granites, which are typically associated with mantle plume activity or hotspots. A2-type granites, on the other hand, have a crustal origin and are generally linked to post-collisional settings. This set of geochemical characteristics indicates that the studied plutons are composed of A$_2$-type granites.

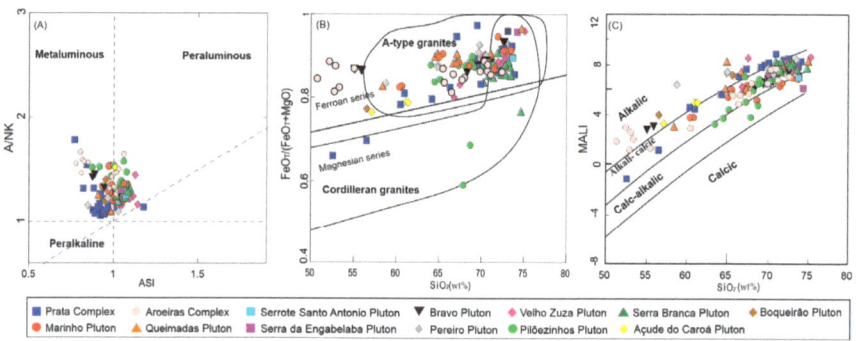

Figure 3. Geochemical characteristics of studied granitoids. (**A**) Alumina saturation index diagram; (**B**) FeO$_T$/(FeO$_T$ + MgO) versus SiO$_2$ diagram; (**C**) SiO$_2$ versus Mali (modified alkali lime index) diagram.

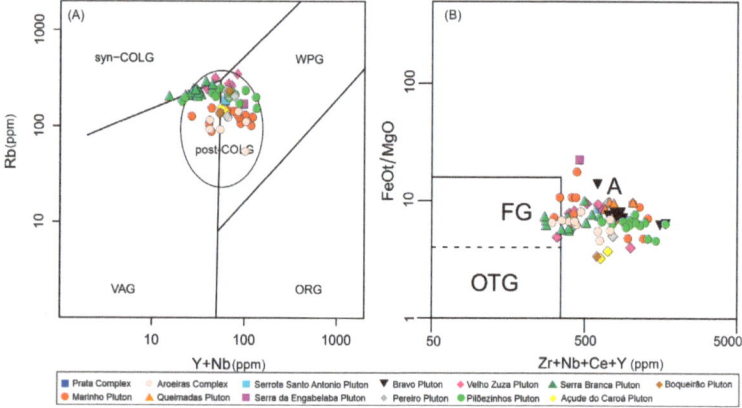

Figure 4. Tectonic setting discrimination diagrams for studied granitoids. (**A**) Pearce et al. [86]: WPG: within-plate granites; syn-COLG: syn-collisional granites; post-COLG: post-collisional; ORG: ocean ridge granites; VAG: volcanic arc granite; (**B**) diagram from Whalen et al. [87]: FG: fractionated granite field; OTG: unfractionated granite field; A: A-type granites.

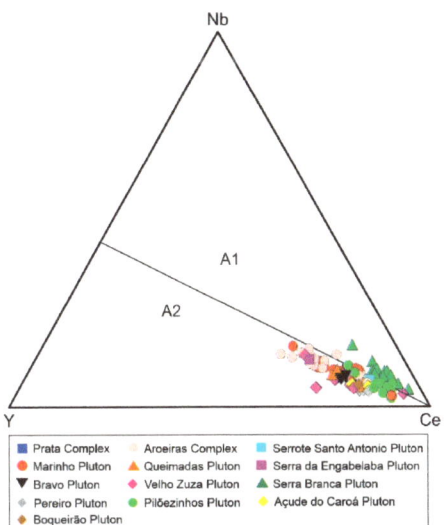

Figure 5. Trace elements of the studied granitoids in the tectonic discriminant diagrams of Eby [34]. A1: Non-orogenic granite; A2: Post-collisional/post-orogenic granite.

5. Mineral Chemistry

The chemical characterization of the mineral phases in the studied granitoids was primarily based on a compilation of data from the scientific literature (Aroerias Pluton [35]; Bravo Pluton [39]; Pilõezinhos Pluton [88]; Serra Branca Suite [35,89]; Queimadas Pluton [36]; and Prata Complex [37]) and a doctoral thesis (Açude do Caroá, Boqueirão, Velho Zuza, and Pereiro Plutons [84]). Additionally, we included unpublished chemical data for amphibole, biotite, and plagioclase from the Marinho Pluton granitoids, along with new trace element data for zircon from the Queimadas, Marinho, Pereiro, Velho Zuza, and Serrote Santo Antônio plutons. All data are shown in Supplementary Table SII.

5.1. Amphibole

The amphibole structural formulas for the studied samples were calculated based on 23 oxygen atoms and the data are available in the Supplementary Table SII. Amphibole data from the Serra da Engabelaba, Santo Antônio, and Pereiro plutons were absent. Amphiboles were classified using Locock's [90] Excel spreadsheet, which follows the guidelines from the International Mineralogical Association for the Nomenclature of the Amphibole Supergroup [91]. Amphibole analyses show Na + K + Ca values between 1.81 and 3.10 a.p.f.u. and Si^{IV} between 5.95 and 7.10 a.p.f.u. (Figure 6A), which are typically reported in magmatic amphiboles [92]. Ca^{2+} is the dominant constituent in the B site, with concentrations much higher than other ions occupying the B site, displaying $^B(Ca + \Sigma M^{2+})/^B\Sigma$ between 0.85 and 1.00, $^BCa/\Sigma B = 0.64$–1.00 and $^B\Sigma M^{2+}/\Sigma B = 0.00$–0.21. The calculated values show that the studied amphiboles have $^B(Ca + \Sigma M^{2+})/\Sigma B \geq 0.75$ and $^BCa/\Sigma B \geq {}^B\Sigma M^{2+}/\Sigma B$, indicating that they belong to the calcium subgroup. The analyses show that in the A site, A (Na + K + 2Ca) values are typically higher than 0.5 and in the C site, C (Al + Fe^{3+} + 2Ti) mostly ranges between 0.5 and 1.5. Therefore, most amphiboles plot in the pargasite–hastingsite field (Figure 6B). In the C site, Fe^{2+} and Fe^{3+} are more abundant than other cations, with the Fe# (Fet/Fet + Mg) ranging from 0.50 to 0.93, also highlighting the high iron concentrations in these minerals. Additionally, K is typically the most abundant constituent in the A site. Therefore, the studied amphiboles are mostly classified as hastingsite and potassic-hastingsite, although, ferro-pargasite and potassic-ferro-pargasite (Boqueirão, Açude do Caroá, Bravo, Pilõezinhos, Prata, and Queimadas plutons), ferro-hornblende and ferro-ferri-hornblende (Açude do Caroá, Marinho, Prata, and Queimadas plutons), and potassic-ferro-ferri-sadanagaite (Marinho pluton) are also recognized in the studied samples. It is noteworthy that four analyses each of amphibole from the Prata Complex and Queimadas Pluton, as well as two analyses from the Açude do Caroá diorite, show Mg–hornblende compositions.

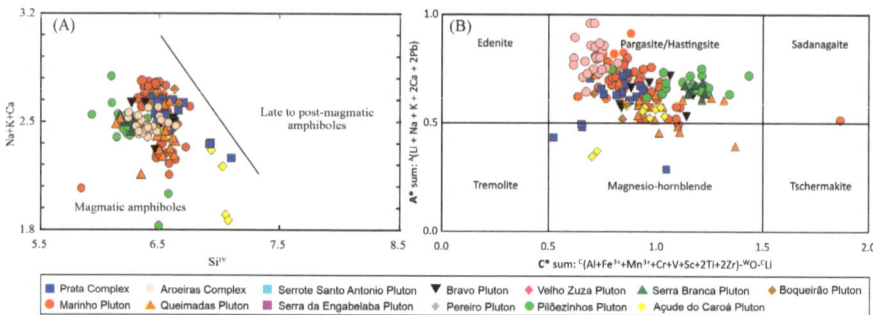

Figure 6. (**A**) Composition of amphibole crystals plotted in the $Si^{IV} \times (Na + K + Ca)$ diagram [92] indicating a magmatic origin for the studied amphiboles; (**B**) studied amphiboles in the A* sum: $^A(Li + Na + K + 2Ca + 2Pb)$ versus C* sum: $^C(Al + Fe^{3+} + Mn^{3+} + Cr + V + Sc + 2Ti + 2Zr)$-$^WO$-$^CLi$ calcium amphiboles classification diagram [91], revealing compositions mainly within the pargasite–hastingsite range.

5.2. Biotite

The structural formula was calculated on the basis of 22 oxygen atoms (Supplementary Table SII) and it was assumed that all iron is in the Fe^{2+} state. The compiled analyses of biotite crystals from the studied plutons are consistent with those observed in primary to re-equilibrated primary biotites; no analyses fall in the secondary biotite field (Figure 7A). In the classification diagram by Foster [93], most of the studied biotites are classified as Fe-biotite to siderophyllite. However, five analyses of biotites from the less evolved facies of the Prata Complex show a more magnesian composition (Figure 7B).

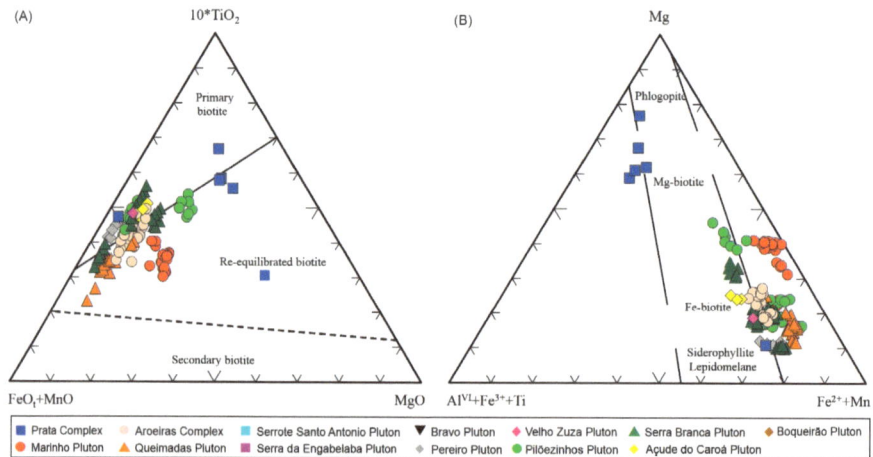

Figure 7. Chemical characteristics of the studied biotite crystals; ternary diagrams produced from Gündüz and Asan [94]. (**A**) $MgO \times 10TiO_2 \times FeO_t + MnO$ ternary diagram [95], showing the studied biotites straddling between the primary and re-quilibrated fields; (**B**) Classification of biotite crystals in the $Al^{IV} + Fe^{3+} + Ti \times Fe^{2+} + Mn \times Mg$ ternary diagram [93], showing predominantly Fe-rich compositions.

The Fe# [Fe/(Fe + Mg)] versus SI (solidification index) diagram (Figure 8A) has proven effective in studies involving several plutons, as the chemical trend in biotite is directly linked to the evolutionary character of the host granitoids. The SI is calculated using the formula [SI = 100 × MgO/(MgO + FeO + Fe$_2$O$_3$ + Na$_2$O + K$_2$O)], where higher SI values indicate less evolved rocks, while lower SI values indicate more evolved rocks. This correlation between SI values and the degree of evolution in biotite-bearing granites underscores the utility of the SI metric in petrological studies. The biotite crystals of the Prata Complex project at extreme positions on the Fe# versus SI diagram, highlighting the broad compositional spectrum of this intrusion. The biotite analyses from gabbroic rocks yield SI values > 60.0. According to the Speer [95] classification, they are eastonite (Figure 8B). One analysis from a more evolved sample shows SI = 22.50 showing a siderophyllite-rich composition (Figure 8B). The biotites from the Pereiro Pluton yield high Fe# values (0.86–0.88) and low SI (20.78–23.66), consistent with a high degree of magmatic evolution for these host granitoids. The biotites from the Pilõezinhos Pluton have SI values ranging from 28.25 to 53.28. Biotites from dioritic enclaves have SI = 49.03–53.28 and are rich in the eastonite molecule (Figure 8B), whereas the biotites from the more evolved facies have SI from 28.25 to 38.56 and are rich in the siderophyllite molecule. Biotites from all other plutons show siderophyllite-rich compositions (Figure 8B).

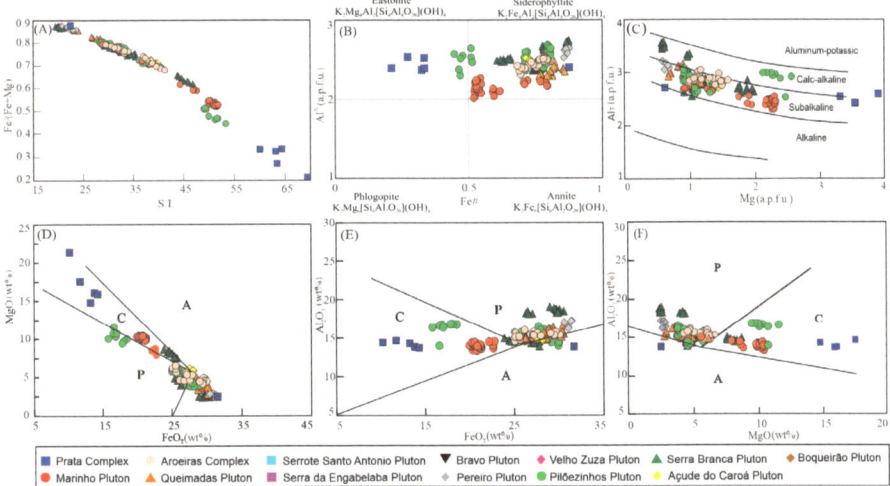

Figure 8. Chemical classification of the studied biotites. (**A**) Fe/(Fe + Mg) versus S.I. [100MgO/(MgO + FeO + Fe$_2$O$_3$ + Na$_2$O + K$_2$O)] diagram; (**B**) Fe# × Al^{IV} diagram; (**C**) Mg × Al$_T$ diagram after Nachit et al. [96]; (**D**–**F**) discriminant diagrams after Abdel-Rahman [97]. Legend: A: alkaline anorogenic; C: calc-alkaline; P: peraluminous.

The Al$_T$ contents of the studied biotite crystals range from 2.31 to 3.58 a.p.f.u. In the Nachit et al. [96] classification, the analyses plot in the sub-alkaline series field (Figure 8C), except for a few analyses from the Pilõezinhos Pluton, the Serra Branca Suite and the less evolved lithotypes of the Prata Complex, instead scattering in the calc-alkaline biotite field. In the discriminant diagrams of Abdel Rahman [97] (Figure 8D–F), biotite analyses from the less evolved facies of the Prata Complex, Pilõezinhos Pluton, and a few analyses from the Serra Branca and Marinho plutons, plot into the calc-alkaline series field, whereas biotites from the remaining plutons straddle between the peraluminous and alkaline fields.

5.3. Feldspars

The chemical study of feldspars was conducted using 167 analyses sourced from the literature, along with nine new analyses performed on feldspars from the Marinho

Pluton (Supplementary Table SII). The structural formulae were calculated on the basis of eight oxygen atoms. The plagioclases of the studied granitoids are sodic with albite contents ranging from 41% to 99%. The compositions range mainly from albite to andesine, with two plagioclase analyses from the Prata Complex classified as labradorite (Figure 9). Most of the studied plagioclase crystals exhibit core-to-rim profiles with compositional zoning characterized by an increase in sodium and a decrease in calcium, typical of normal zoning, which is commonly associated with fractionation processes. However, reverse zoning is also observed in some plagioclase crystals from the Prata Complex and the Queimadas Pluton, indicating additional partial melting processes during the magmatic evolution of these plutons or a new influx of more primitive magma into the magma chamber. The Potassic feldspars are mainly orthoclase (Figure 9), with compositions of 58%–98%.

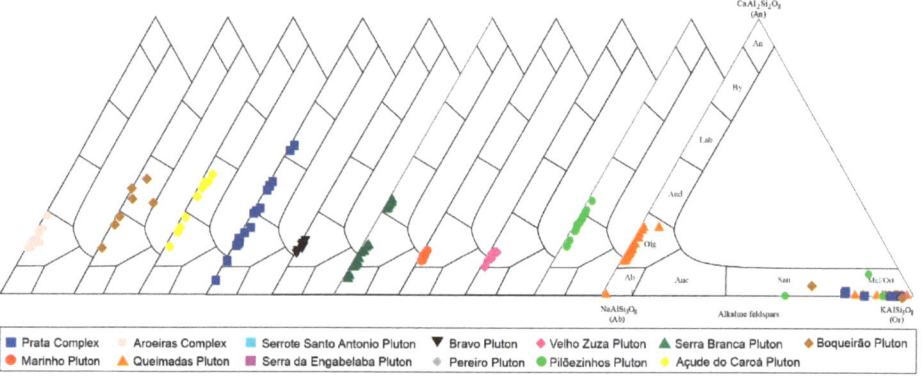

Figure 9. An-Ab-Or ternary diagram for the classification of the studied plagioclase and K-feldspar [98].

5.4. Accessory Minerals

Mineral chemistry data for accessory mineral phases are not available in the literature for all the studied plutons. In the Pilõezinhos Pluton, Lima et al. [88] analyzed opaque minerals which showed a predominance of ilmenite, which is locally surrounded by secondary titanite coronas. The high Al_2O_3 content in secondary titanite ($Al_2O_3 > 6\%$; [88]) was interpreted as a distinguishing feature from primary titanite crystals, which exhibit euhedral habit and lower Al_2O_3 content (3.6% to 4.0%).

Santos et al. [89] analyzed apatite crystals from the Serra Branca granitoids and showed structural formulas characterized by contents of Ca at 9.4–9.6, of P at 5.7–5.8, and F at 2.3–2.4, classifying them as fluorapatite [89]. High cerium values also occur in these apatites (1.1%–2.1%).

5.5. Zircon Trace Element Composition

In situ zircon trace element abundances were measured in samples from five plutons (Serrote Santo Antonio, Marinho, Queimadas, Serra do Velho Zuza, and Pereiro) using an Elan 6100DRC ICP-MS instrument coupled to a nanosecond New Wave Research UP-213 laser ablation system, at the Geosciences Center of the São Paulo University. The measurements were undertaken in previously dated zircon grains to avoid inherited grains, using a laser spot of 30 μm, a 4 Hz repetition rate, and an energy fluence of ~10 J/cm^2. The total acquisition time was 120 s, equally divided between background and laser ablation ion signals. The NIST SRM 612 glass wafer was employed as the external calibration standard [99], while an averaged SiO_2 abundance of 31.6% was chosen as the internal standard, following S.R.F. Vlach (pers. communication). The (version 4.0) *Glitter* software [100] was used for drift correction, data reduction, and elemental abundance determinations. Trace element values for the investigated zircon grains are given in Supplementary Table SIII. Analyses yielding anomalous P, Th, Ti, and LREE were excluded and were interpreted to represent the analysis of microinclusions (e.g., apatite, titanite, or rutile) interfering with the analysis [101–104].

Zircon is an important accessory phase in most granitic samples, except for the highly differentiated ones [105]. Zircon is resistant to alteration and metamorphic processes under a wide range of conditions. Therefore, it is widely used in petrogenetic geochronological studies. Despite its simple chemical composition ($ZrSiO_4$), zircon can accept many minor and trace elements into its crystal lattice, which provides information about the chemical composition of the melt from which the granite crystallized [106–112]. However, concentrations of many nonformula elements, such as LREEs (light rare earth elements), can be modified by hydrothermal and low-temperature fluids [113–115].

Zircon crystals in the studied granites form prismatic euhedral to subhedral grains, ranging from 50 to 250 μm in length and showing {101} pyramid endings, features common to zircons crystallized in granites of the alkaline series [105]. The zircon crystals from the Serrote Santo Antonio granite, however, rarely show pyramidal endings. Cathodoluminescence (CL) images show ubiquitous concentric oscillatory zoning, mainly in the crystal rims. Embayments and narrow overgrowths are recorded mainly in the late dike of the Marinho Pluton (FMJ-55) and the Serra Branca dikes [35], indicating local resorption and reprecipitation potentially associated with hydrothermal processes. Apatite and titanite form microinclusions in zircon from most analyzed samples.

The analyzed zircon grains show high but variable Th/U ratios, reflecting high Th contents. The highest values were recorded in the zircons from the Queimadas granites (0.44–3.36, mean 0.98), whilst the lowest values were recorded in the zircons from the Marinho Pluton (0.24–2.76, mean 0.64). Th/U ratios > 0.2 recorded in zircon grains from all studied plutons are typically associated with magmatic origin [103,116].

Hafnium contents in the studied zircons from all studied plutons range from 7941 to 13,507 ppm (Figure 10). The Y abundances are high in the studied samples, ranging from 465 to 3478 ppm (mean 1568 ppm). Zircons from the Serra do Velho Zuza Pluton yield the highest Y values (956–3478 ppm, mean 1807 ppm). The Nb abundance and Nb/Ta ratios vary widely across all samples. However, the Nb/Ta ratios for most samples are <10, except for the analyses from the Serra do Velho Zuza Pluton, in which Nb/Ta values distinguish two zircon populations: (i) Nb/Ta ratios > 10 due to high Nb abundances (74–228 ppm), and (ii) Nb/Ta ratios < 10 (Figure 10A). Zircon grains from all studied granites show negative correlations between Hf contents and Zr/Hf and Nb/Ta ratios (Figure 10A,B).

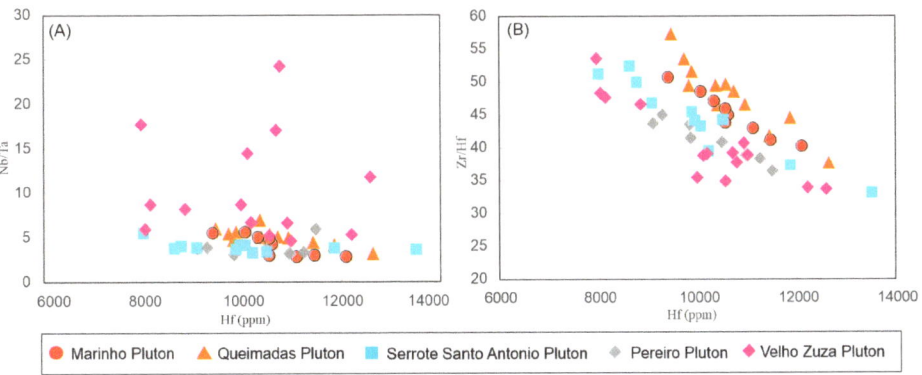

Figure 10. Variation diagrams for studied zircon crystals. (**A**) Hf (ppm) × Nb/Ta; (**B**) Hf (ppm) × Zr/Hf.

Chondrite-normalized REE patterns (Figure 11A–E) of the zircon grains from all studied granitic plutons are, with some exceptions, similar. The patterns are characterized by a steeply rising slope due to HREE enrichment relative to LREEs, positive Ce, and negative Eu anomalies, with Ce/Ce* [Ce$_N$/(La$_N$ × Pr$_N$)] = 1.2–23.6 and Eu/Eu* [Eu$_N$/(Sm$_N$ × Gd$_N$)] = 0.02–0.38. These patterns are typical of igneous zircon [104]. Zircon grains from the Pereiro Pluton show lower LREE abundances (24.7–74.8 ppm), deeper Eu (Eu/Eu* = 0.04–0.09), and higher Ce anomalies (Ce/Ce* = 1.88–23.58) than the other studied granites (Figure 11E). Zircon grains from the rapakivi syenogranite of the Marinho Pluton are characterized, in general, by higher LREE abundance (24.6–1136 ppm) resulting in less steep-rising slope patterns (Figure 11B). Most of the analyzed zircon grains from the Marinho granitoids exhibit slightly positive Ce anomalies [(Ce/Ce*)$_N$ = 1.20–2.65], with only one grain displaying a more pronounced Ce anomaly [(Ce/Ce*)$_N$ = 7.98].

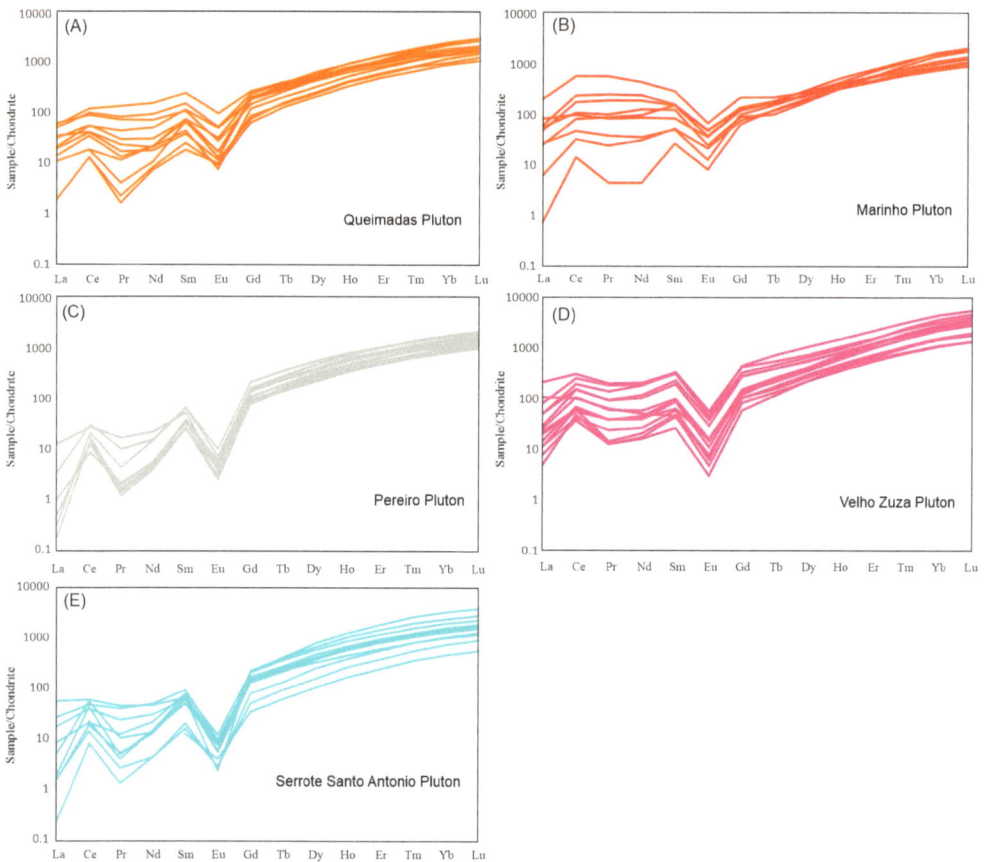

Figure 11. Chondrite-normalized REE patterns [117] of the zircon crystals from the studied plutons. (**A**) Serrote Santo Antônio Pluton; (**B**) Marinho Pluton; (**C**) Queimadas Pluton; (**D**) Velho Zuza Pluton; (**E**) Pereiro Pluton.

6. Discussion

6.1. Mineralogical Characterization, Chemical Affinity, and Petrogenetic Implications

Although granites exhibit relatively simple mineralogy, they are notably recognized for their significant modal and chemical variability, which reflects the diverse modes of occurrence and genesis [18]. The mineral composition of granite is a crucial factor in distinguishing the chemical signature and source of the parental magma, offering essential insights into the magmatic evolution processes.

In the studied granitoids, the mafic mineralogy is characterized by Fe-rich calcium amphibole [91] and biotite of siderophyllite composition, which, together with quartz and perthitic feldspars, form the essential minerals. Titanite, allanite, apatite, zircon, and Fe-Ti oxides, predominantly ilmenite, are the most frequently observed accessory minerals. The composition of the mafic minerals, associated with the presence of primary ilmenite crystals, is typical of ilmenite series granites [85], supporting their classification as ferroan A-type granites [23,28]. The analyzed zircons yield high Y, Th, and U consistent with values reported for zircons from A-type granites [118].

According to Xie et al. [119], zircons from aluminous A-type granites exhibit ThO_2 contents of less than 1 wt%, unlike zircons from peralkaline A-type granites, which can reach up to 10 wt%. In the latter, microinclusions of thorite are common due to thorium concentrations exceeding the solubility limit of Th in the zircon structure. However, late alteration of zircon can occur via fluids accumulated at the end of the magmatic evolution of an A-type granite. Most of the zircon grains analyzed from the Queimadas, Marinho, Pereiro, and Serrote Santo Antônio plutons exhibit Th concentrations below 300 ppm. In contrast, in the Velho Zuza Pluton, which is the most aluminous granite (Figure 3A), only one zircon grain shows a Th contents below 300 ppm, while the other analyzed grains display Th concentrations ranging from 335 to 1025 ppm. Santos [120] reported a wide Th variation (60–1250 ppm) in zircon grains from the aluminous A-type Serra Branca granites. Thus, zircon Th contents do not discriminate aluminous A-type from normal A-type granites.

In general, the majority of analyzed zircon grains present high levels of Th and LREEs, in addition, Figure 12 shows positive correlation of LREEs with Th and Y in zircon from of the most plutons. The data presented show that, at least for the studied A-type granites, Th abundance in zircon is not a simple function of the granite composition. It appears to depend on the Th content in the source of the magma, the time of zircon crystallization and the composition of the late hydrothermal fluids. Early crystallized inclusions of allanite and monazite may also influence the available thorium content in the magma. However, microinclusions of these minerals were not detected during the zircon grain analyses, and no other features suggesting the crystallization of thorium-rich early mineral phases were identified.

The chemical composition of biotite can be a valuable indicator of the chemical affinity of the magma from which it crystallized [96,97]. In the studied granites, the biotites chemical compositions reflect an intermediate chemical affinity between peraluminous and alkaline (Figure 8D–F). However, in the discriminant diagram by Nachit et al. [96], the biotite compositions indicate a subalkaline chemical signature (Figure 8C), consistent with the whole-rock chemical compositions of alkali-calcic granitoids (Figure 3C) as well as most A-type granitoids worldwide. The biotites from the more magnesian facies plot within the field of biotites from calc-alkaline granitoids, with some biotites from the Prata Complex showing compositions similar to high Mg-Ti biotites (phlogopite). The granitoids of the Prata Complex exhibit many field features suggestive of interaction with more mafic magmas. Both chemical and field data indicate that mafic melts derived from the lithospheric mantle contributed to the formation of the Prata Complex magma.

The zircon trace element signatures of the studied granites closely align with the findings from the biotite chemical compositions and the whole-rock chemistry. According to Vilalva et al. [121], zircons from rocks of the alkaline association yield Hf < 10,000 ppm, whereas zircons from rocks of the subalkaline rock association yield Hf > 10,000 ppm. However, this limit cannot be applied to classify the studied rocks, because values both above and below this threshold have been shown in the same sample. On the other hand, it could suggest granite compositions in the transition between subalkaline and alkaline associations.

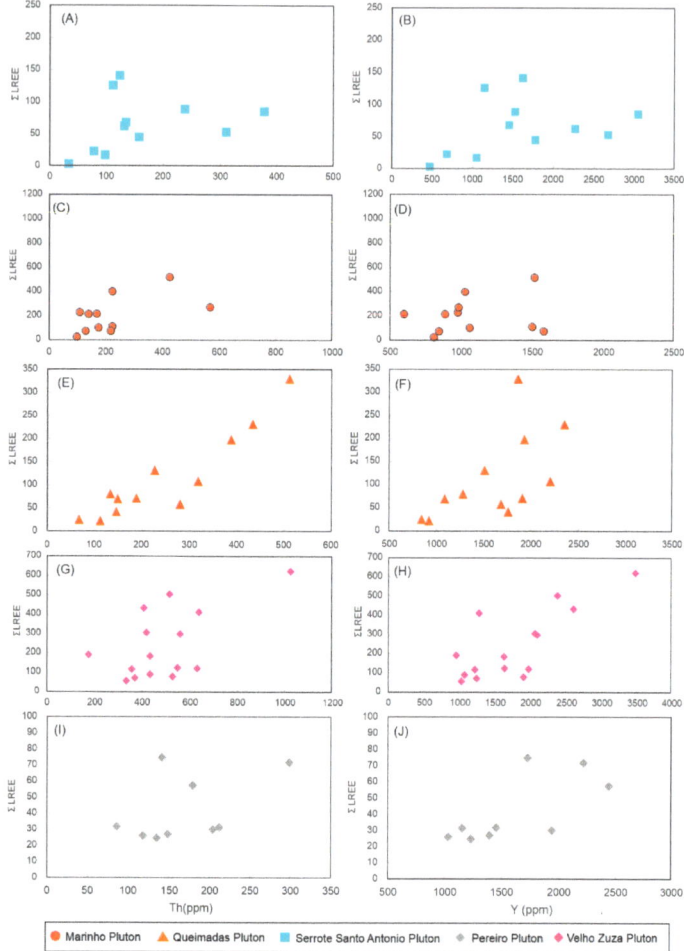

Figure 12. Th × Σ LREE and Y × Σ LREE plot for studied zircon crystals. (**A**,**B**) Serrote Santo Antônio Pluton; (**C**,**D**) Marinho Pluton; (**E**,**F**) Queimadas Pluton; (**G**,**H**) Velho Zuza Pluton; (**I**,**J**) Pereiro Pluton. A zircon grain from the Marinho Pluton (FMJ-55-Zr-12) has anomalous Th and LREE concentrations (Supplementary Table SIII) and, therefore, was not included in (**C**,**D**).

Breiter et al. [118] used zircon trace elements from an extensive dataset to discriminate zircon from highly evolved, evolved, and normal A-type and S-type granites. They found that zircon Zr/Hf ratios < 25 are typical of strongly evolved granites, Zr/Hf ratios in the 25–55 intervals characterize evolved granites, while normal granites have Zr/Hf ratios > 55. Except for one grain of the Queimadas Pluton, zircon grains from all studied plutons exhibit Zr/Hf ratios between 33 and 54, these being typical for evolved A-type granites, according to Breiter et al. [118]. Hawkesworth and Kemp [8] used zircon Th/U and Nb/Hf ratios to discriminate between I-type and peralkaline A-type granites. The zircon of the studied granites plots within the A-type granitoid field (Figure 13A). The correspondence between the studied zircon grains and the A-type field in the Nb/Hf vs. Th/U plot by Hawkesworth and Kemp [8] suggests that, in addition to distinguishing between zircon from peralkaline A-type granites and I-type granites, this diagram is also effective in discriminating between zircon crystals from I-type granites and non-peralkaline A-type granites.

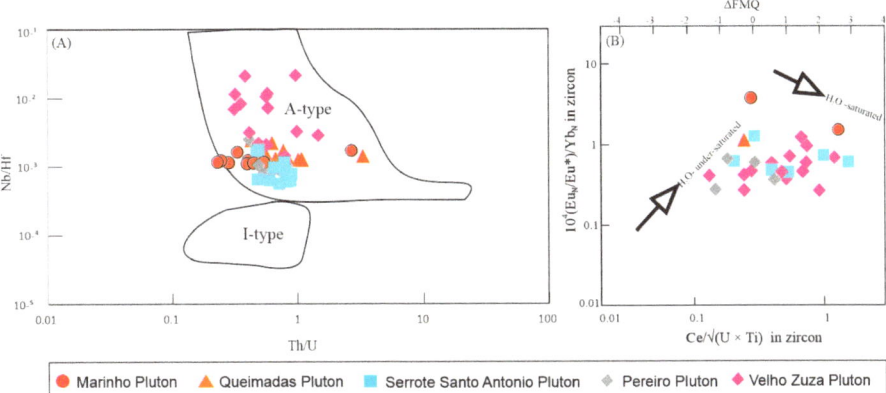

Figure 13. Discriminant diagrams based on the trace element chemistry of the analyzed zircons. (**A**) An Nb/Hf vs. Th/U plot for the studied zircon crystals. The A-type and I-type fields are from Hawkesworth and Kemp [8] using zircon data of peralkaline A-type granites akin to those of Nigeria and I-type granites from the Lachlan Fold Belt (SE Australia), respectively; (**B**) the $10^4(Eu/Eu^*)_N/Yb_N$ vs. $Ce/\sqrt{(U \times Ti)}$ plot [112] for the studied zircon crystals.

The water content in A-type granites is a frequently debated topic, although it is traditionally accepted that these granitoids evolve from anhydrous to undersaturated magmas. The chemical composition of zircon can provide valuable insights into this issue. According to Loucks et al. [122], during the ascent of a volatile-rich magma, zircon and plagioclase coprecipitate, while hydrothermal fluid is exsolving and segregating from the decompressing melt. These processes produce a negative correlation between zircon $(Eu/Eu^*)_N/Yb_N$ (a useful melt hygrometer) and zircon $Ce/\sqrt{(U \times Ti)}$ (a useful indicator of melt oxidation state). Most zircon grains from the Marinho, Queimadas, and Pereiro plutons exhibit Ti concentrations below the detection limit, making it difficult to establish a reliable correlation between $(Eu/Eu^*)_N/Yb_N$ and $Ce/\sqrt{(U \times Ti)}$ ratios (Figure 13B). Two grains from the Marinho Pluton show Ti concentrations above the detection limit along with high $(Eu/Eu^*)_N/Yb_N$ values, forming a negative trend in Figure 13B, suggesting that these zircon grains may have crystallized from a more hydrous melt. The four analyses from the Pereiro granitoids and the six analyses from the Serrote Santo Antonio Pluton, which show Ti concentrations above the detection limit, form flat trends. Zircon grains from the Velho Zuza Pluton display a weak positive correlation between $(Eu/Eu^*)_N/Yb_N$ and $Ce/\sqrt{(U \times Ti)}$ ratios (Figure 13B), indicating crystallization from an H_2O-undersaturated melt.

6.2. Estimation of Crystallization Parameters

6.2.1. Temperature

In previous studies [39,84,88,89], the crystallization temperatures of the studied A-type magmas were determined using the classic amphibole–plagioclase geothermometer, which is based on the plagioclase–amphibole equilibrium [123]. This geothermometer yielded temperatures (Table 1) ranging from 810 to 730 °C for the Aroeiras Complex, 790 to 740 °C for the Pilõezinhos Pluton, 710 to 560 °C for the Serra Branca Suite, 780 to 580 °C for the Bravo Pluton, and 810 to 780 °C for the Açude do Caroá Pluton.

Amphiboles are complex hydrous silicate minerals that can incorporate many elements into their structure, such as aluminum, iron, magnesium, and titanium. The temperature and pressure conditions strongly influence these compositional variations during crystallization. Based on the concentrations of major oxides in amphiboles, Ridolfi et al. [124] and Ridolfi and Rezulli [125] developed geothermometers with an associated error of ≥ 50 °C. The amphibole–liquid thermometer proposed by Putirka [126] enhanced the precision of Ridolfi and Renzulli's [125] geothermometer (± 30 °C). Putirka's thermometer [126] was applied to the studied granitoids; the lowest temperatures were recorded in the Bravo Pluton (730–650 °C), while the Açude do Caroá Pluton granitoids exhibited temperatures as high as 880 °C, the highest calculated by this geothermometer among the studied plutons.

Zr saturation thermometry was performed to infer the liquidus temperatures of the studied magmatic systems. The solubility of zircon in crustal melts is influenced by both melt composition and temperature [127–129], making the concentration of Zr in whole-rock compositions a valuable tool in thermometric studies of magmatic melts. The temperatures obtained using the calibration proposed by Watson and Harrison [126] for the studied plutons are presented in Table 1. Temperatures range from 790 to 690 °C in the Aroeiras Complex and from 950 to 770 °C in the Pilõezinhos Pluton. For the other granitoids, the temperature values fall within the range covered by these two plutons (950–690 °C).

During the zircon analyses, apatite microinclusions were identified within zircon, suggesting that using the apatite saturation thermometer [130] may be more appropriate for determining the liquidus temperatures of the studied magmatic systems. The temperatures obtained using this geothermometer are shown in Table 1. The less evolved granitoids and those enclosing large number of mafic enclaves, with evidence for magma mixing and mingling yielded the highest temperatures (1160–740 °C). On the other hand, granitoids free of mafic enclaves or containing only rare mafic enclaves (the Velho Zuza, Boqueirão, Serra Branca, Serrote Santo Antonio and Pereiro plutons) exhibit a more restricted and somewhat lower temperature range (1000–780 °C).

The bar graph (Figure 14) illustrates the temperature range at which the studied granitic magmas crystallized. The highest temperatures (liquidus) were obtained using the apatite saturation thermometer [130], while the lowest temperatures correspond to those determined by the amphibole–liquid thermometer of Putirka [126]. Because amphibole data are not available for the Serra do Velho Zuza, Serrote Santo Antônio, Serra da Engabelada, and Pereiro plutons, the minimum crystallization temperatures for these granitoids were inferred from the lowest temperature, provided by zircon saturation thermometer with calibration by Watson and Harrison [129].

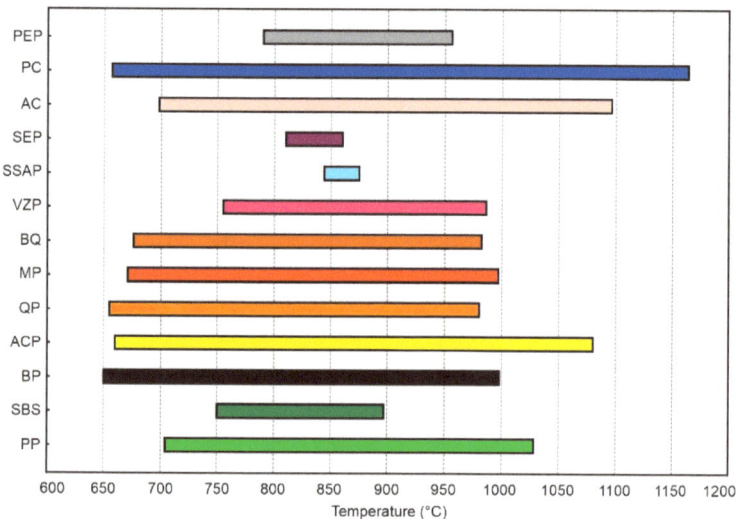

Figure 14. Bar graph illustrating the temperature range at which the studied magmas crystallized. Colors as in the other figures. Abbreviations: PP—Pilõezinhos Pluton; SBS—Serra Branca Suite; BP—Bravo Pluton; ACP—Açude do Caroá Pluton; QP—Queimadas Pluton; MP—Marinho Pluton; BP—Boqueirão Pluton; VZP—Velho Zuza Pluton; SSAP—Serrote Santo Antônio Pluton; SEP—Serra da Engabelaba Pluton; AC—Aroeiras Complex; PC—Prata Complex and PEP—Pereiro Pluton.

6.2.2. Pressure

To estimate the pressure in granitic magmatic systems, the total aluminum-in-hornblende barometer is commonly used [131–133]. However, the proposed calculations do not apply to the granitic rocks from this study, as they exhibit higher temperature ranges than those used in the original calibration (655–700 °C). Furthermore, the studied A-type granites have iron-rich amphiboles with Fe/(Fe + Mg) > 0.65 associated with low fO_2 conditions which typically result in an increase in Al substitution producing unreliable geobarometer results [132]. New equations for the aluminum-in-hornblende geobarometer have been proposed over the past decades, incorporating increasingly refined natural and experimental data with reduced uncertainties [124,125,134]. However, Putirka [126] highlights that most of these aluminum-in-hornblende geobarometers, including the more recent calibrations, must adhere to the conditions outlined by Anderson and Smith [132]. This requirement makes it difficult to apply geobarometers to rocks that crystallized under low fO_2, such as A-type granites.

Mutch et al. [134] developed a geobarometer that combines selected experimental data with a wide range of new experimental results. The calibration, performed at near-solidus temperatures and covering a broad pressure range (0.8 to 10 kbar), applies to a wider range of granitic rocks. We applied this geobarometer to determine the crystallization pressures of the studied granitoids (Table 1). Figure 15 illustrates the pressure range for the crystallization of the studied granitic magmas. The lowest pressures were recorded in the granitoids of the Açude do Caroá Pluton (2.5–4.6 kbar), while the Pilõezinhos Pluton granitoids exhibited the highest pressures (5.2–7.0 kbar). The geobarometer developed by Mutch et al. [134] applies to granitoids containing the mineral assemblage amphibole + plagioclase + biotite + alkali feldspar + quartz + ilmenite/titanite + apatite + magnetite. Considering this mineral assemblage is common in the studied granitoids, the obtained pressures are considered reliable. Additionally, the Al^{IV} and Al^{VI} values of the amphiboles analyzed in this study fall within the pressure range obtained through the geobarometer of Mutch et al. [134]. The large pressure variations observed within some individual plutons can be explained by the crystallization of amphibole grains at greater depths, subsequently transported by the magma to shallower levels where the rest of the pluton crystallized.

Table 1. Estimation of temperature and pressure for the studied plutons. Abbreviations: PP—Pilõezinhos Pluton; SBS—Serra Branca Suite; BP—Bravo Pluton; ACP—Açude do Caroá Pluton; QP—Queimadas Pluton; MP—Marinho Pluton; BP—Boqueirão Pluton; VZP—Velho Zuza Pluton; SSAP—Serrote Santo Antônio Pluton; SEP—Serra da Engabelaba Pluton; AC—Aroeiras Complex; PC—Prata Complex and PEP—Pereiro Pluton.

Plutons	PP		SBS		BP		ACP		QP		MP		BQ	
Temperature (°C)	Min	Max	Min	Max	Min	Max	Min	Max	Min	Max	Min	Max	Min	Max
Blund and Holland [123]	745	795	560	711	581	785	776	812	-	-	-	-	-	-
Putirka [126]	704	758	750	789	650	730	660	876	655	736	671	802	676	691
Watson and Harrison [129]	772	946	743	838	847	893	804	826	752	903	795	930	857	857
Harrison and Watson [130]	916	1028	784	897	882	998	1037	1080	757	980	865	997	982	982
Pressure (kbar)	Min	Max	Min	Max	Min	Max	Min	Max	Min	Max	Min	Max	Min	Max
Mutch et al. [134]	5.2	7.0	5.98	6.45	4.53	6.39	2.5	4.6	3.85	6.72	3.94	5.38	4.29	5.21

Plutons	VZP		SSAP		SEP		AC		PC		PEP	
Temperature (°C)	Min	Max	Min	Max	Min	Max	Min	Max	Min	Max	Min	Max
Blund and Holland [123]	-	-	-	-	-	-	733	809	-	-	-	-
Putirka [126]	-	-	-	-	-	-	698	740	657	769	-	-
Watson and Harrison [129]	755	895	844	844	810	810	687	790	804	916	790	878
Watson and Harrison [130]	790	986	875	875	860	860	804	1096	714	1164	851	956
Pressure (kbar)	Min	Max	Min	Max	Min	Max	Min	Max	Min	Max	Min	Max
Mutch et al. [134]	-	-	-	-	-	-	4.01	5.75	-	-	-	-

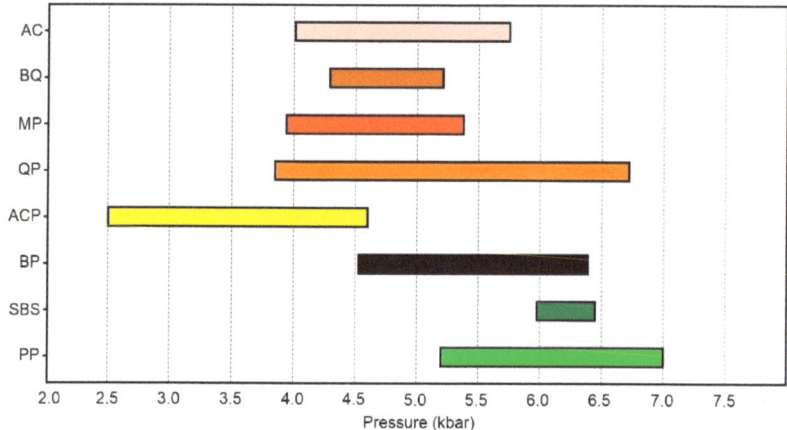

Figure 15. Bar graph illustrating the crystallization pressures of the studied granitoids. PP—Pilõezinhos Pluton; SBS—Serra Branca Suite; BP—Bravo Pluton; ACP—Açude do Caroá Pluton; QP—Queimadas Pluton; MP—Marinho Pluton; BQ—Boqueirão Pluton and AC—Aroeiras Complex.

6.2.3. Oxygen Fugacity

The compositions of biotite, amphibole, and Fe-Ti oxides were also analyzed to estimate the oxygen fugacity (fO_2) of the studied granitoids, as these minerals are highly sensitive to variations of fO_2. Previous studies have demonstrated that the composition of these phases can be effectively used as proxies for assessing oxygen fugacity in magmatic systems [29,135,136]. The amphiboles in the studied granitoids show a slight variation in Fe# [Fe/(Fe + Mg)], with most falling within the range of 0.7 to 0.9. The amphiboles are thus iron-rich, and their Fe# values are characteristic of crystallization under low to intermediate fO_2 conditions (Figure 16A). The Fe/(Fe + Mg) values of the studied biotites follow the same patterns observed in the amphibole compositions (Figure 16B). Most of the studied biotite have Fe# > 0.65, similar to those reported by Anderson et al. [137] for biotite of granites that originated from magmas crystallizing under low fO_2. The composition of the mafic minerals aligns with the overall geochemical signature of these rocks, which predominantly belong to the ferroan series of Frost et al. [19]. This series is typically associated with magmas that evolved under reducing conditions, characteristic of the ilmenite series granites described by Ishihara [85]. It is important to note that some of the analyzed biotites and amphiboles exhibit low Fe# values; these were obtained from Açude do Caroá diorites and the less evolved facies of the Prata Complex, as well as the Pilõezinhos and Marinho plutons. As they are more magnesian, these samples plot in the field associated with minerals crystallized under more oxidizing conditions, or that crystallized from a magma of more basaltic composition (dioritic) and higher temperature, which agree with the conspicuous presence of mafic enclaves. The granitoids of the Pilõezinhos and Serra Branca plutons are the only granitoids among the ones studied, with reported detailed studies of opaque minerals [88,89]. In these granitoids, the Fe-Ti oxides are predominantly primary ilmenite crystals, which agree with the mineral and whole-rock chemistry data reported in the literature. These integrated data suggest that the A-type granites of the Central Subprovince of the Borborema Province crystallized from magmas that evolved under low fO_2 conditions, similar to most A-type granites worldwide.

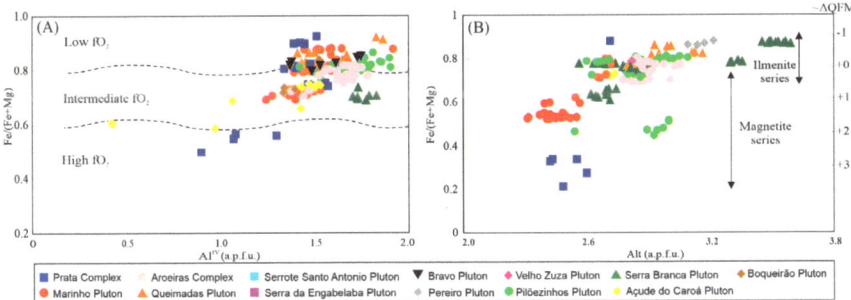

Figure 16. Chemical diagrams for inferring fO$_2$ conditions. (**A**) Al^{IV} × Fe/(Fe + Mg) diagram for the studied amphiboles, with fields according to Anderson and Smith [132]; (**B**) Al$_T$ (total aluminum) × Fe/(Fe + Mg) diagram for the studied biotites.

6.3. Post-Magmatic Processes

The zircon REE patterns characterized by positive Ce anomalies and negative Eu anomalies and high Th/U ratios recorded in zircon grains from all studied plutons are characteristics of igneous zircon. However, relatively high LREE abundances were recorded in zircon grains from Marinho, Serra do Velho Zuza, and Queimadas plutons (Figures 11 and 12), associated with high Y abundances and a positive correlation between ΣLREE and Y, suggesting zircon alteration by LREE- and HFSE-enriched late hydrothermal fluids or the precipitation of zircon from these fluids. The presence of hydrothermal zircon is evident in the projection of the zircon compositions in the (Sm/La)$_N$ versus (Ce/Ce*)$_N$ bivariate discriminant plot (Figure 17) with magmatic and hydrothermal zircon fields after Hoskin [115]. This bivariate plot shows clearly that the zircon grains from the Pereiro granites and some from the Serrote Santo Antonio granites are the least hydrothermaly altered zircon grains, while most of the analyzed zircon grains fall within the hydrothermal zircon field (Figure 17). The presence of zircon grains with both higher- and lower-LREE abundances in the same rock, suggests that the zircon LREE and HFSE enrichment involved both alteration of previously crystallized grains by hydrothermal fluids and zircon crystallization from these fluids. Our data agree with earlier studies showing that hydrothermal alteration of zircon is a common process in evolved to highly evolved A$_2$-type granites [118,119,121,138].

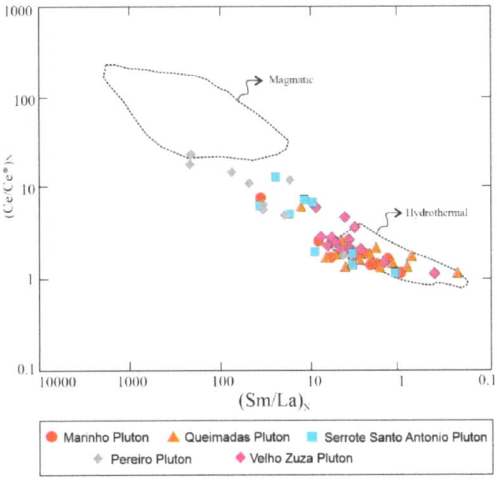

Figure 17. Bivariate discriminant plot (Sm/La)$_N$ × (Ce/Ce*)$_N$ of the analyzed zircon crystals.

7. Conclusions

Based on a thorough review of available mineral chemistry and whole-rock geochemistry data from 13 plutons and a dike swarm, including some complementary data, along with new zircon trace element data from five of those plutons, crystallization conditions and post-magmatic processes and their implications for the genesis of Ediacaran–Cambrian A-Type granites from the Central Subprovince of the Borborema Province were evaluated.

These granitoids form small bodies, except for the larger Prata Complex, and were intruded into Paleoproterozoic gneisses and migmatites or Neoproterozoic supracrustals between approximately 580 and 525 Ma, succeeding large intrusions of post-collisional high-K granitoids (590–580 Ma). Several intrusions contain numerous mafic enclaves indicating magma mixing and mingling as an important component in their formation. This is supported by the wide compositional variation in mafic minerals within some of the plutons, revealing that interactions between felsic and mafic magmas played a significant role in the magmatic evolution of those plutons. The mineral assemblages of these granitoids, the chemical composition of the mafic minerals, including siderophyllite-rich biotite and iron-rich calcic amphiboles, the sub-alkaline signature of the biotites, and the zircon trace element composition, align with the whole-rock chemistry. These characteristics classify them as ferroan, alkali-calcic, and dominantly metaluminous granitoids, which are typical features of A-type granites. The crystallization parameters derived from mineral chemistry show temperatures ranging from 1160 to 650 °C in the granitoids containing numerous mafic enclaves, and from 990 to 680 °C in the more evolved granitoids. Crystallization pressures, determined using the Al-in-amphibole geobarometer, mostly range from ~4.0 to 7.0 kbar, corresponding to crustal depths of ~12 to 21 km. The mineral assemblages characterized by Fe-rich mafic phases and ilmenite show that the studied A-type granites crystallized under low fO_2 conditions, akin to classical ferroan A-type granites. We highlight that the zircon trace elements indicate hydrothermal alteration in some of the studied granites, which appears to be a common feature of high-SiO_2 A-type granites. Most of the studied A-type granites intruded within or near shear zones. Thus, the post-magmatic hydrothermal alteration observed in zircon crystals may be associated with the reactivation of these shear zones.

Supplementary Materials: The following supporting information can be downloaded at: https://www.mdpi.com/article/10.3390/min14101022/s1. Table SI: Whole-rock geochemistry data; Table SII: Mineral chemistry data; Table SIII: Zircon trace element composition.

Author Contributions: Conceptualization, J.V.d.L. and I.d.P.G.; methodology, J.V.d.L. and I.d.P.G.; software, J.V.A.d.A., C.C.G.B., and L.d.S.; validation, J.V.d.L., I.d.P.G., J.V.A.d.A., and A.F.d.S.F.; formal analysis, I.d.P.G. and J.V.d.L.; investigation, J.V.d.L., I.d.P.G., J.V.A.d.A., and C.C.G.B.; resources, I.d.P.G.; data curation, J.V.d.L. and C.C.G.B.; writing—original draft preparation, J.V.d.L. and I.d.P.G.; writing—review and editing, J.V.d.L., I.d.P.G., J.V.A.d.A., L.d.S., and A.F.d.S.F.; supervision, J.V.d.L. and I.d.P.G.; project administration, I.d.P.G. All authors have read and agreed to the published version of the manuscript.

Funding: This research received no external funding.

Data Availability Statement: The original contributions presented in the study are included in the article/supplementary material, further inquiries can be directed to the corresponding authors.

Acknowledgments: We would like to express our heartfelt gratitude to the three anonymous reviewers for their insightful comments and constructive feedback, which significantly enhanced the quality of our manuscript.

Conflicts of Interest: The authors declare that they have no conflicts of interest.

References

1. Wedepohl, K.H. Chemical Composition and Fractionation of the Continental Crust. *Geol. Rundsch.* **1991**, *80*, 207–223. [CrossRef]
2. Rudnick, R.L.; Gao, S. Composition of the Continental Crust. In *Treatise on Geochemistry*; Elsevier: Amsterdam, The Netherlands, 2003; Volume 3, pp. 1–64.
3. Taylor, S.R.; McLennan, S.M. The Geochemical Evolution of the Continental Crust. *Rev. Geophys.* **1995**, *33*, 241–265. [CrossRef]

4. Kemp, A.I.S.; Hawkesworth, C.J.; Foster, G.L.; Paterson, B.A.; Woodhead, J.D.; Hergt, J.M.; Gray, C.M.; Whitehouse, M.J. Magmatic and Crustal Differentiation History of Granitic Rocks from Hf-O Isotopes in Zircon. *Science* **2007**, *315*, 980–983. [CrossRef]
5. Barbarin, B. A Review of the Relationships between Granitoid Types, Their Origins and Their Geodynamic Environments. *Lithos* **1999**, *46*, 605–626. [CrossRef]
6. Brown, M. Granite: From Genesis to Emplacement. *Bull. Geol. Soc. Am.* **2013**, *125*, 1079–1113. [CrossRef]
7. Chappell, B.W.; White, A.J.R. I- and S-Type Granites in the Lachlan Fold Belt. *Earth Environ. Sci. Trans. R. Soc. Edinb.* **1992**, *83*, 1–26. [CrossRef]
8. Hawkesworth, C.J.; Kemp, A.I.S. Using Hafnium and Oxygen Isotopes in Zircons to Unravel the Record of Crustal Evolution. *Chem. Geol.* **2006**, *226*, 144–162. [CrossRef]
9. Iles, K.A.; Hergt, J.M.; Woodhead, J.D.; Ickert, R.B.; Williams, I.S. Petrogenesis of Granitoids from the Lachlan Fold Belt, Southeastern Australia: The Role of Disequilibrium Melting. *Gondwana Res.* **2020**, *79*, 87–109. [CrossRef]
10. Jacob, J.-B.; Moyen, J.-F.; Fiannacca, P.; Laurent, O.; Bachmann, O.; Janoušek, V.; Farina, F.; Villaros, A. Crustal Melting vs. Fractionation of Basaltic Magmas: Part 2, Attempting to Quantify Mantle and Crustal Contributions in Granitoids. *Lithos* **2021**, *402–403*, 106292. [CrossRef]
11. Moyen, J.-F.; Janoušek, V.; Laurent, O.; Bachmann, O.; Jacob, J.-B.; Farina, F.; Fiannacca, P.; Villaros, A. Crustal Melting vs. Fractionation of Basaltic Magmas: Part 1, Granites and Paradigms. *Lithos* **2021**, *402–403*, 106291. [CrossRef]
12. Vigneresse, J.L. A New Paradigm for Granite Generation. *Earth Environ. Sci. Trans. R. Soc. Edinb.* **2004**, *95*, 11–22. [CrossRef]
13. Sawyer, E.W.; Cesare, B.; Brown, M. When the Continental Crust Melts. *Elements* **2011**, *7*, 229–234. [CrossRef]
14. Hawkesworth, C.; Cawood, P.; Kemp, T.; Storey, C.; Dhuime, B. A Matter of Preservation. *Science* **2009**, *323*, 49–50. [CrossRef]
15. Cawood, P.A.; Chowdhury, P.; Mulder, J.A.; Hawkesworth, C.J.; Capitanio, F.A.; Gunawardana, P.M.; Nebel, O. Secular Evolution of Continents and the Earth System. *Rev. Geophys.* **2022**, *60*, e2022RG000789. [CrossRef]
16. Laurent, O.; Martin, H.; Moyen, J.F.; Doucelance, R. The Diversity and Evolution of Late-Archean Granitoids: Evidence for the Onset of "Modern-Style" Plate Tectonics between 3.0 and 2.5 Ga. *Lithos* **2014**, *205*, 208–235. [CrossRef]
17. Condie, K.C.; Pisarevsky, S.A.; Puetz, S.J.; Roberts, N.M.W.; Spencer, C.J. A-Type Granites in Space and Time: Relationship to the Supercontinent Cycle and Mantle Events. *Earth Planet Sci. Lett.* **2023**, *610*, 118125. [CrossRef]
18. Bonin, B.; Janoušek, V.; Moyen, J.-F. Chemical Variation, Modal Composition and Classification of Granitoids. *Geol. Soc. Lond. Spec. Publ.* **2020**, *491*, 9–51. [CrossRef]
19. Frost, B.R.; Barnes, C.G.; Collins, W.J.; Arculus, R.J.; Ellis, D.J.; Frost, C.D. A Geochemical Classification for Granitic Rocks. *J. Petrol.* **2001**, *42*, 2033–2048. [CrossRef]
20. Gao, P.; Zheng, Y.; Zhao, Z. Experimental Melts from Crustal Rocks: A Lithochemical Constraint on Granite Petrogenesis. *Lithos* **2016**, *266–267*, 133–157. [CrossRef]
21. Loiselle, M.C.; Wones, D.R. Characteristics and Origin of Anorogenic Granites. *Geol. Soc. Am. Abstr. Program* **1979**, *11*, 468.
22. Frost, C.D.; Frost, B.R. On Ferroan (A-Type) Granitoids: Their Compositional Variability and Modes of Origin. *J. Petrol.* **2011**, *52*, 39–53. [CrossRef]
23. Bonin, B. A-Type Granites and Related Rocks: Evolution of a Concept, Problems and Prospects. *Lithos* **2007**, *97*, 1–29. [CrossRef]
24. Johansson, Å. A Tentative Model for the Origin of A-Type Granitoids. *Minerals* **2023**, *13*, 236. [CrossRef]
25. Collins, W.J.; Huang, H.-Q.; Bowden, P.; Kemp, A.I.S. Repeated S–I–A-Type Granite Trilogy in the Lachlan Orogen and Geochemical Contrasts with A-Type Granites in Nigeria: Implications for Petrogenesis and Tectonic Discrimination. *Geol. Soc. Lond. Spec. Publ.* **2020**, *491*, 53–76. [CrossRef]
26. Collins, W.J.; Beams, S.D.; White, A.J.R.; Chappell, B.W. Nature and Origin of A-Type Granites with Particular Reference to Southeastern Australia. *Contrib. Miner. Petrol.* **1982**, *80*, 189–200. [CrossRef]
27. Creaser, R.A.; Price, R.C.; Wormald, R.J. A-Type Granites Revisited: Assessment of a Residual-Source Model. *Geology* **1991**, *19*, 163–166. [CrossRef]
28. Eby, G.N. The A-Type Granitoids: A Review of Their Occurrence and Chemical Characteristics and Speculations on Their Petrogenesis. *Lithos* **1990**, *26*, 115–134. [CrossRef]
29. Dall'Agnol, R.; de Oliveira, D.C. Oxidized, Magnetite-Series, Rapakivi-Type Granites of Carajás, Brazil: Implications for Classification and Petrogenesis of A-Type Granites. *Lithos* **2007**, *93*, 215–233. [CrossRef]
30. King, P.L.; White, A.J.R.; Chappell, B.W.; Allen, C.M. Characterization and Origin of Aluminous A-Type Granites from the Lachlan Fold Belt, Southeastern Australia. *J. Petrol.* **1997**, *38*, 371–391. [CrossRef]
31. Sylvester, P.J. *Post-Collisional Strongly Peraluminous Granites*; Elsevier: Amsterdam, The Netherlands, 1998; Volume 45.
32. Ferré, E.C.; Caby, R.; Peucat, J.J.; Capdevila, R.; Monié, P. Pan-African, Post-Collisional, Ferro-Potassic Granite and Quartz–Monzonite Plutons of Eastern Nigeria. *Lithos* **1998**, *45*, 255–279. [CrossRef]
33. Goodenough, K.M.; Lusty, P.A.J.; Roberts, N.M.W.; Key, R.M.; Garba, A. Post-Collisional Pan-African Granitoids and Rare Metal Pegmatites in Western Nigeria: Age, Petrogenesis, and the 'Pegmatite Conundrum'. *Lithos* **2014**, *200–201*, 22–34. [CrossRef]
34. Eby, G.N. Chemical Subdivision of the A-Type Granitoids:Petrogenetic and Tectonic Implications. *Geology* **1992**, *20*, 641. [CrossRef]
35. Amorim, J.V.A.d.; de Pinho Guimarães, I.; Farias, D.J.S.; Lima, J.V.d.; Santos, L.; Ribeiro, V.B.; Brainer, C. Late-Neoproterozoic Ferroan Granitoids of the Transversal Subprovince, Borborema Province, NE Brazil: Petrogenesis and Geodynamic Implications. *Int. Geol. Rev.* **2019**, *61*, 1745–1767. [CrossRef]

36. Almeida, C.N.; Guimarães, I.P.; Silva Filho, A.F. A-Type Post-Collisional Granites in the Borborema Province—NE Brazil: The Queimadas Pluton. *Gondwana Res.* **2002**, *5*, 667–681. [CrossRef]
37. Guimarães, I.P.; Silva Filho, A.F.; Melo, S.C.; Macambira, M.B. Petrogenesis of A-Type Granitoids from the Alto Moxoto and Alto Pajeu Terranes of the Borborema Province, Ne Brazil: Constraints from Geochemistry and Isotopic Composition. *Gondwana Res.* **2005**, *8*, 347–362. [CrossRef]
38. Lima, J.V.; Guimarães, I.d.P.; Santos, L.; Amorim, J.V.A.; Farias, D.J.S. Geochemical and Isotopic Characterization of the Granitic Magmatism along the Remígio—Pocinhos Shear Zone, Borborema Province, NE Brazil. *J. S. Am. Earth Sci.* **2017**, *75*, 116–133. [CrossRef]
39. Lages, G.d.A.; Marinho, M.d.S.; Nascimento, M.A.L.d.; Medeiros, V.C.d.; Dantas, E.L. Geocronologia e Aspectos Estruturais e Petrológicos Do Pluton Bravo, Domínio Central Da Província Borborema, Nordeste Do Brasil: Um Granito Transalcalino Precoce No Estágio Pós-Colisional Da Orogênese Brasiliana. *Braz. J. Geol.* **2016**, *46*, 41–61. [CrossRef]
40. Hollanda, M.H.B.M.; Archanjo, C.J.; Souza, L.C.; Armstrong, R.; Vasconcelos, P.M. Cambrian Mafic to Felsic Magmatism and Its Connections with Transcurrent Shear Zones of the Borborema Province (NE Brazil): Implications for the Late Assembly of the West Gondwana. *Precambrian Res.* **2010**, *178*, 1–14. [CrossRef]
41. Van Schmus, W.R.; de Brito Neves, B.B.; Hackspacher, P.; Babinski, M. U-Pb and Sm-Nd Geochronologic Studies of the Eastern Borborema Province, Northeastern Brazil: Initial Conclusions. *J. S. Am. Earth Sci.* **1995**, *8*, 267–288. [CrossRef]
42. Caxito, F.d.A.; Santos, L.C.M.d.L.; Ganade, C.E.; Bendaoud, A.; Fettous, E.-H.; Bouyo, M.H. Toward an Integrated Model of Geological Evolution for NE Brazil-NW Africa: The Borborema Province and Its Connections to the Trans-Saharan (Benino-Nigerian and Tuareg Shields) and Central African Orogens. *Braz. J. Geol.* **2020**, *50*. [CrossRef]
43. Van Schmus, W.R.; Oliveira, E.P.; Silva Filho, A.F.; Toteu, S.F.; Penaye, J.; Guimarães, I.P. Proterozoic Links between the Borborema Province, NE Brazil, and the Central African Fold Belt. *Geol. Soc. Lond. Spec. Publ.* **2008**, *294*, 69–99. [CrossRef]
44. Toteu, S.F.; Van Schmus, W.R.; Penaye, J.; Michard, A. New U–Pb and Sm–Nd Data from North-Central Cameroon and Its Bearing on the Pre-Pan African History of Central Africa. *Precambrian Res.* **2001**, *108*, 45–73. [CrossRef]
45. Van Schmus, W.R.; Kozuch, M.; Brito Neves, B.B. Precambrian History of the Zona Transversal of the Borborema Province, NE Brazil: Insights from Sm–Nd and U–Pb Geochronology. *J S. Am. Earth Sci.* **2011**, *31*, 227–252. [CrossRef]
46. Ferreira, A.C.D.; Dantas, E.L.; Fuck, R.A.; Nedel, I.M. Arc Accretion and Crustal Reworking from Late Archean to Neoproterozoic in Northeast Brazil. *Sci. Rep.* **2020**, *10*, 7855. [CrossRef] [PubMed]
47. Lira Santos, L.C.M.; Lages, G.A.; Caxito, F.A.; Dantas, E.L.; Cawood, P.A.; Lima, H.M.; da Cruz Lima, F.J. Isotopic and Geochemical Constraints for a Paleoproterozoic Accretionary Orogen in the Borborema Province, NE Brazil: Implications for Reconstructing Nuna/Columbia. *Geosci. Front.* **2022**, *13*, 101167. [CrossRef]
48. Dantas, E.L.; Hackspacker, P.C.; Van Schmus, W.R.V.; Brito Neves, B.B. Archean Accretion in the Sao Jose Do Campestre Massif, Borborema Province, Northeast Brazil. *Rev. Bras. Geociências* **1998**, *28*, 221–228. [CrossRef]
49. Guimarães, I.P.; Van Schmus, W.R.; Brito Neves, B.B.; Bretas Bittar, S.M.; Silva Filho, A.F.; Armstrong, R. U–Pb Zircon Ages of Orthogneisses and Supracrustal Rocks of the Cariris Velhos Belt: Onset of Neoproterozoic Rifting in the Borborema Province, NE Brazil. *Precambrian Res.* **2012**, *192–195*, 52–77. [CrossRef]
50. Neves, S.P. Constraints from Zircon Geochronology on the Tectonic Evolution of the Borborema Province (NE Brazil): Widespread Intracontinental Neoproterozoic Reworking of a Paleoproterozoic Accretionary Orogen. *J. S. Am. Earth Sci.* **2015**, *58*, 150–164. [CrossRef]
51. Hollanda, M.H.B.M.; Archanjo, C.J.; Bautista, J.R.; Souza, L.C. Detrital Zircon Ages and Nd Isotope Compositions of the Seridó and Lavras Da Mangabeira Basins (Borborema Province, NE Brazil): Evidence for Exhumation and Recycling Associated with a Major Shift in Sedimentary Provenance. *Precambrian Res.* **2015**, *258*, 186–207. [CrossRef]
52. Oliveira, E.P.; McNaughton, N.J.; Windley, B.F.; Carvalho, M.J.; Nascimento, R.S. Detrital Zircon U–Pb Geochronology and Whole-Rock Nd-Isotope Constraints on Sediment Provenance in the Neoproterozoic Sergipano Orogen, Brazil: From Early Passive Margins to Late Foreland Basins. *Tectonophysics* **2015**, *662*, 183–194. [CrossRef]
53. Araujo, C.E.G.; Cordani, U.G.; Basei, M.A.S.; Castro, N.A.; Sato, K.; Sproesser, W.M. U–Pb Detrital Zircon Provenance of Metasedimentary Rocks from the Ceará Central and Médio Coreaú Domains, Borborema Province, NE-Brazil: Tectonic Implications for a Long-Lived Neoproterozoic Active Continental Margin. *Precambrian Res.* **2012**, *206–207*, 36–51. [CrossRef]
54. Santos, T.J.S.; Fetter, A.H.; Hackspacher, P.C.; Van Schmus, W.R.; Nogueira Neto, J.A. Neoproterozoic Tectonic and Magmatic Episodes in the NW Sector of Borborema Province, NE Brazil, during Assembly of Western Gondwana. *J. S. Am. Earth Sci.* **2008**, *25*, 271–284. [CrossRef]
55. Sial, A.N.; Ferreira, V.P. Magma Associations in Ediacaran Granitoids of the Cachoeirinha–Salgueiro and Alto Pajeú Terranes, Northeastern Brazil: Forty Years of Studies. *J. S. Am. Earth Sci.* **2016**, *68*, 113–133. [CrossRef]
56. Oliveira, E.P.; Bueno, J.F.; McNaughton, N.J.; Silva Filho, A.F.; Nascimento, R.S.; Donatti-Filho, J.P. Age, Composition, and Source of Continental Arc- and Syn-Collision Granites of the Neoproterozoic Sergipano Belt, Southern Borborema Province, Brazil. *J. S. Am. Earth Sci.* **2015**, *58*, 257–280. [CrossRef]
57. Nascimento, M.A.L.; Galindo, A.C.; Medeiros, V.C. Ediacaran to Cambrian Magmatic Suites in the Rio Grande Do Norte Domain, Extreme Northeastern Borborema Province (NE of Brazil): Current Knowledge. *J. S. Am. Earth Sci.* **2015**, *58*, 281–299. [CrossRef]

58. Guimarães, I.P.; Silva Filho, A.F.; Almeida, C.N.; Van Schmus, W.R.; Araújo, J.M.M.; Melo, S.C.; Melo, E.B. Brasiliano (Pan-African) Granitic Magmatism in the Pajeú-Paraíba Belt, Northeast Brazil: An Isotopic and Geochronological Approach. *Precambrian Res.* **2004**, *135*, 23–53. [CrossRef]
59. Silva Filho, A.F.; Guimarães, I.d.P.; Santos, L.; Armstrong, R.; Van Schmus, W.R. Geochemistry, U–Pb Geochronology, Sm–Nd and O Isotopes of ca. 50 Ma Long Ediacaran High-K Syn-Collisional Magmatism in the Pernambuco Alagoas Domain, Borborema Province, NE Brazil. *J. S. Am. Earth Sci.* **2016**, *68*, 134–154. [CrossRef]
60. Ferreira, V.P.; Silva, T.R.d.; Lima, M.M.C.d.; Sial, A.N.; Silva Filho, A.F.; Neves, C.H.F.S.; Tchouankoue, J.P.; Lima, S.S. Tracing a Neoproterozoic Subduction to Collision Magmatism in the Eastern Pernambuco–Alagoas Domain, Northeastern Brazil. *J. S. Am. Earth Sci.* **2023**, *128*, 104457. [CrossRef]
61. Neves, S.P.; Tommasi, A.; Vauchez, A.; Carrino, T.A. The Borborema Strike-Slip Shear Zone System (NE Brazil): Large-Scale Intracontinental Strain Localization in a Heterogeneous Plate. *Lithosphere* **2021**, *2021*, 6407232. [CrossRef]
62. Caby, R. *Precambrian Terranes of Benin-Nigeria and Northeast Brazil and the Late Proterozoic South Atlantic Fit*; Terranes in the Circum-Atlantic Paleozoic Orogens: Boulder, CO, USA, 1989; pp. 145–158.
63. Vauchez, A.; Neves, S.P.; Caby, R.; Corsini, M.; Egydio-Silva, M.; Arthaud, M.; Amaro, V. The Borborema Shear Zone System, NE Brazil. *J. S. Am. Earth Sci.* **1995**, *8*, 247–266. [CrossRef]
64. Araujo, C.E.G.; Weinberg, R.F.; Caxito, F.A.; Lopes, L.B.L.; Tesser, L.R.; Costa, I.S. Decratonization by Rifting Enables Orogenic Reworking and Transcurrent Dispersal of Old Terranes in NE Brazil. *Sci. Rep.* **2021**, *11*, 5719. [CrossRef]
65. Archanjo, C.J.; Hollanda, M.H.B.M.; Rodrigues, S.W.O.; Brito Neves, B.B.B.; Armstrong, R. Fabrics of Pre- and Syntectonic Granite Plutons and Chronology of Shear Zones in the Eastern Borborema Province, NE Brazil. *J. Struct. Geol.* **2008**, *30*, 310–326. [CrossRef]
66. Santos, L.C.M.d.L.; Dantas, E.L.; Cawood, P.A.; Lages, G.d.A.; Lima, H.M.; dos Santos, E.J.; Caxito, F.A. Early to Late Neoproterozoic Subduction-Accretion Episodes in the Cariris Velhos Belt of the Borborema Province, Brazil: Insights from Isotope and Whole-Rock Geochemical Data of Supracrustal and Granitic Rocks. *J. S. Am. Earth Sci.* **2019**, *96*, 102384. [CrossRef]
67. Santos, L.C.M.d.L.; Dantas, E.L.; Cawood, P.A.; Lages, G.d.A.; Lima, H.M.; dos Santos, E.J. The Cariris Velhos Tectonic Event in Northeast Brazil. *J. S. Am. Earth Sci.* **2010**, *29*, 61–76. [CrossRef]
68. Lira Santos, L.C.M.; Dantas, E.L.; Cawood, P.A.; Lages, G.d.A.; Lima, H.M.; Santos, E.J. Accretion Tectonics in Western Gondwana Deduced From Sm-Nd Isotope Mapping of Terranes in the Borborema Province, NE Brazil. *Tectonics* **2018**, *37*, 2727–2743. [CrossRef]
69. Santos, L.C.M.D.L.; de Oliveira, R.G.; Lages, G.d.A.; Dantas, E.L.; Caxito, F.; Cawood, P.A.; Fuck, R.A.; Lima, H.M.; Santos, G.L.; Neto, J.F.d.A. Evidence for Neoproterozoic Terrane Accretion in the Central Borborema Province, West Gondwana Deduced by Isotopic and Geophysical Data Compilation. *Int. Geol. Rev.* **2022**, *64*, 1574–1593. [CrossRef]
70. Brito Neves, B.B.; Santos, E.J.d.; Fuck, R.A.; Santos, L.C.M.L. A Preserved Early Ediacaran Magmatic Arc at the Northernmost Portion of the Transversal Zone Central Subprovince of the Borborema Province, Northeastern South America. *Braz. J. Geol.* **2016**, *46*, 491–508. [CrossRef]
71. Neves, S.P. Proterozoic History of the Borborema Province (NE Brazil): Correlations with Neighboring Cratons and Pan-African Belts and Implications for the Evolution of Western Gondwana. *Tectonics* **2003**, *22*. [CrossRef]
72. Mariano, G.; Neves, S.P.; Siva Filho, A.F.; Guimaraes, I.P. Diorites of the High-K Calc-Alkalic Association: Geochemistry and Sm-Nd Data and Implications for the Evolution of the Borborema Province, Northeast Brazil. *Int. Geol. Rev.* **2001**, *43*, 921–929. [CrossRef]
73. Lima, J.V.d.; Guimarães, I.d.P.; Neves, S.P.; Basei, M.A.S.; Silva Filho, A.F.; Brainer, C.C.G. Post-Collisional, High-Ba-Sr Teixeira Batholith Granites: Evidence for Recycling of Paleoproterozoic Crust in the Alto Pajeú Domain, Borborema Province—NE-Brazil. *Lithos* **2021**, *404–405*, 106469. [CrossRef]
74. Hollanda, M.H.B.M.; Pimentel, M.M.; Jardim de Sá, E.F. Paleoproterozoic Subduction-Related Metasomatic Signatures in the Lithospheric Mantle beneath NE Brazil: Inferences from Trace Element and Sr–Nd–Pb Isotopic Compositions of Neoproterozoic High-K Igneous Rocks. *J. S. Am. Earth Sci.* **2003**, *15*, 885–900. [CrossRef]
75. Neves, S.P.; Bruguier, O.; Vauchez, A.; Bosch, D.; Silva, J.M.R.d.; Mariano, G. Timing of Crust Formation, Deposition of Supracrustal Sequences, and Transamazonian and Brasiliano Metamorphism in the East Pernambuco Belt (Borborema Province, NE Brazil): Implications for Western Gondwana Assembly. *Precambrian Res.* **2006**, *149*, 197–216. [CrossRef]
76. Santos, T.J.S.; Fetter, A.H.; Neto, J.A.N. Comparisons between the Northwestern Borborema Province, NE Brazil, and the Southwestern Pharusian Dahomey Belt, SW Central Africa. *Geol. Soc. Lond. Spec. Publ.* **2008**, *294*, 101–120. [CrossRef]
77. Araújo, M.N.C.; Vasconcelos, P.M.; Alves da Silva, F.C.; Jardim de Sá, E.F.; Sá, J.M. 40Ar/39Ar Geochronology of Gold Mineralization in Brasiliano Strike-Slip Shear Zones in the Borborema Province, NE Brazil. *J. S. Am. Earth Sci.* **2005**, *19*, 445–460. [CrossRef]
78. Araújo Neto, J.F.d.; Santos, L.C.M.d.L.; Viegas, G.; Souza, C.P.d.; Miggins, D.; Cawood, P.A. Structural and Geochronological Constraints on the Portalegre Shear Zone: Implications for Emerald Mineralization in the Borborema Province, Brazil. *J. Struct. Geol.* **2023**, *174*, 104921. [CrossRef]
79. Lira Santos, L.C.M.; Dantas, E.L.; Cawood, P.A.; José dos Santos, E.; Fuck, R.A. Neoarchean Crustal Growth and Paleoproterozoic Reworking in the Borborema Province, NE Brazil: Insights from Geochemical and Isotopic Data of TTG and Metagranitic Rocks of the Alto Moxotó Terrane. *J. S. Am. Earth Sci.* **2017**, *79*, 342–363. [CrossRef]

80. Santos, L.; Guimarães, I.P.; Silva Filho, A.F.; Farias, D.J.S.; Lima, J.V.; Antunes, J. V Magmatismo Ediacarano Extensional Na Província Borborema, NE Brasil: Pluton Serra Branca. *Comun. Geológicas* **2014**, *101*, 199–203.
81. Miranda, A.W.A. Evolução Estrutural das zonas de Cisalhamento Dúcteis na Porção Centro-Leste do Domínio da Zona Transversal na Província Borborema. Ph.D. Thesis, Universidade do Estado do Rio de Janeiro, RiodeJaneiro, Brazil, 2010.
82. Guimarães, I.P.; Silva Filho, A.F.; Lima, J.V.; Farias, D.J.S.; Amorim, J.V.A.; Silva, F.M.J. V Geocronologia U-Pb e Caracterização Geoquímica Dos Granitoides Do Pluton Marinho, Província Borborema. In Proceedings of the 46° Congresso Brasileiro de Geologia Anais Proceedings, Santos, Brazil, 5 October 2012.
83. Silva, J.M.R.D.; Mariano, G. Geometry and Kinematics of the Afogados Da Ingazeira Shear Zone, Northeast Brazil. *Int. Geol. Rev.* **2000**, *42*, 86–95. [CrossRef]
84. Melo, E.B. de Petrologia e Geoquimica de Granitoides a Sudeste da zona de Cisalhamento Afogados da Ingazeira—PE Nordeste do Brasil. Ph.D. Thesis, Universidade Federal de Pernambuco, Recife, Brazil, 2004.
85. Ishihara, S. The Magnetite-Series and Ilmenite-Series Granitic Rocks. *Min. Geol.* **1977**, *27*, 293–305.
86. Pearce, J.A.; Harris, N.B.W.; Tindle, A.G. Trace Element Discrimination Diagrams for the Tectonic Interpretation of Granitic Rocks. *J. Petrol.* **1984**, *25*, 956–983. [CrossRef]
87. Whalen, J.B.; Currie, K.L.; Chappell, B.W. A-Type Granites: Geochemical Characteristics, Discrimination and Petrogenesis. *Contrib. Miner. Petrol.* **1987**, *95*, 407–419. [CrossRef]
88. Lima, J.V.d.; Guimaraes, I.D.P.; Santos, L.; Farias, D.J.; Antunes, J.V. Química Mineral e Condições de Cristalização de Granitos Intrudidos Ao Longo Da Zona de Cisalhamento Remígio—Pocinhos, NE Do Brasil: Plúton Pilõezinhos. *Pesqui. Em Geociências* **2017**, *44*, 123. [CrossRef]
89. Santos, L.; Guimaraes, I.P.; Farias, D.J.S.; Lima, J. V Geologia, petrografia e quimica mineral de plutons contrastantes, Serra Branca e Coxixola, ao longo da Zona de Cisalhamento Coxixola, Provicia Borborema, NE Brasil. *Estud. Geológicos* **2013**, *23*, 111–142.
90. Locock, A.J. An Excel Spreadsheet to Classify Chemical Analyses of Amphiboles Following the IMA 2012 Recommendations. *Comput. Geosci.* **2014**, *62*, 1–11. [CrossRef]
91. Hawthorne, F.C.; Oberti, R.; Harlow, G.E.; Maresch, W.V.; Martin, R.F.; Schumacher, J.C.; Welch, M.D. Nomenclature of the Amphibole Supergroup. *Am. Miner.* **2012**, *97*, 2031–2048. [CrossRef]
92. Czamanske, G.K.; Wones, D.R. Oxidation During Magmatic Differentiation, Finnmarka Complex, Oslo Area, Norway: Part 2, The Mafic Silicates1. *J. Petrol.* **1973**, *14*, 349–380. [CrossRef]
93. Foster, M.D. Interpretation of the Composition of Trioctahedral Micas. *US Geol. Surv. Prof. Pap. B* **1960**, *354*, 1–49.
94. Gündüz, M.; Asan, K. MagMin_PT: An Excel-based mineral classification and geothermobarometry program for magmatic rocks. *Mineral. Mag.* **2023**, *87*, 1–9. [CrossRef]
95. Speer, J.A. Micas in Igneous Rocks. In *Micas*; Bailey, S.W., Ed.; De Gruyter: Berlin, Germany, 1984; Volume 13, pp. 299–356.
96. Nachit, H.; Razafimahefa, N.; Stussi, J.-M.; Carron, J.-P. Composition Chimique Des Biotites et Typologie Magmatique Des Granitoïdes. *Comtes Rendus Hebd. De L'academie Des Sci.* **1985**, *301*, 813–818.
97. Abdel-Rahman, A.-F.M. Nature of Biotites from Alkaline, Calc-Alkaline, and Peraluminous Magmas. *J. Petrol.* **1994**, *35*, 525–541. [CrossRef]
98. Deer, W.A.; Howie, R.; Zussman, J. *An Introduction to the Rock—Forming Minerals*, 2nd ed.; Pearson Education Limited: London, UK, 1992; pp. 1–712.
99. Jochum, K.P.; Weis, U.; Stoll, B.; Kuzmin, D.; Yang, Q.; Raczek, I.; Jacob, D.E.; Stracke, A.; Birbaum, K.; Frick, D.A.; et al. Determination of Reference Values for NIST SRM 610-617 Glasses Following ISO Guidelines. *Geostand. Geoanalyt. Res.* **2011**, *35*, 397–429. [CrossRef]
100. Van Achterbergh, E.; Ryan, C.G.; Jackson, S.E.; Griffin, W.L. Data Reduction Software for LA-ICP-MS. In *Laser-Ablation-ICPMS in the Earth Sciences; Principles and Applications*; Sylvester, P.J., Ed.; Mineralogical Association of Canada: Ottawa, ON, Canada, 2001; pp. 239–243.
101. Burnham, A.D. Key Concepts in Interpreting the Concentrations of the Rare Earth Elements in Zircon. *Chem. Geol.* **2020**, *551*, 119765. [CrossRef]
102. El-Bialy, M.Z.; Ali, K.A. Zircon Trace Element Geochemical Constraints on the Evolution of the Ediacaran (600–614Ma) Post-Collisional Dokhan Volcanics and Younger Granites of SE Sinai, NE Arabian-Nubian Shield. *Chem. Geol.* **2013**, *360–361*, 54–73. [CrossRef]
103. Rubatto, D. Zircon Trace Element Geochemistry: Partitioning with Garnet and the Link between U-Pb Ages and Metamorphism. *Chem. Geol.* **2002**, *184*, 123–138. [CrossRef]
104. Hoskin, P.W.O.; Schaltegger, U. The Composition of Zircon and Igneous and Metamorphic Petrogenesis. *Rev. Miner. Geochem.* **2003**, *53*, 27–62. [CrossRef]
105. Pupin, J.P. Zircon and Granite Petrology. *Contrib. Miner. Petrol.* **1980**, *73*, 207–220. [CrossRef]
106. Harley, S.L.; Kelly, N.M. Zircon Tiny but Timely. *Elements* **2007**, *3*, 13–18. [CrossRef]
107. Smythe, D.J.; Brenan, J.M. Magmatic Oxygen Fugacity Estimated Using Zircon-Melt Partitioning of Cerium. *Earth Planet Sci. Lett.* **2016**, *453*, 260–266. [CrossRef]
108. Valley, J.W. Oxygen Isotopes in Zircon. *Rev. Miner. Geochem.* **2003**, *53*, 343–385. [CrossRef]
109. Kirkland, C.L.; Smithies, R.H.; Taylor, R.J.M.; Evans, N.; McDonald, B. Zircon Th/U Ratios in Magmatic Environs. *Lithos* **2015**, *212–215*, 397–414. [CrossRef]

110. Trail, D.; Watson, E.B.; Tailby, N.D. Ce and Eu Anomalies in Zircon as Proxies for the Oxidation State of Magmas. *Geochim. Cosmochim. Acta* **2012**, *97*, 70–87. [CrossRef]
111. Schiller, D.; Finger, F. Application of Ti-in-Zircon Thermometry to Granite Studies: Problems and Possible Solutions. *Contrib. Miner. Petrol.* **2019**, *174*, 51. [CrossRef] [PubMed]
112. Loucks, R.R.; Henríquez, G.J.; Fiorentini, M.L. Zircon and Whole-Rock Trace Element Indicators of Magmatic Hydration State and Oxidation State Discriminate Copper Ore-Forming from Barren Arc Magmas. *Econ. Geol.* **2024**, *119*, 511–523. [CrossRef]
113. Schaltegger, U. Hydrothermal Zircon. *Elements* **2007**, *203*, 183–203. [CrossRef]
114. Fu, B.; Mernagh, T.P.; Kita, N.T.; Kemp, A.I.S.; Valley, J.W. Distinguishing Magmatic Zircon from Hydrothermal Zircon: A Case Study from the Gidginbung High-Sulphidation Au-Ag-(Cu) Deposit, SE Australia. *Chem. Geol.* **2009**, *259*, 131–142. [CrossRef]
115. Hoskin, P.W.O. Trace-Element Composition of Hydrothermal Zircon and the Alteration of Hadean Zircon from the Jack Hills, Australia. *Geochim. Cosmochim. Acta* **2005**, *69*, 637–648. [CrossRef]
116. Williams, I.S.; Claesson, S. Isotopic Evidence for the Precambrian Provenance and Caledonian Metamorphism of High Grade Paragneisses from the Seve Nappes, Scandinavian Caledonides. *Contrib. Miner. Petrol.* **1987**, *97*, 205–217. [CrossRef]
117. Taylor, S.R.; McLennan, S.H. *The Continental Crust: Its Composition and Evolution*; Blackwell Scientific: Oxford, UK, 1985.
118. Breiter, K.; Lamarão, C.N.; Borges, R.M.K.; Dall'Agnol, R. Chemical Characteristics of Zircon from A-Type Granites and Comparison to Zircon of S-Type Granites. *Lithos* **2014**, *192–195*, 208–225. [CrossRef]
119. Xie, L.; Wang, R.; Chen, X.; Qiu, J.; Wang, D. Th-Rich Zircon from Peralka Line A-Type Granite: Mineralogical Features and Petrological Implications. *Chin. Sci. Bull.* **2005**, *50*, 809–817. [CrossRef]
120. Santos, L. Caracterização petrológica e geoquímica dos granitoides intrudidos ao longo da zona de cisalhamento Coixixola, Província Borborema, NE Brasil: Plutons Serra Branca e Coixixola. Master's Thesis, Universidade Federal de Pernambuco, Recife, Brazil, 2013.
121. Vilalva, F.C.J.; Simonetti, A.; Vlach, S.R.F. Insights on the Origin of the Graciosa A-Type Granites and Syenites (Southern Brazil) from Zircon U-Pb Geochronology, Chemistry, and Hf and O Isotope Compositions. *Lithos* **2019**, *340–341*, 20–33. [CrossRef]
122. Loucks, R.R.; Fiorentini, M.L.; Henríquez, G.J. New Magmatic Oxybarometer Using Trace Elements in Zircon. *J. Petrol.* **2020**, *61*, egaa034. [CrossRef]
123. Blundy, J.D.; Holland, T.J.B. Calcic Amphibole Equilibria and a New Amphibole-Plagioclase Geothermometer. *Contrib. Miner. Petrol.* **1990**, *104*, 208–224. [CrossRef]
124. Ridolfi, F.; Renzulli, A.; Puerini, M. Stability and Chemical Equilibrium of Amphibole in Calc-Alkaline Magmas: An Overview, New Thermobarometric Formulations and Application to Subduction-Related Volcanoes. *Contrib. Miner. Petrol.* **2010**, *160*, 45–66. [CrossRef]
125. Ridolfi, F.; Renzulli, A. Calcic Amphiboles in Calc-Alkaline and Alkaline Magmas: Thermobarometric and Chemometric Empirical Equations Valid up to 1130 °C and 2.2 GPa. *Contrib. Miner. Petrol.* **2012**, *163*, 877–895. [CrossRef]
126. Putirka, K. Amphibole Thermometers and Barometers for Igneous Systems and Some Implications for Eruption Mechanisms of Felsic Magmas at Arc Volcanoes. *Am. Miner.* **2016**, *101*, 841–858. [CrossRef]
127. Watson, E.B.; Wark, D.A.; Thomas, J.B. Crystallization Thermometers for Zircon and Rutile. *Contrib. Miner. Petrol.* **2006**, *151*, 413–433. [CrossRef]
128. Boehnke, P.; Watson, E.B.; Trail, D.; Harrison, T.M.; Schmitt, A.K. Zircon Saturation Re-Revisited. *Chem. Geol.* **2013**, *351*, 324–334. [CrossRef]
129. Watson, E.B.; Harrison, T.M. Zircon Saturation Revisited: Temperature and Composition Effects in a Variety of Crustal Magma Types. *Earth Planet Sci. Lett.* **1983**, *64*, 295–304. [CrossRef]
130. Harrison, T.M.; Watson, E.B. The Behavior of Apatite during Crustal Anatexis: Equilibrium and Kinetic Considerations. *Geochim. Cosmochim. Acta* **1984**, *48*, 1467–1477. [CrossRef]
131. Schmidt, M.W. Amphibole Composition in Tonalite as a Function of Pressure: An Experimental Calibration of the Al-in-Hornblende Barometer. *Contrib. Miner. Petrol.* **1992**, *110*, 304–310. [CrossRef]
132. Anderson, J.L.; Smith, D.R. The Effects of Temperature and FO2 on the Al-in-Hornblende Barometer. *Am. Miner.* **1995**, *80*, 549–559. [CrossRef]
133. Hammarstrom, J.M.; Zen, E. Aluminum in Hornblende: An Empirical Igneous Geobarometer. *Am. Miner.* **1986**, *71*, 1297–1313.
134. Mutch, E.J.F.; Blundy, J.D.; Tattitch, B.C.; Cooper, F.J.; Brooker, R.A. An Experimental Study of Amphibole Stability in Low-Pressure Granitic Magmas and a Revised Al-in-Hornblende Geobarometer. *Contrib. Miner. Petrol.* **2016**, *171*, 85. [CrossRef]
135. Wones, D.R. Significance of the Assemblage Titanite+magnetite+quartz in Granitic Rocks. *Am. Miner.* **1989**, *74*, 744–749.
136. Frost, B.R.; Lindsley, D.H. Equilibria among Fe-Ti Oxides, Pyroxenes, Olivine, and Quartz: Part II. Application. *Am. Miner.* **1992**, *77*, 1004–1020.
137. Anderson, J.L.; Barth, A.P.; Wooden, J.L.; Mazdab, F. Thermometers and Thermobarometers in Granitic Systems. *Rev Miner. Geochem.* **2008**, *69*, 121–142. [CrossRef]
138. Nardi, L.V.S.; Formoso, M.L.L.; Müller, I.F.; Fontana, E.; Jarvis, K.; Lamarão, C. Zircon/Rock Partition Coefficients of REEs, Y, Th, U, Nb, and Ta in Granitic Rocks: Uses for Provenance and Mineral Exploration Purposes. *Chem. Geol.* **2013**, *335*, 1–7. [CrossRef]

Disclaimer/Publisher's Note: The statements, opinions and data contained in all publications are solely those of the individual author(s) and contributor(s) and not of MDPI and/or the editor(s). MDPI and/or the editor(s) disclaim responsibility for any injury to people or property resulting from any ideas, methods, instructions or products referred to in the content.

Article

Mineralogical and Geochemical Response to Fluid Infiltration into Cambrian Orthopyroxene-Bearing Granitoids and Gneisses, Dronning Maud Land, Antarctica

Ane K. Engvik [1,*], Fernando Corfu [2], Ilka C. Kleinhanns [3], Heinrich Taubald [3] and Synnøve Elvevold [4]

[1] Geological Survey of Norway, N-7491 Trondheim, Norway
[2] Department of Geosciences, University of Oslo, N-0316 Oslo, Norway; fernando.corfu@geo.uio.no
[3] Department of Geosciences, University of Tübingen, Schnarrenbergstrasse 94-96, D-72076 Tübingen, Germany; kleinhanns@ifg.uni-tuebingen.de (I.C.K.); taubald@uni-tuebingen.de (H.T.)
[4] Norwegian Polar Institute, N-9296 Tromsø, Norway; synnove.elvevold@npolar.no
* Correspondence: ane.engvik@ngu.no

Abstract: Fluid infiltration into Proterozoic and Early Palaeozoic dry, orthopyroxene-bearing granitoids and gneisses in Dronning Maud Land, Antarctica, has caused changes to rock appearance, mineralogy, and rock chemistry. The main mineralogical changes are the replacement of orthopyroxene by hornblende and biotite, ilmenite by titanite, and various changes in feldspar structure and composition. Geochemically, these processes resulted in general gains of Si, mostly of Al, and marginally of K and Na but losses of Fe, Mg, Ti, Ca, and P. The isotopic oxygen composition ($\delta^{18}O_{SMOW} = 6.0‰–9.9‰$) is in accordance with that of the magmatic precursor, both for the host rock and infiltrating fluid. U-Pb isotopes in zircon of the altered and unaltered syenite to quartz-monzonite indicate a primary crystallization age of 520.2 ± 1.0 Ma, while titanite defines alteration at 485.5 ± 1.4 Ma. Two sets of gneiss samples yield a Rb-Sr age of 517 ± 6 Ma and a Sm-Nd age of 536 ± 23 Ma. The initial Sr and Nd isotopic ratios suggest derivation of the gneisses from a relatively juvenile source but with a very strong metasomatic effect that introduced radiogenic Sr into the system. The granitoid data indicate instead a derivation from Mid-Proterozoic crust, probably with additions of mantle components.

Keywords: Antarctica; Dronning Maud Land; fluid; orthopyroxene granitoid; metasomatism; U-Pb; O-isotopes; H-isotopes; Rb-Sr

Citation: Engvik, A.K.; Corfu, F.; Kleinhanns, I.C.; Taubald, H.; Elvevold, S. Mineralogical and Geochemical Response to Fluid Infiltration into Cambrian Orthopyroxene-Bearing Granitoids and Gneisses, Dronning Maud Land, Antarctica. *Minerals* **2024**, *14*, 772. https://doi.org/10.3390/min14080772

Academic Editors: Ignez de Pinho Guimarães and Jefferson Valdemiro De Lima

Received: 24 May 2024
Revised: 23 July 2024
Accepted: 23 July 2024
Published: 29 July 2024

Copyright: © 2024 by the authors. Licensee MDPI, Basel, Switzerland. This article is an open access article distributed under the terms and conditions of the Creative Commons Attribution (CC BY) license (https:// creativecommons.org/licenses/by/ 4.0/).

1. Introduction

Fluid infiltration into the middle and lower crust has been an important research topic over the last few decades as it triggers metamorphic reactions and deformation and changes rock properties (e.g., [1–3]). Since the milestone paper of Austrheim [4] on fluid-induced eclogitization in the lower crust, the infiltration of volatiles and its importance for metamorphic and metasomatic reactions has been reported world-wide (e.g., [5–7]). Volatiles released during prograde metamorphism can affect magmatic crystallization (e.g., [8]) and can infiltrate and cause mineral reactions and replacements in dry crust (e.g., [9,10]). Fluid infiltration can cause local effects in single narrow zones but can also be regionally important, affecting crustal domains on the order of >100 km (e.g., [11]). Fluid–rock interaction results in petrographic and petrological responses, changing not only mineral compositions but also the texture, appearance, and color of rocks.

A key question when working on fluid petrology concerns the composition of the fluid. The volatile composition is the sum of the different compounds of the elements H, C, and N, usually as mixtures of H_2O, CO_2, N_2, and different hydrocarbons. There is then the associated question of the sources of the fluid, whether it has a magmatic, metamorphic, mantle, or meteoric origin. Through fluid flow and fluid–rock interactions, rock elements can be transported as ligands in metasomatic fluids, changing the rock's chemical composition (e.g., [12,13]).

The orthopyroxene-bearing granitoids and high-grade gneisses in Dronning Maud Land, Antarctica (DML; Figure 1), have been extensively infiltrated by fluids, forming bleached alteration zones [14,15]. The field expression of the alteration zones, as discordant light bands flanking pegmatite or aplite veins, was the basis for pointing to a late-magmatic fluids source [16]. Engvik and Stöckhert [17] described its volatile H_2O-CO_2 composition and Engvik et al. [18] modeled time scales concerning the fluid flow, heat transfer, and temperature-dependent decay of interconnected porosity.

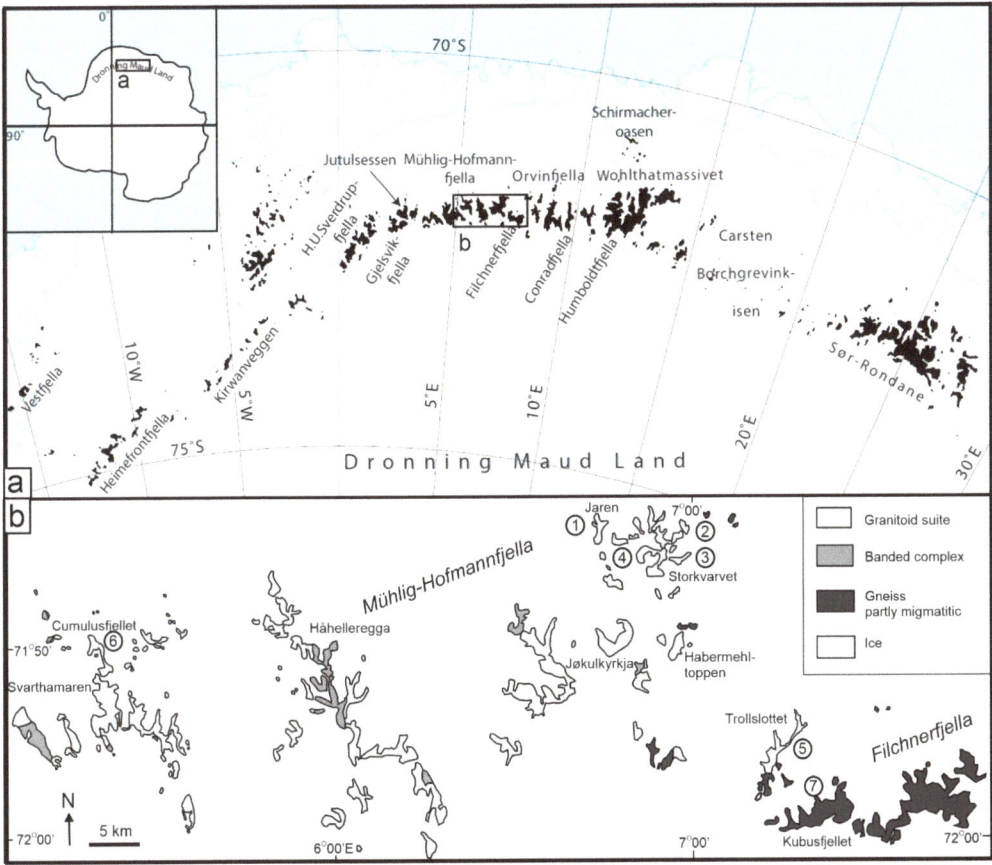

Figure 1. (**a**) Geographical situation map of central Dronning Maud Land. (**b**) Geological map of the studied part of Mühlig–Hofmannfjella and Orvinfjella. The numbers indicate the sampled localities (Table 1) (modified from [16]).

In this paper, we present and discuss petrographic, geochemical, and isotopic responses to fluid infiltration into Proterozoic and Early Palaeozoic dry gneisses and granitoids in DML. The U-Pb isotopes in zircon and titanite define the age relationships between voluminous orthopyroxene-bearing intrusions and discordant alteration zones. Stable H- and O-isotopes inform of the rock and fluid origins. Bulk rock chemical and Sr, Nd isotope compositions test the geochemical response to volatile infiltration and are discussed with respect to element transfer and mineral reactions.

2. Geological Setting

The mountain range of DML (Figure 1) was formed by multi-plate collisions of various parts of East and West Gondwana during the Pan-African event in the Late Neoproterozoic-Early Palaeozoic and represents the southern part of the East African-Antarctic Orogen in Gondwana reconstructions (e.g., [19,20]). The central DML formed during two main tectono-thermal events, the first in the late Mesoproterozoic and the second in the late Neoproterozoic to Cambrian. The late Mesoproterozoic rocks have protolith ages of about 1.2–1.0 Ga [21–24].

The dominant structures of the region formed between 640 and 500 Ma, deformation and tectonism accompanying high-grade metamorphism and voluminous magmatism [21,25,26]. The gneisses reached peak granulite facies metamorphic conditions of about 800–900 °C, at intermediate pressures (e.g., [27,28]). Combined structural studies, geochronology, and petrology have revealed an early compressional event followed by near-isothermal decompression and extension, and a final isobaric cooling [26,29–31].

In the late stages of the Pan-African event widespread post-tectonic granitoid intrusions (530–485 Ma) were emplaced in the high-grade gneisses [15,21,32]. This voluminous igneous suite extends between 2° and 15° E comprising large plutons of syenite, quartz-monzonite, granite, monzonite, diorite, gabbro, and several generations of granitic to mafic dykes. Relatively dry orthopyroxene and fayalite-bearing granitoids are an important component of the suite.

3. Material and Methods

Access conditions are poor in DML, and the only sampling was performed during the Norwegian Antarctic Research Expedition 1996/1997. The main properties of key samples used in this paper are listed in Table 1. Detailed petrographic studies were performed via optical microscopy and scanning electron microscopy (SEM) using a LEO1450 VP instrument at the Geological Survey of Norway (NGU, Trondheim, Norway). Mineral identification was performed with an energy-dispersive spectrometer (EDS) mounted on a scanning electron microscope. The degree of rock alteration was estimated based on petrographic observations.

The U-Pb analyses on zircon and titanite (Table 2) were carried out via ID-TIMS [33] at the University of Oslo (Norway). Zircon grains were selected under a binocular microscope and chemically abraded [34] before dissolution in HF at 195 °C. Titanite was air-abraded [35] and dissolved in HF on a hotplate. The spike was a mixture of ^{202}Pb, ^{205}Pb, and ^{235}U. The purification of U and Pb was performed with one-stage procedures in anion exchange resin, zircon with HCl, and titanite with mixed HBr–HCl. The blank correction was $\leq$2 pg for Pb and 0.1 pg for U. A more detailed description of the procedure in the Oslo laboratory is given in Corfu [36]. The decay constants are those of Jaffey et al. [37]. The data were calculated and plotted using the program Isoplot [38].

Table 1. List of samples, assemblages, and petrographic characteristics (abbreviations from Whitney and Evans) [39].

Sample	Locality (Loc.)	Rock Type	Main Assemblage	Access/Minor Minerals
AHA91	Loc. 1—Jaren	Opx-diorite	Qz Pl perthitic-Kfs Opx Bt Amp	Opq Ap Zrc
AHA92	Loc. 1—Jaren	Granodiorite (light)	Qz Pl Mc Bt Amp	Opq Ap Zrc
AHA106	Loc. 2—Storskvarvet	Opx-quartz monzonite	Qz Pl Perthitic-Kfs Opx Bt Amp	Opq Ap Zrc
AHA107	Loc. 2—Storskvarvet	Diorite (light)	Qz Pl Mc Myrmekite Bt Amp	Opq Ap Zrc
AHA119	Loc. 3—Storskvarvet	Opx-diorite	Qz Pl Opx Bt Amp	Perthite Opq Ap Zrc
AHA120	Loc. 3—Storskvarvet	Quartz monzonite (light)	Qz Pl and anti-perthite Bt Amp	Opq Ap Zrc
AHA144	Loc. 4—Storskvarvet	Opx-quartz monzonite	Qz Pl Perthitic-Kfs Opx Bt Amp Myrmekite	Opq Ap Zrc Aln
AHA145	Loc. 4—Storskvarvet	Quartz monzonite (light)	Qz Pl Mc Bt Amp	Opq Ap Zrc Aln Ttn

Table 1. Cont.

Sample	Locality (Loc.)	Rock Type	Main Assemblage	Access/Minor Minerals
AHA197	Loc. 5—Trollslottet	Opx-syenite/quartz monzonite	Qz Pl perthitic-Kfs Opx Bt Amp Myrmekite	Opq Ap Zrc
AHA199	Loc. 5—Trollslottet	Quartz monzonite (light)	Qz Pl Mc Bt Amp Myrmekite	Opq Ap Zrc Ttn
AHA200 I	Loc. 5—Trollslottet	Quartz monzonite (light)	Qz Pl Mc Bt Amp Myrmekite	Opq Ap Zrc Ttn
AHA200 II	Loc. 5—Trollslottet	Aplite	Qz Pl Mc	Amf Bt Zrc Ap
AHA217	Loc. 6—Cumulusfjellet	Opx-granite	Qz Pl perthitic-Kfs Bt Opx Myrmekite	Opq Ap Zrc
AHA218	Loc. 6—Cumulusfjellet	Granite (light)	Qz Pl perthite/Mc Opx-remnants Bt Myrmekite	Opq Ap Zrc Aln
AHA 193A	Loc. 7—Kubusfjellet	Opx gneiss (dark)	Qz Pl perthitic-Kfs antiperthite Opx Bt Grt	Opq Ap Zrc
AHA 193C	Loc. 7—Kubusfjellet	Gneiss (light)	Qz Pl perthite/Mc Opx-remnents Bt Grt	Opq Ap Zrc
AHA 205A	Loc. 7—Kubusfjellet	Opx gneiss (dark)	Qz Pl perthitic-Kfs Opx Bt Grt	Myrmekite Opq Ap Zrc
AHA 205GHJ	Loc. 7—Kubusfjellet	Gneiss (light)	Qz Pl perthite/Mc Bt Myrmekite Opx-remnants Grt	Opq Ap Zrc

Sample	Replacement Minerals
AHA91	Weak alteration
AHA92	Pl microveining sericitization; perthite => Mc as microveining and subgrains
AHA106	Weak alteration; Opx => Bt; Pl microveined sericitization; perthite microveined
AHA107	Fine symplectite of Amp+Qz and Bt+Qz; Pl microveined and sericitization; perthite => Mc microveined and replaced to finer subgrains
AHA119	Weak alteration; minor fine-grained Bt+Qz symplectite
AHA120	Weak alteration; Fld microveining and sericitization; Bt + Qz symplectite
AHA144	Weak alteration
AHA145	Pl microveining sericitization; perthite => Mc as microveining and subgrains; Amp => Bt
AHA197	Weak alteration; perthite microveining
AHA199	Pl microveined and sericitized; perthite => Mc and microveined; Bt+Qz fine symplectite; Opq => Ttn
AHA200 I	Pl microveined and sericitized; perthite => Mc and microveined; Bt Amp replaced with finer grained/symplectite; Opq => Ttn
AHA200 II	
AHA217	Weak alteration perthite microvein
AHA218	Pl sericitization and microveining; perthite => Mc microveining and subgrains; Opx => Bt and Amp
AHA 193A	
AHA 193C	Opx => Bt; Pl sericitization microveins; perthite => Mc microveins
AHA 205A	
AHA 205GHJ	Opx => Bt; Pl sericitization microveins; perthite => Mc microveins

Sample	Alteration [%]	Characteristic Textures
AHA91	<5	Equigranular; coarse-grained Qz Pl perthite Bt Amp; few finer grains
AHA92	50	Heterogranular; Coarse Qz Fld; medium Amph Bt
AHA106	15	Equigranular; coarse-grained Qz Pl perthite Bt Amp; few finer grains
AHA107	50	Heterogranular; coarse Qz Fld and few Amph Bt; fine symplectite of Amp+Qz and Bt+Qz
AHA119	<5	Equigranular; coarse Qz Pl; medium Bt Amp; Pl phenocrysts
AHA120	20	Equigranular; coarse Qz Pl; medium Bt Amp; fine Bt-Qz symplectites
AHA144	<5	Heterogranular; coarse to large Qz Pl perthite; coarse Amf Bt
AHA145	60–70	Heterogranular; coarse Qz Pl Amp; fine replacements
AHA197	<5	Equigranular; coarse Qz Pl perthite; coarse to medium Bt Amp
AHA199	70	Heterogranular; coarse Qz and Fld; fine sympectites of Amp+Qz and Bt+Qz
AHA200 I	80	Heterogranular; coarse Qz and Fld; medium to fine Amp Bt Qz; fine symplectite Bt-Qz
AHA200 II		Strong heterogranular varying coarse to fine grained; embayed grain boundaries; dusty Qz and Fld
AHA217	<5	Heterogranular coarse Fld Qz Bt; phenocryst perthite; fine myrmeckite Qz Fld along grain boundaries; some fine Bt aggregates
AHA218	50	Heterogranular; coarse Qz Fld; medium Amph Bt
AHA 193A	<1	Equigranular medium grained; triple-point and embayed grain boundaries; Grt anhedral rims
AHA 193C	40	Heterogranular; medium Qz Pl perthite/Mc; medium to fine Bt
AHA 205A	10	Equigranular medium grained; triple-point and embayed grain boundaries; Grt anhedral rims
AHA 205GHJ	30–50	Equigranular, medium grained, triple-point grain boundaries, undulating grain boundaries; grt subhedral rims

Stable isotopic data are presented in Table 3. The oxygen isotope compositions (^{16}O, ^{18}O) of handpicked mineral separates of quartz, K-feldspar, biotite, and amphibole were measured at the University of Tübingen (Germany) using a method similar to that of Sharp [41] and Rumble and Hoering [42] and described in more detail by Kasemann et al. [43]. Between 2 and 4 mg of sample was loaded onto small Pt sample holders, which were pumped to a vacuum of about 10^{-6} mbar. After prefluorination of the sample chamber overnight, the samples were heated with a CO_2-laser in 50 mbar of pure F_2. Excess F_2 was separated from the O_2 using KCl at 150 °C by producing KF and releasing Cl_2. The extracted O_2 was collected quantitatively via adsorption on a molecular sieve (13X) at liquid nitrogen temperature in a sample vial. Subsequently, the vial was removed from the line and heated to room temperature, releasing O_2 as a gas, which was analyzed isotopically using a Finnigan MAT 252 isotope ratio mass spectrometer. Oxygen isotope compositions are given in the standard δ-notation, expressed relative to VSMOW (Vienna Standard Mean Ocean Water) in permil (‰). Replicate oxygen isotope analyses of the standards, using NBS-28 quartz and UWG-2 garnet [43], have an average precision of ±0.1‰ for $\delta^{18}O$. The accuracy of $\delta^{18}O$ values is better than 0.2‰ compared to the accepted $\delta^{18}O$ values for NBS-28 of 9.64‰ and UWG-2 of 5.8‰.

Table 2. U-Pb zircon and titanite data.

Mineral, Characteristics [1]	Weight [µg] [2]	U [ppm] [2]	Th/U [3]	Pbi [4] [ppm]	Pbc [5] [pg]	$^{206}Pb/^{204}Pb$ [6]	$^{207}Pb/^{235}U$ [7]	2 s [abs]	$^{206}Pb/^{238}U$ [7]	2 s [abs]	rho
AHA 197											
Z sp-fr [1]	13	269	0.44	0.0	1.5	12,376	0.668	0.002	0.08394	0.00017	0.90
Z lp [1]	15	187	0.59	0.0	0.8	17,721	0.666	0.002	0.08371	0.00018	0.88
Z lp-flat [1]	11	183	0.53	0.0	1.3	8352	0.666	0.002	0.08370	0.00017	0.83
Z lp-fr [1]	2	575	0.64	0.0	0.8	7102	0.660	0.002	0.08329	0.00019	0.82
AHA199											
Z lp [1]	1	771	0.62	0.0	0.3	13,970	0.669	0.003	0.08414	0.00025	0.88
Z fr [1]	6	261	0.42	0.0	1.9	4447	0.669	0.002	0.08399	0.00020	0.83
T pb [3]	8	45	1.38	2.0	17.9	117	0.621	0.019	0.07855	0.00033	0.22
T pb [2]	1	167	1.05	9.3	11.4	90	0.610	0.025	0.07805	0.00040	0.23
T pb [5]	1	217	0.90	14.9	16.9	80	0.591	0.027	0.07786	0.00045	0.14

Mineral, Characteristics [1]	$^{207}Pb/^{206}Pb$ [7]	2 s [abs]	$^{206}Pb/^{238}U$ [7] [Ma]	$^{207}Pb/^{235}U$ [7] [Ma]	$^{207}Pb/^{206}Pb$ [7] [Ma]	2 s [abs]	D [8] [%]
AHA 197							
Z sp-fr [1]	0.05776	0.00006	519.6	519.8	520.8	2.4	0.2
Z lp [1]	0.05769	0.00007	518.2	518.2	518.1	2.8	0.0
Z lp-flat [1]	0.05769	0.00009	518.2	518.1	517.9	3.4	−0.1
Z lp-fr [1]	0.05751	0.00011	515.7	514.9	511.0	4.3	−1.0
AHA199							
Z lp [1]	0.05769	0.00011	520.8	520.3	518.2	4.1	−0.5
Z fr [1]	0.05778	0.00012	519.9	520.2	521.6	4.5	0.3
T pb [3]	0.05738	0.00173	487.4	490.7	506.1	64.9	3.8
T pb [2]	0.05667	0.00224	484.5	483.5	478.8	85.1	−1.2
T pb [5]	0.05510	0.00250	483.3	471.8	416.3	98.3	−16.7

[1] Z = zircon; T = titanite; lp = long prismatic; sp = short prismatic; fr = fragment; pb = pale brown; [N] = number of grains in fraction. [2] Weight and concentrations are known to be better than 10%, except for those near and below the ca. 1 µg limit of resolution of the balance. [3] Th/U model ratio inferred from the 208/206 ratio and age of the sample. [4] Pbi = initial common Pb. [5] Pbc = total common Pb in sample (initial + blank). [6] Raw data corrected for fractionation and spike. [7] Corrected for fractionation, spike, blank, and initial common Pb [40]; error calculated by propagating the main sources of uncertainty. [8] D = degree of discordancy.

For the D/H analysis of biotite, an extraction line as described in Vennemann and O'Neil's study [44] was used. Depending on the water content, sufficient amounts of hydrous minerals were loaded into 12 cm long quartz tubes in order to obtain >1 mg H_2O. Water was released by heating the minerals in the tubes using a torch and was then converted to H_2 using Zn (see [44] for further details). H_2 was subsequently measured on a Finnigan MAT 252 Mass Spectrometer, using the dual inlet device. External precision is typically ±2‰; all values are reported relative to VSMOW.

Table 3. Stable isotopic data.

Sample	Rock Type	Mineral	$\delta^{18}O_{SMOW}$ (‰)	δD_{SMOW} (‰)
AHA 197	Opx-syenite—quartz monzonite	Quartz	9.1	
AHA 200	Quartz monzonite (light)	Quartz	9.9	
AHA 197	Opx-syenite—quartz monzonite	K-feldspar	8.8	
AHA 200	Quartz monzonite (light)	K-feldspar	8.8	
AHA 197	Opx-syenite—quartz monzonite	Amphibole	6.5	
AHA 200	Quartz monzonite (light)	Amphibole	5.7	
AHA 197	Opx-syenite—quartz monzonite	Biotite	6.4	−101.4
AHA 200	Quartz monzonite (light)	Biotite	6.0	−288.8

Samples used for geochemical analyses were selected as being representative and homogenous, with good control of mineralogy and petrography. Whole-rock major and trace element analyses (Table 4) were carried out at the NGU (Trondheim, Norway). Major elements were measured on fused glass beads prepared via 1:7 dilution with lithium tetraborate. Trace elements were measured from pressed tablets on a PANalytical Axios XRF spectrometer equipped with a 4 kW Rh X-ray end-window tube, using synthetic and international standards for calibration.

Table 4. Whole-rock major and trace element geochemistry.

	Major Elements (wt%)											
Sample	SiO_2	Al_2O_3	Fe_2O_3	TiO_2	MgO	CaO	Na_2O	K_2O	MnO	P_2O_5	LOI	Sum
AHA91	62.3	14.0	9.07	1.86	1.18	4.36	2.73	4.08	0.103	0.662	0.015	100.0
AHA92	64.3	14.9	6.64	1.28	0.784	3.80	2.92	4.27	0.080	0.494	−0.039	99.5
AHA 106	65.3	15.3	5.89	1.14	0.658	3.45	2.91	5.10	0.072	0.448	0.070	100.0
AHA 107	61.8	15.3	7.59	1.67	1.04	4.61	3.10	3.37	0.100	0.828	0.204	99.6
AHA 119	60.3	14.6	10.1	2.07	1.31	4.73	3.14	3.11	0.114	0.929	0.029	100.0
AHA 120	63.9	15.3	6.64	1.28	0.759	3.91	3.11	4.32	0.084	0.515	0.054	99.9
AHA 144	62.7	14.5	8.34	1.55	0.990	3.58	2.53	5.37	0.097	0.589	0.010	100.0
AHA 145	65.4	14.4	6.43	1.20	0.686	2.88	2.85	4.98	0.081	0.471	0.223	99.6
AHA 197	61.6	16.2	6.93	1.31	1.41	3.58	3.51	5.50	0.083	0.417	0.069	101.0
AHA 199	63.1	15.7	6.64	1.05	1.13	2.67	3.39	5.66	0.081	0.334	0.120	99.9
AHA 217	71.4	14.9	2.69	0.358	0.375	1.47	3.52	5.79	0.036	0.158	0.303	101.0
AHA 218	74.6	13.6	1.58	0.213	0.140	1.12	2.95	5.81	0.017	0.038	0.346	100.0
AHA 193A	71.8	13.2	5.73	0.712	0.493	2.23	2.25	4.24	0.084	0.188	−0.143	101.0
AHA 193C	71.2	13.9	3.87	0.448	0.517	1.33	1.84	6.82	0.044	0.161	0.010	100.0
AHA 205A	71.6	13.8	4.38	0.532	0.374	1.98	2.23	5.17	0.057	0.139	−0.035	100.0
AHA 205H	71.2	14.5	2.37	0.599	0.632	1.54	2.33	6.99	0.010	0.248	0.157	101.0

	Trace Elements (mg/kg)										
Sample	Ba	Ce	Co	Cr	Cu	Ga	La	Mo	Nb	Nd	Ni	Pb
AHA91	2170	384	12.7	15.9	9.5	26.2	181	5.1	39.5	183	6.4	33.8
AHA92	2450	260	9.2	4.2	6.1	26.9	132	3.2	29.2	96	5.7	35.8
AHA 106	2770	247	8.0	4.6	6.5	27.3	133	2.7	26.9	100	6.2	40.8
AHA 107	1770	255	11.4	19.6	6.7	29.1	122	1.9	43.5	113	6.8	102
AHA 119	1440	322	15.4	9.1	9.1	28.9	143	4.5	48.3	148	5.3	37.5
AHA 120	2170	323	8.9	6.6	6.9	28.2	174	3.5	31.2	131	5.6	42.0
AHA 144	2600	352	10.8	14.7	9.5	26.1	155	2.8	37.5	147	5.4	39.1
AHA 145	2340	235	8.5	23.7	<2	25.1	114	2.7	35.4	100	6.2	42.8
AHA 197	2390	238	15.0	13.6	6.2	24.4	108	2.3	28.3	119	6.3	37.5
AHA 199	2190	202	11.4	4.5	<2	26.1	98	1.9	39.0	96	7.8	48.8
AHA 217	925	79	<4	6.6	<2	21.7	48	2.9	14.4	41	6.3	40.0
AHA 218	684	118	<4	<4	<2	19.0	39	1.7	8.7	20	5.0	41.7
AHA 193A	896	127	6.7	86.3	8.3	19.3	65	1.6	17.2	58	9.2	19.7
AHA 193C	992	136	5.4	22.1	7.7	20.1	64	<1	9.5	56	9.3	30.0
AHA 205A	1040	86	6.0	16.0	5.9	20.0	44	<1	14.4	35	7.5	25.7
AHA 205H	1240	34	5.1	14.9	5.6	20.7	<20	1.1	10.0	16	8.2	39.2

Table 4. Cont.

Sample	Rb	Sc	Sr	Th	U	V	W	Y	Yb	Zn	Zr
AHA91	140	22.3	515	34.5	5.9	58.3	5.8	108	5.8	187	640
AHA92	147	12.0	599	32.5	6.0	29.6	<5	60.6	<5	136	477
AHA 106	165	11.9	601	24.8	5.9	29.8	<5	63.8	<5	121	500
AHA 107	191	12.3	587	28.6	6.9	40.5	10.9	97.1	6.4	207	534
AHA 119	176	17.7	521	26.7	5.6	56.1	5.9	95.3	5.3	269	604
AHA 120	158	11.7	546	37.1	8.1	31.6	<5	79.2	6.7	145	566
AHA 144	155	16.8	536	23.9	5.4	43.5	<5	90.3	<5	170	672
AHA 145	257	8.3	466	24.0	7.0	25.5	<5	66.3	<5	146	561
AHA 197	131	14.3	588	15.2	5.1	71.1	<5	67.3	<5	117	566
AHA 199	298	12.2	466	18.4	7.7	56.9	5.8	88.2	5.3	140	536
AHA 217	225	7.6	224	12.4	4.1	20.0	<5	30.5	<5	72.0	259
AHA 218	132	<5	191	18.4	3.1	6.4	<5	7.9	<5	34.0	183
AHA 193A	136	12.9	127	18.8	2.2	20.8	5.9	113	8.3	78.5	410
AHA 193C	254	9.3	133	23.6	2.4	15.6	<5	75.6	<5	80.2	226
AHA 205A	170	9.9	140	16.8	2.4	16.1	<5	80.9	<5	67.8	315
AHA 205H	267	<5	181	4.3	2.3	29.2	<5	19.4	<5	57.1	213

Sample	Cl	F	Cs	Ge	Hf	Sm	Te
AHA91	<0.05	0.54	<10	6.0	17.4	30	20
AHA92	<0.05	0.44	<10	6.1	15.3	14	16
AHA 106	<0.05	0.46	<10	6.4	17.1	24	17
AHA 107	0.227	0.47	<10	6.8	15.9	25	19
AHA 119	0.093	0.81	<10	6.1	14.4	23	19
AHA 120	<0.05	0.54	<10	6.6	20.8	20	21
AHA 144	0.056	0.49	<10	6.2	20.2	24	<10
AHA 145	0.058	0.51	<10	6.5	15.5	15	19
AHA 197	<0.05	0.45	<10	6.0	17.0	17	<10
AHA 199	0.060	0.51	<10	6.5	14.4	20	21
AHA 217	<0.05	0.38	<10	6.5	13.6	<10	16
AHA 218	<0.05	0.32	<10	6.6	12.2	<10	<10
AHA 193A	<0.05	0.32	<10	6.9	16.1	<10	19
AHA 193C	<0.05	0.36	<10	6.4	12.0	12	<10
AHA 205A	<0.05	0.31	<10	6.8	15.4	<10	<10
AHA 205H	<0.05	0.40	<10	6.4	13.8	<10	<10

Rubidium-Sr and Sm-Nd isotopic analyses (Table 5) were performed on the Finnigan MAT262 mass spectrometer located ats the Isotope Geochemistry group, Eberhard Karls University of Tübingen, Germany. Prior to digestion, the samples were mixed with tracer solutions enriched in the ^{87}Rb-^{84}Sr and ^{149}Sm-^{150}Nd isotopes. Rubidium, Sr, and REE were separated from one single rock digest using standard cation exchange procedures followed by the separation of Sm and Nd using reverse ion chromatography with HDEHP resin. Reproducibility for NBS SRM 987 measurements during the study was 0.710256 ± 16 (2sd; n = 9) and 0.056500 ± 08 (2sd) for ^{87}Sr/^{86}Sr and ^{84}Sr/^{86}Sr, respectively. Analytical mass bias was corrected relative to ^{88}Sr/^{86}Sr = 0.1194 using exponential law. Analytical mass bias correction for Rb measurements was based on repeated analyses of NBS SRM 984 yielding ^{85}Rb/^{87}Rb$_{raw}$ = 2.5952 ± 13 (2sd, n = 7), resulting in an exponential mass bias of 0.4‰/amu. Repeated measurements of the La Jolla standard (n = 6) gave 0.511831 ± 26 (2sd) and 0.348404 ± 14 (2sd) for ^{143}Nd/^{144}Nd and ^{145}Nd/^{144}Nd, respectively. Analytical mass bias was corrected with ^{146}Nd/^{144}Nd = 0.7219 using an exponential law. Total procedural blanks were consistently below 0.5 ng for all four elements and were thus negligible.

Table 5. WR Rb-Sr and Sm-Nd isotope data.

Sample ID	Rb [µg/g]	Sr [µg/g]	^{87}Rb/^{86}Sr	^{87}Sr/^{86}Sr	±2 S.E.	TDM [a] [Ga]	^{87}Sr/^{86}Sr [@ 0.5 Ga]
AHA 106	179	611	0.822	0.714957	0.000010	1.09	0.70910
AHA 107	207	613	0.941	0.715830	0.000012	1.01	0.70913
AHA 119	192	540	0.992	0.716184	0.000010	0.99	0.70912
AHA 120	174	574	0.851	0.715260	0.000009	1.08	0.70920
AHA 144	168	551	0.848	0.715209	0.000010	1.07	0.70916
AHA 145	291	488	1.667	0.720615	0.000011	0.77	0.70874
AHA 197	141	613	0.643	0.713208	0.000010	1.20	0.70863
AHA 199	334	483	1.919	0.721822	0.000012	0.71	0.70815
AHA 217	255	238	2.996	0.730783	0.000009	0.67	0.70944
AHA 218	148	201	2.054	0.724052	0.000010	0.74	0.70941
AHA 193A	150	131	3.211	0.772378	0.000011	1.53	0.74950
AHA 193C	295	139	5.926	0.792361	0.000011	1.06	0.75014
AHA 205A	187	145	3.606	0.775184	0.000010	1.41	0.74949
AHA 205H	317	193	4.618	0.766434	0.000010	0.97	0.73353

Sample ID	Sm [µg/g]	Nd [µg/g]	^{147}Sm/^{144}Nd	^{143}Nd/^{144}Nd	±2 S.E.	eNd [b] [today]	TDM [c] [Ga]	eNd [b] [@ 0.5 Ga]
AHA 106	20.1	115	0.1081	0.511927	0.000007	−13.7	1.76	−8.1
AHA 107	25.9	134	0.1190	0.511892	0.000008	−14.4	2.02	−9.5
AHA 119	28.5	154	0.1143	0.511900	0.000006	−14.2	1.91	−9.0
AHA 120	24.4	139	0.1086	0.511942	0.000007	−13.4	1.75	−7.8
AHA 144	31.6	175	0.1115	0.511882	0.000008	−14.6	1.88	−9.2
AHA 145	20.9	119	0.1085	0.511861	0.000007	−15.0	1.86	−9.4
AHA 197	24.7	133	0.1149	0.512130	0.000007	−9.80	1.57	−4.6
AHA 199	22.5	108	0.1279	0.512105	0.000007	−10.2	1.85	−5.9
AHA 217	9.77	40.9	0.1474	0.512219	0.000008	−8.00	2.13	−4.9
AHA 218	3.90	24.5	0.0988	0.511950	0.000007	−13.3	1.59	−7.1
AHA 193A	15.5	92.9	0.1028	0.512049	0.000009	−11.3	1.51	−5.4
AHA 193C	16.0	86.0	0.1149	0.512088	0.000006	−10.6	1.64	−5.4
AHA 205A	9.67	51.5	0.1160	0.512087	0.000010	−10.6	1.65	−5.5
AHA 205H	6.82	24.3	0.1731	0.512293	0.000008	−6.60	3.20	−5.1

[a] TDM calculated assuming linear DM evolution with present-day values of ^{87}Rb/^{86}Sr of 0.022 and ^{87}Sr/^{86}Sr of 0.7025 [45] using l ^{87}Rb of $1.42 \times 10^{-11} a^{-1}$. [b] calculated with chondritic values of ^{147}Sm/^{144}Nd of 0.1960 and ^{143}Nd/^{144}Nd of 0.512630 [46]. [c] TDM calculated assuming linear DM evolution with present-day values of ^{147}Sm/^{144}Nd of 0.2137 and ^{143}Nd/^{144}Nd of 0.51315 [47] using l^{147}Sm of $6.54 \times 10^{-12} a^{-1}$ and chondritic values [46].

4. Results

4.1. Field Relations

In Mühlig–Hofmannfjella and Filchnerfjella (5–8° E), two major rock types are distinguished, gneisses and an extensive igneous suite, with the latter including a banded complex (Figure 1).

The sequence of gneisses comprises granitic gneisses, metapelites, and mafic rocks, all partly migmatitic. Garnet and orthopyroxene are the principal minerals of the granitic gneisses, together with perthite, plagioclase, and quartz and including cordierite and sillimanite in the metapelites. The gneisses reached granulite facies conditions of 850–885 °C and 0.55–0.70 GPa. Their metamorphic, mineral chemistry, and structural evolution was reported by Engvik and Elvevold [30] and Elvevold and Engvik [28]. The migmatitic gneisses have leucosome quartz-feldspar layers, where layers enriched in mafic minerals define a foliation.

The igneous suite is part of a larger complex intruding the gneisses and extending between 5 and 15° E, often constituting larger nunataks (Figure 2a). In the study area, the intrusions are mainly quartz-monzonite and granite. Orthopyroxene is usually a part of the mineral assemblage in pristine parts of the granitoid, dominating the bedrock with a brownish outcrop color. These rocks are medium- to coarse-grained and contain typical megacrysts of mesoperthitic K-feldspar. A banded complex (Figure 1), as described by Engvik et al. [48], is characterized by dark reddish, light, and darker bands that represent orthopyroxene and olivine-bearing medium-grained charnockite, coarse granitic material, and fine-grained doleritic rocks, respectively.

Figure 2. Field photos. (**a**) Nunatak of orthopyroxene-bearing granitoid with the characteristic dark brownish outcrop color (Håhelleregga). (**b**) Alteration halo around discordant aplitic vein crosscutting migmatitc gneiss (locality 7, Kubusfjellet). The vein (arrow) is about 5 cm thick with an alteration halo extending 0.5 m into the host rock. The field of view is 1.5 m. (**c**) Alteration halo around the pegmatitic vein cutting dark brownish-colored orthopyroxene-bearing syenite (locality 5, Trollslottet). The field of view is about 1 m. (**d**) A high density of crisscrossing veins causes heavy alteration of the dark orthopyroxene-bearing syenite of the Trollslottet nunatak (Locality 5, the cliff face is about 70 m high).

Both the gneisses and the igneous suite are discordantly crosscut by conspicuous light bands with a central pegmatitic or aplitic vein of granitoid composition (Figure 2b–d). The veins have a typical thickness ranging between a few mm and 15 cm and are surrounded by alteration halos showing a marked change in the color of the host rocks. The halos are roughly symmetric on both sides of the central vein and display sharp boundaries to pristine host rock. The widths of the alteration halos range from some centimeters to more than a meter. In addition, light bleaching is locally developed, with more diffuse relations to the dark host rock.

4.2. Petrography

Petrography has been investigated in sample pairs of brownish outcrop-colored pristine gneiss and granitoids and their altered light-colored equivalent crossing the alteration zone, with samples collected within distances on a meter scale.

4.2.1. Orthopyroxene Gneiss

Migmatitic orthopyroxene gneiss is variably garnet bearing and dominated by quartz, feldspar as perthitic K-feldspar ($An_{<1}Ab_{20}Kfs_{79}$), plagioclase ($An_{35}Ab_{64}Kfs_{1}$), locally with antiperthitic texture, besides the orthopyroxene ($Fs_{68-74}En_{24-29}Wo_{1}$), biotite and minor myrmekite (Figure 3a). Opaque phases, apatite, and zircon occur as accessories. The gneiss is equigranular and medium-grained. Biotite (Ti = 0.44–0.58 a.p.f.u.; F = 0.31–0.68 a.p.f.u.; $Fe^{2+}/(Fe^{2+} + Mg)$ = 0.66–0.70) defines with its crystal-preferred orientation the variable foliation. Quartz and feldspars show embayed-grained boundaries and, together with orthopyroxene, triple-point grain boundaries. Garnet ($Alm_{78-84}Prp_{5-9}Grs_{9-12}Sps_{2}$) occurs as anhedral, rounded grains or with resorbed rims.

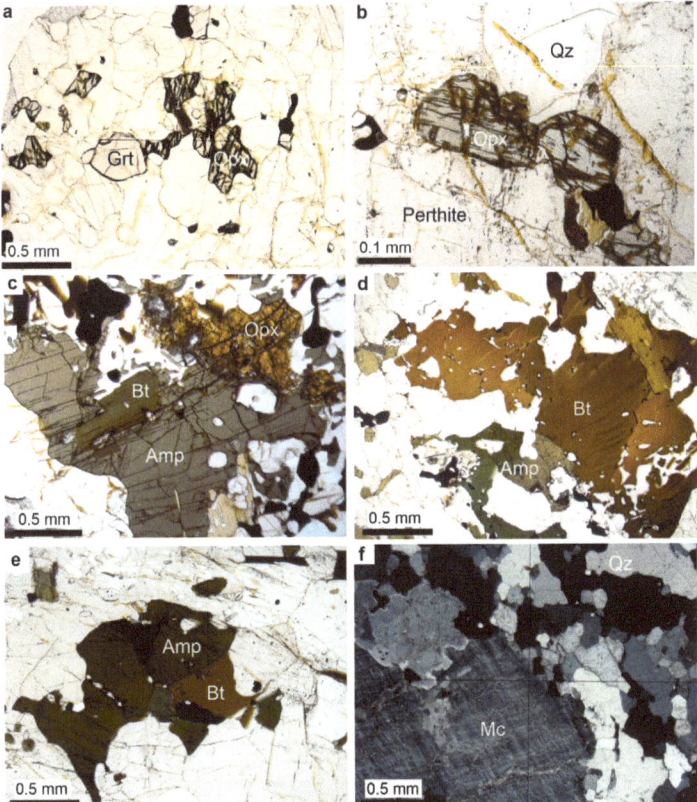

Figure 3. Micrographs of pristine orthopyroxene-bearing gneiss and granitoid intrusions (mineral abbreviations following Whitney and Evans [38]. (**a**) Garnet-orthopyroxene gneiss with major quartz, perthitic K-feldspar, and plagioclase occur equigranular with embayed and triple-point grain boundaries (sample AHA193A, plane light). (**b**) Euhedral orthopyroxene in the quartz and feldspar matrix (sample AHA217, orthopyroxene granite, plane light). (**c**) Coarse subhedral amphibole with medium-grained biotite and orthopyroxene (sample AHA144, orthopyroxene-bearing quartz-monzonite, plane light). (**d**) Amphibole and biotite in the matrix of quartz and perthitic K-feldspar; note the coarse crystals of biotite (sample AHA197, orthopyroxene-bearing granite, plane light). (**e**) Amphibole and biotite in the matrix of perthite, plagioclase, and quartz; note the well-developed coarse crystals of amphibole and biotite (sample AHA197, orthopyroxene-bearing granite, plane light). (**f**) Fine-grained quartz and feldspars of aplite but with a strong heterogeneity including some coarse grains. Remark dusty appearance of quartz and feldspar (sample AHA200II, crossed polarizers).

4.2.2. Orthopyroxene-Bearing Granitoids

The investigated unaltered granitoids occur as orthopyroxene-bearing granite, syenite, quartz monzonite, and diorite. In general, they are relatively equigranular with respect to their major assemblage of medium to coarse grains of quartz, perthitic K-feldspar, plagioclase, biotite, and amphibole, and with fine- to medium-grained orthopyroxene. Feldspars occur partly as phenocrysts. Mafic minerals constitute 5%–10% of the rock, including orthopyroxene, amphibole, and biotite. Orthopyroxene has euhedral to subhedral grain boundaries (Figure 3b,c) and shows a weak alteration texture along microfractures or a brownish surface alteration. For the quartz monzonite of locality 5, with the mineral chemistry described by Engvik and Elvevold [49], perthitic K-feldspar shows a total chemistry of $An_{0-2}Ab_{16-45}Kfs_{53-84}$ and plagioclase are oligoclase ($An_{29}Ab_{69}Kfs_2$). Orthopyroxene is $Fs_{69-71}En_{25-26}Wo_2$. Amphibole is hastingsite; Ti = 0.19–0.20 a.p.f.u.; F = 0.61–0.62 a.p.f.u.; $Fe^{2+}/(Fe^{2+}+Mg)$ = 0.64–0.65). The medium- to coarse-grained crystals of biotite (Ti = 0.45–0.50 a.p.f.u.; F = 0.60–0.87 a.p.f.u.); $Fe^{2+}/(Fe^{2+} + Mg)$ = 0.64–0.66) are subhedral (Figure 3c–e). Myrmekite is present, and apatite, zircon, opaque phases (Fe-Ti oxides), and allanite occur as accessories.

4.2.3. Granitoid Pegmatitic and Aplitic Veins

Pegmatites and aplites are mainly composed of quartz, plagioclase, microcline and perthite, with minor biotite and amphibole and accessory apatite, zircon, Fe-Ti oxides, carbonate and white mica. Pegmatites are coarse-grained with a grain size of up to several cm. Aplites have highly heterogranular grain sizes varying from fine to coarse grains between <0.1 mm and up to 5 mm (Figure 3f). Quartz and feldspars contain a high density of dust represented by fluid inclusions and tiny phases such as sericite and opaque minerals.

4.2.4. Alteration Textures

Orthopyroxene, in general, disappears in the altered samples, both in gneiss and granitoids, although remnants can be preserved locally within its replacements of biotite + quartz-symplectite (Figure 4a) or amphibole + quartz-symplectite (Figure 4b). A replacement of amphibole occurs along grain boundaries and cleavage planes and on sub-grain boundaries. Where amphibole is fractured, the microveins are filled with biotite and trails of opaque phases (Figure 4c,d). Both amphibole and biotite are replaced by finer grains in the altered samples (Figure 4a,b). In the central part of the altered zones, amphibole disappears, while biotite is replaced only by finer-grained spread grains. Ti of the biotites shows a decrease (down to 0.13 a.p.f.u.) compared to the unaltered rocks, while F increases and Cl is present (<1.08 and 0.18 a.p.f.u., respectively). The $Fe^{2+}/(Fe^{2+} + Mg)$ ratio of the biotite decreases in the gneiss (>0.63) but shows an increase in the quartz monzonite (<0.69). Similar mineral chemical changes are shown in the hastingsite of quartz monzonite where Ti decreases (down to 0.10 a.p.f.u.), while F, Cl and the $Fe^{2+}/(Fe^{2+}+ Mg)$-ratio increase (up to 0.37 a.p.f.u., 0.16 p.f.u. and 0.77, respectively). In the most intensely altered halos, titanite has grown at the expense of ilmenite (Figure 4e).

In the altered zones, conspicuous changes have affected the feldspars. In thin sections, they reveal abundant sealed micro-cracks or microveins cutting across grain boundaries and boundaries between different minerals (Figure 5c–f). The micro-cracks in feldspar are sealed by albite, whereas those in quartz are generally healed and marked by trails of secondary fluid inclusions. The formation of microveins in feldspars is connected to a replacement of perthitic K-feldspar to microcline (Figure 5c,d). The Kfs-component of perthite increases along the micro-cracks into microcline ($An_{<1}Ab_{7-10}Kfs_{90-93}$). In strongly altered samples, perthite is totally replaced by microcline (Figure 5a,b, sample AHA145). Plagioclase undergoes albitization and sericitization ($An_{1-4}Ab_{96}Kfs_{0-2}$; Figure 5e,f). The dusty appearance on the microscale is caused by the fluid inclusions, with tiny hydrous silicates such as sericite and tiny needles of opaque phases widespread throughout the crystals (Figure 5c,d).

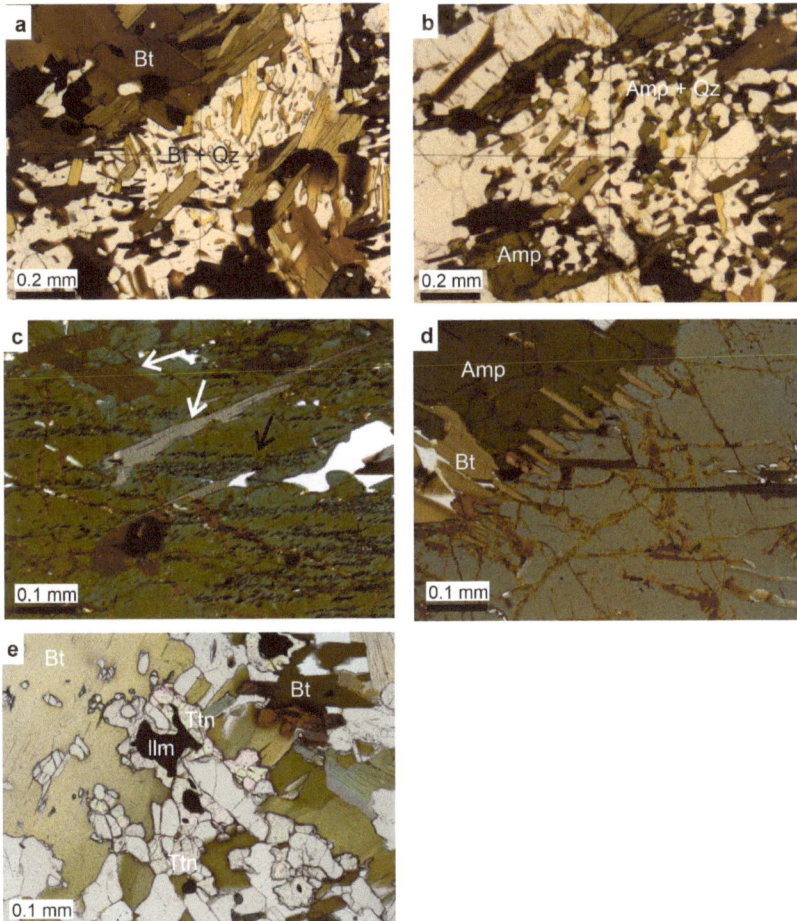

Figure 4. Replacement of mafic minerals in alteration zones (micrographs, plane light). (**a**) Biotite + quartz fine-grained symplectites (sample AHA107, altered quartz-monzonite). (**b**) Amphibole + quartz fine-grained symplectites (sample AHA107, altered quartz-monzonite). (**c**) Replacement of amphibole along cleavage planes and microfractures to biotite (white arrows) and Fe-oxide (black arrow; sample AHA145 altered quartz-monzonite). (**d**) Replacement of amphibole to biotite along cleavage planes, micro-cracks, and sub-grain boundaries (sample AHA145, altered quartz-monzonite). (**e**) Replacement of coarse biotite to finer biotite grains and of ilmenite to titanite (sample AHA199, altered quartz monzonite).

4.3. Geochemical Composition

The granitoids analyzed in this study range in composition from diorite to quartz monzonite and granite (Figure 6). Alteration caused a consistent decrease in Ca, whereas for the alkalies, the effect is inconsistent, indicating both increases and decreases. Silica shows an increase in almost all sample pairs, whereas Fe and Mn consistently decrease. For the other elements, the changes vary for each sample pair.

Figure 5. Feldspars in alteration zones. (**a**,**b**) Replacement of original feldspar to subgrains and production of a high density of micropores, fluid inclusions, and tiny grains of sericite and biotite. Plane light (**a**) and crossed polarizers (**b**) (sample AHA145, altered quartz-monzonite). (**c**) Replacement of perthitic K-feldspar to microcline (crossed polarizers, sample AHA 193C, altered gneiss). (**d**) Alteration of perthitic K-feldspar along microfractures (arrows) and replacement to microcline (crossed polarizers, sample AHA199, altered syenite). (**e**,**f**) Sericitization (arrows) and growth of biotite (brown phase) and titanite along microfractures in plagioclase. Plane light (**e**) and crossed polarizers (**f**) (sample AHA145, altered quartz-monzonite).

4.4. U-Pb Geochronology on Zircon and Titanite

Two samples were processed for U-Pb dating. Sample AHA 197 is a coarse-grained orthopyroxene-bearing granite sampled 1.8 m from the central vein of the alteration halo, whereas sample AHA199 is the altered equivalent sample, 20 cm from the central vein. Both contain zircon, but the altered sample also contains secondary titanite.

The zircon populations from the two samples display similar morphological characteristics. They consist of short prismatic grains with recognizable euhedral shapes, but somewhat subrounded and irregular, and in general with common inclusions of other minerals. Two zircon grains from sample AHA199 yield concordant data, defining a Concordia age of 520.2 ± 1.0 Ma (Table 2; Figure 7). The four analyses from the other sample AHA197 instead show some spread along the Concordia curve; one analysis is identical to those from AHA199, whereas the other three are between 2 and 5 million years younger. The pattern suggests that the two samples are coeval and the shift in the AHA197 data is interpreted as the result of partial Pb loss.

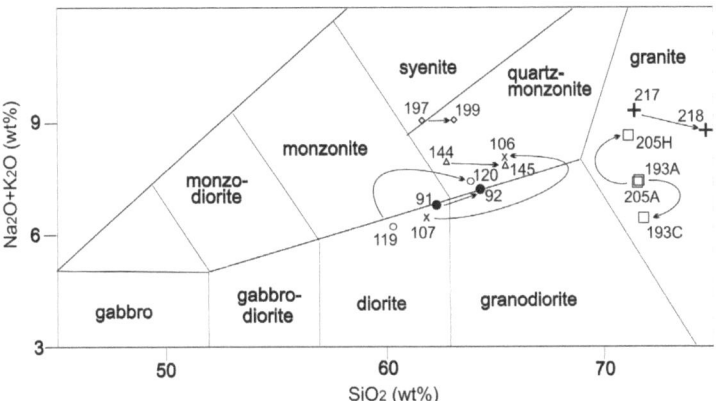

Figure 6. TAS-plot. Arrows link the unaltered and altered samples with their direction pointing to the alteration. Symbols are the same as in Figures 8 and 9.

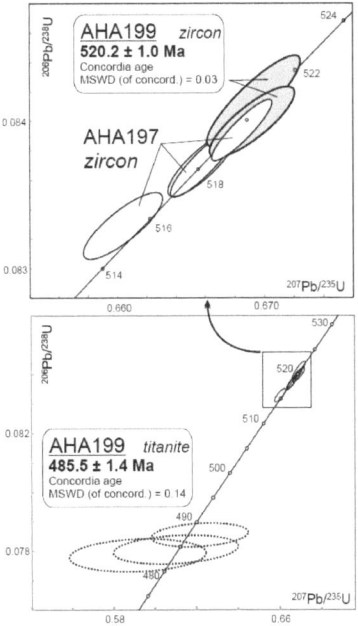

Figure 7. U-Pb analyses of zircon and titanite in orthopyroxene-bearing granite (sample AHA197) and alteration zone (sample AHA199). Ellipses (full lines for zircon and dashed lines for titanite) indicate 2 sigma uncertainty.

Titanite from sample AHA199 consists mostly of pale brown fragments, in part still containing a core of ilmenite. Three analyses indicate U contents of 45–220 ppm and Th/U of 0.9–1.4, which is interesting because it contradicts the widespread, but simplistic notion that secondary titanite always has low Th/U. The three titanite analyses overlap within error defining a Concordia age of 485.5 ± 1.4 Ma (Figure 7).

4.5. O and H Stable Isotopes

Mineral separates of the orthopyroxene-bearing granite of sample AHA197 and the related altered granitoid (sample AHA200) were analyzed for the stable isotopes of O (δ^{18}O) and H (δD), (Table 3). Oxygen was analyzed from both samples using quartz, K-feldspar, amphibole, and biotite. In addition, the δD composition was obtained on biotite separate.

The δ^{18}O$_{SMOW}$ values show a relatively good consistency for the specific minerals within the pristine orthopyroxene-bearing granite and altered sample. Quartz gives δ^{18}O$_{SMOW}$ of 9.1 and 9.9‰, respectively, K-feldspar gives 8.8‰ for both samples, amphibole gives 6.5 and 5.7‰, respectively, and biotite gives 6.4 and 6.0‰. In contrast, the δD$_{SMOW}$ of biotite is very different between the two samples, with a value of −101‰ for the orthopyroxene-bearing granite and −289‰ for the altered granitoid.

4.6. Rb-Sr and Sm-Nd Isotopes

Isotopic ratios for Rb/Sr and Sm/Nd were obtained for five pairs of granitoid rocks and two pairs of gneisses, with each pair including the pristine rock and its altered counterpart (Table 5; Figure 8a).

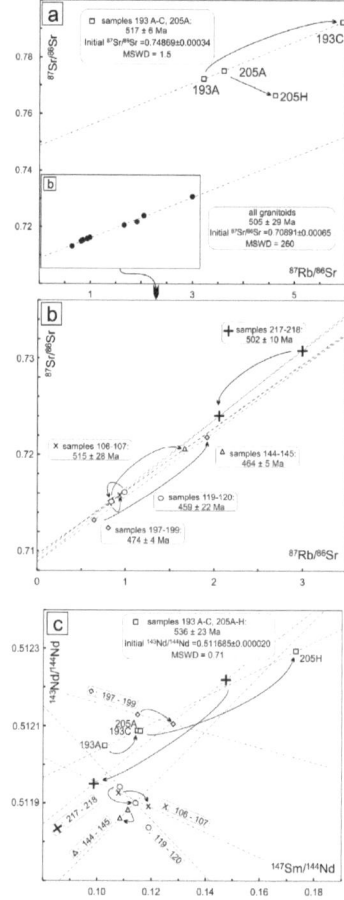

Figure 8. Radiogenic isotopic plot. Arrows link the unaltered to the altered sample and their direction points to the alteration. (**a**,**b**) ^{87}Rb/^{86}Sr vs. ^{87}Sr/^{86}Sr. (**c**) ^{147}Sm/^{144}Nd vs. ^{143}Nd/^{144}Nd. See Section 5.

For the Rb-Sr isotopic data, the two sets of rocks fall into distinct groupings. They plot along linear arrays with very distinct initial ratios of >0.74 for the gneisses and <0.71 for the granitoids (Figure 8a).

Three of the gneisses, including one of the altered rocks, fit on an isochron with an age of 517 ± 6 Ma (MSWD = 1.5). The analysis for the second altered gneiss (205H) is significantly displaced from the isochron, halfway to the array of the granitoids, indicating that alteration brought material from the granitoid magma into the original gneiss.

The array of the granitoids shows a large scatter defining an errorchron with a large MSWD of 260. In spite of the scatter, the errorchron age of 505 ± 29 Ma overlaps within uncertainty with the ages obtained from zircon and titanite (520.2 ± 1.0 and 485.5 ± 1.4 Ma, respectively; Figure 7). Calculations of pristine and altered sets separately yield ages of 516 ± 26 Ma (MSWD = 222) and 479 ± 110 Ma (MSWD = 256). Although these ages overlap the zircon and titanite ages for the primary crystallization and the alteration, they have considerable internal uncertainties. The excess scatter can be related to both initial variations in the Sr isotopic composition of different units and by shifts linked to the alteration process, as can be deduced from the two-point isochron dates calculated from each pair of unaltered and altered rock (Figure 8b).

The Sm-Nd data show a somewhat similar picture (Figure 8c). The results for the four gneiss samples (altered and unaltered) define a true isochron (MSWD = 0.71) with an age of 536 ± 23 Ma, which is within uncertainty of the Rb-Sr age of three of the same samples (Figure 8a) and also overlaps with the zircon age for the granitoids. In contrast, the Sm-Nd data for the granitoids are scattered unsystematically and the slopes of lines through pairs of unaltered and altered counterparts have no consistent trend.

5. Discussion

5.1. Mineralogical Effects of Alteration

A fluid-induced alteration of the pristine magmatic rocks is supported by (1) the light alteration halos with spatial vicinity to a pegmatite or an aplite, (2) the mineral replacements in the light rocks, including formation of the water-bearing minerals amphibole, biotite, and sericite, and (3) the microfabric showing a high density of microveins, in accordance with the results of Engvik et al. [16]. In addition, there is a direct connection between fluid inclusions having volatile H_2O-CO_2 compositions and micro-cracking [17].

The marked change in macroscopic rock surface color correlates with the change in mineral assemblage and microfabric. The light color resulting from alteration contrasts with the areas dominated by the dark brownish weathering color of the host rock and bedrock in Mühlig–Hofmannfjella. The feldspar-dominated orthopyroxene-bearing granitoids have clear crystals of pristine feldspar and quartz with a dark brownish luster. The microfabric evolution with micro-cracking and replacements of coarse crystals to finer grains, together with a dusty spread of mica, opaque phases, and fluid inclusions, cause the bleaching and dull appearance of the rocks. The related replacement of perthitic K-feldspar to microcline contributes to the bleaching. The transition in the luster and color of the rocks is also coordinated by the replacement of orthopyroxene with biotite and amphibole. The phenomenon is observed over a minimum distance of 150 km, from the western limb of Mühlig–Hofmannfjella [15] to the east of Orvinfjella [14], which illustrates its regional importance.

The veins in the center of the alteration halos have a granitoid composition and occur as aplite or pegmatite. The mineral assemblage in the veins indicates formation from a hydrous granitic melt, for which the solidus temperature is typically about 650–700 °C at a pressure of about 0.3 GPa (e.g., [50]). Myrmekite is a common feature indicating subsolidus ionic reactions between the feldspar and a fluid phase. Based on the study of field relations, mineralogical transformations, microstructures, and fluid inclusions, the level of emplacement is considered to have been the middle crust near the crustal-scale brittle-ductile transition zone at temperatures of about 300–400 °C [16,17].

5.2. Interpretation of Fluid Source—Stable Isotopes

The fluid composition of the alteration zones has been studied by Engvik and Stöckhert [17], revealing a dominance of H_2O-CO_2, with variable and mostly subordinate N_2. The variable proportions of N_2-inclusions in the granitic veins are presumably derived from a source in the host rock. Aqueous inclusions-bearing silicate daughter crystals indicate a very high solute content and correspondingly high temperatures of formation. Based on the geological setting, combined with the fluid inclusion study, the infiltration was interpreted as late-magmatic derived from the hydrous granitic melt intruding the fractures represented by the central pegmatitic and aplitic veins. This observation suggests vigorous fluid circulation and fluid mixture.

The stable isotopic composition of silicate mineral separates can reflect the origin of both the rocks and the infiltrating fluid [51,52]. It will retain information from the protolith phases, but, depending on the degree of alteration and replacement, the isotopic composition will undergo a shift during fluid infiltration. The measurements of the pristine orthopyroxene-bearing syenite (sample AHA197) should give constraints on the source and origin of the rock. For this rock, the $\delta^{18}O_{SMOW}$ composition ranges from 6.4 to 9.1‰ for quartz, K-feldspar, amphibole, and biotite (Table 3) and is in accordance with a magmatic precursor and original values from such protoliths [53].

In altered granitoids, oxygen is already present in significant concentrations in the silicate minerals, and shifting its $\delta^{18}O$ composition will require large amounts of infiltrating fluids. The $\delta^{18}O_{SMOW}$ compositions of the minerals in the altered sample AHA200 are within the same range as those in the pristine orthopyroxene syenite, showing 6.0 to 9.9‰ (Table 3). However, the sample is strongly altered; thus, the preservation of the magmatic $\delta^{18}O_{SMOW}$ signature supports a magmatic origin of the fluid responsible for the alteration. This is in accordance with the interpretation from the fluid inclusion study of Engvik and Stöckhert [17].

For δD-values, the measured δD of the biotite of the pristine orthopyroxene-bearing syenite (sample AHA197) is -101‰, while the mineral of the altered sample (AHA200) yielded -289‰. The measured δD-signature of this orthopyroxene-bearing syenite is in accordance with an upper mantle-derived magma [54–56]. Wang et al. [57] found that there is significant hydrogen isotope fractionation between coexisting structural units in a melt and also strong fluid/melt H/D fractionation effects on H_2O dissolved in melts of hydrous magmatic systems. During the strong alteration caused by the fluid infiltration in our study area, biotite is completely replaced (Figure 4b–d). This implies that the H-isotopic signature of the altered sample will be strongly influenced by the infiltrating fluid. The large discrepancy of the δD-values measured for the biotite in the altered rock in comparison to the pristine orthopyroxene-bearing syenite can be explained by variable redox conditions during changing oxygen fugacity in the late-magmatic processes [55].

5.3. Geochemical Response to Fluid Infiltration

Figure 9 illustrates the bulk rock geochemical changes crossing alteration fronts from pristine orthopyroxene-bearing granitoid samples (left) to the altered variant (right). One of the sample pairs (106–107) shows a very distinct chemical behavior, different from the other samples, and this suggests that the altered rock was not an original equivalent of the pristine rock. In the following discussion, the data from this pair are therefore not further considered.

In the granitoids, there is a consistent increase in SiO_2 in the alteration zone, coupled with general decreases for TiO_2, Fe_2O_3, MgO, MnO, CaO and P_2O_3. The contents of Al_2O_3 and Na_2O show both moderate increases and decreases (Figure 9). In the gneisses, there are distinct increases for Al_2O_3 and K_2O, and to some degree for MgO, and distinct decreases for Fe_2O_3, MnO, and CaO but inconsistent changes for the other elements.

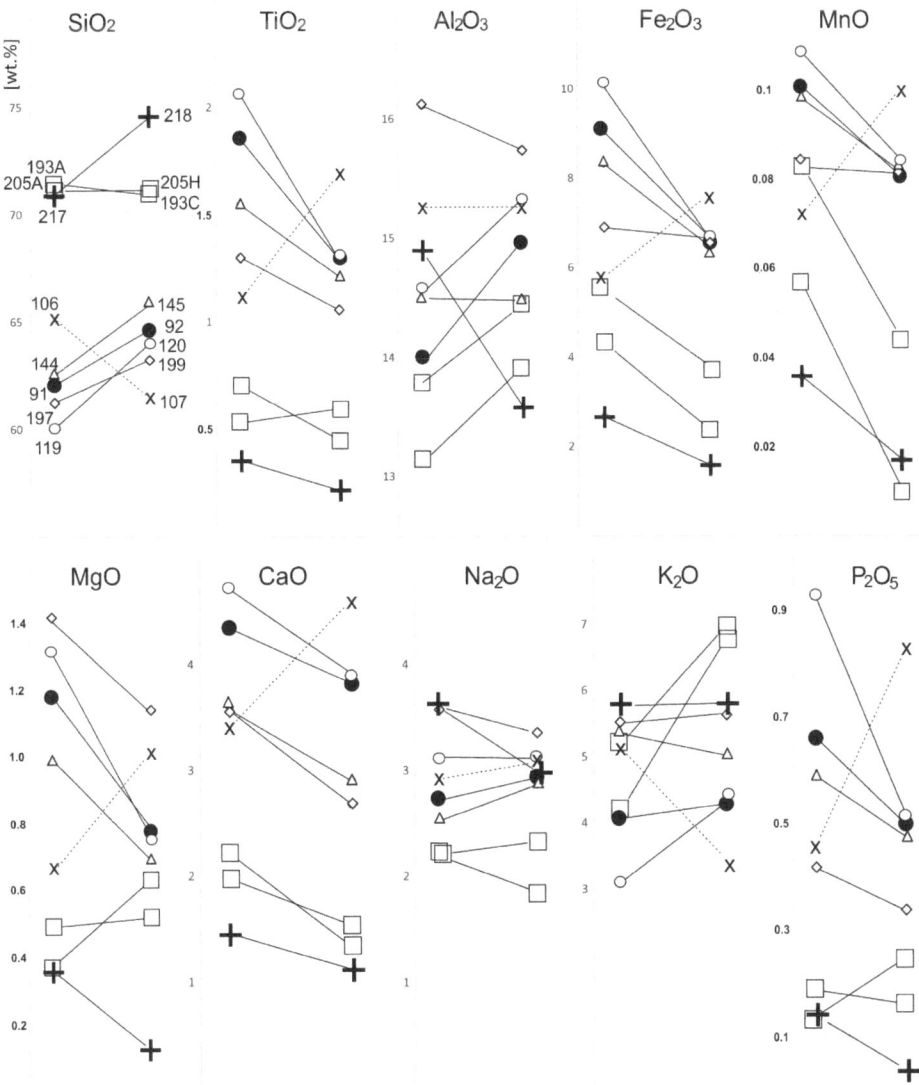

Figure 9. Geochemical variation crossing alteration zones from unaltered rock (**left**) to altered rock (**right**). Symbols are the same as in Figure 6.

The bulk geochemical response to the fluid alteration (Figure 9) can be explained in part by reactions and mass transport connected to the replacement of orthopyroxene and amphibole (Figure 4a–d). Given that TiO_2 is mainly hosted by ilmenite, its general decrease can be due to the breakdown of ilmenite during rock alteration, partly replaced by titanite (Figure 4d,e). However, TiO_2 is also a trace element in amphibole and biotite and may be affected during the replacement processes of these minerals.

The decrease in CaO (Figure 9) can be explained by plagioclase and perthite replacement to albite, as seen by the albite-filled microveins which in the strongest altered samples occur with a high density (Figure 5a–d). Na_2O and K_2O are both highly fluid mobile elements; however, they show smaller or variable changes during alteration. These elements tend to be kept in place by their incorporation both in the feldspar replacement products albite and microcline (Figure 5) and by the replacement product biotite and very fine-spread sericite (Figure 5a,b,e,f). The parallel decrease in P and Ca also suggests a possible connection with apatite, which may have been affected during the alteration process. Apatite is, however, present on both sides of the alteration front, and thus, the hypothesis is difficult to verify. The data also reveal a general decrease in the Zr abundance in the altered samples (Table 5). Given that Zr is mainly carried by zircon, which shows essentially no evidence of having been affected by the alteration event, the reduction in Zr points to a dilution effect rather than loss. This may also explain the reduction in the P abundance.

The changes in the isotopic ratios (Figure 8) mimic in part the changes in the major elements. For example, for pairs 144–145 and 197–199, there is a significant increase in the Rb-Sr ratio, the result of increasing Rb and decreasing Sr (Tables 4 and 5). The Sr decrease mirrors the decreases in Ca, Mg, and Fe, which may be due to the transformation of hornblende to biotite. The increase in Si may reflect its addition through external fluids, supported perhaps by the microveining seen in the feldspar (Table 1) but could also reflect relative enrichment due to the loss of other major components. A similar pattern also characterizes sample pairs 91–92 and 119–120, with the exception of Al, which increases in both cases.

The opposite behavior is observed for the granite pair 217–218. In this case, the Rb-Sr ratio decreases significantly, but apart from the increase in silica, all of the other considered elements also decrease. These changes presumably reflect the replacement of orthopyroxene by amphibole and biotite and dilution by incoming solutions adding more Si, K, and Rb.

5.4. Alteration Effects on the Sr and Nd Isotopic Systems

The Rb-Sr isotopic data for the granitoids (Figure 8a,b) plot along an array, which is scattered, indicating some heterogeneity in the initial Sr composition of the various units and the fluids causing alteration. Nevertheless, the degree of scatter is approximately the same for the altered and unaltered sets, indicating that the fluids affecting the rocks were likely derived from a magmatic system with comparable isotopic signatures.

The Sm-Nd results (Figure 8c) obtained for the same granitoids show a much more disordered behavior, which likely reflects both the initial heterogeneity of the various magma batches and modifications of the Sm-Nd ratios during alteration.

For the two sets of gneisses, the Sm-Nd system yields a true isochron, showing that the fluid causing chemical and mineralogical changes did not disrupt the Nd isotopic composition. This coherence of the Sm-Nd system implies a co-magmatic source of the fluid. That coherence is partially also observed for the Rb-Sr data of the gneisses, which define an isochron, except for one of the altered gneiss samples (205H). The Sr composition of this sample deviates significantly from the isochron, and trends toward less radiogenic compositions such as those in the granitoids. One may therefore conclude that this gneiss location was intruded by material derived from the granitoid magmatic system.

5.5. Timing of the Processes and Implications for the Source of Magmas and Fluids

The new U-Pb ages of a sample pair from the orthopyroxene-bearing syenite to quartz monzonite of Trollslottet (Locality 5) constrain the ages for both the magmatic crystallization and the fluid-induced alteration. The U-Pb zircon data of both the pristine host rock and the altered samples date the magmatic intrusion to 520 ± 1 Ma (Figure 7). During the intense alteration, ilmenite is replaced by titanite; thus, the U-Pb titanite age of 485.5 ± 1.4 Ma (Figure 7) constrains the time of alteration. This age overlaps with the youngest episode of magmatism recorded in the area [21,32], supporting a late-magmatic source of the infiltrating fluid. Although the titanite age documents a 35-million-year discrepancy between the time of magmatic emplacement and alteration at Trollslottet, the concept of an autometasomatic processes, as suggested by Markl and Henjes-Kunst [14] could still be valid in some cases, and autometasomatism may have been responsible for parts of the more diffuse alteration fronts in the area.

The Rb-Sr and Sm-Nd data for the two sets of gneisses from Kubusfjellet define isochrons with ages of 517 ± 6 Ma and 536 ± 23 Ma, respectively, that match the U-Pb zircon age of the granitoids (Figure 8a,c). The initial ratio of the Rb-Sr isochron for the gneisses ($^{87}Sr/^{86}Sr$ = 0.74869) is significantly more radiogenic than that of the granitoids ($^{87}Sr/^{86}Sr$ = 0.70891), in principle suggesting a derivation of the gneisses from older sources than the granitoids. By contrast, for the Sm-Nd system, the difference in the isotopic ratios between gneisses and granitoids is much less substantial and in part non-existent. Moreover, the Nd isotopic ratios and ε_{Nd} of the gneisses are more radiogenic than those of most of the granitoid rocks (ε_{Nd} about −5.5 for the gneisses and −5 to −10 for the granitoids), which as such implies a derivation of the granitoids from older crust than the gneisses, the opposite information to that provided by the Rb-Sr system. Since this behavior is shown by both unaltered and altered samples, the conflicting information cannot be attributed to alteration, although the latter probably added some further complications. The very radiogenic Sr composition of the gneisses, and their moderately negative ε_{Nd} values suggest instead that the rocks originated from an AFC (assimilation and fractional crystallization) process involving juvenile melts interacting with relatively young but highly metasomatized crustal components during alteration, causing its complexity). Alternatively, SCLM (subcontinental lithospheric mantle) could also host these crustal signatures for Sr and Nd isotopes. The contrasting isotopic signature of the granitoids, with lower ε_{Nd} values and relatively non-radiogenic $^{87}Sr/^{86}Sr$, is instead compatible with a derivation principally from the melting of Mid-Proterozoic crust with variable addition of magmas derived from mantle sources affected by more minimal metasomatic effects.

Our new U-Pb data are in accordance with the earlier constrained ages of the voluminous magmatic complex extending between 2° and 13° E of Pan-African post-tectonic granitoid intrusions ranging 530–485 Ma in age [15,21,32,48].

6. Concluding Remarks

Fluids intruding a dry orthopyroxene-bearing granitoid and gneissic complex in Dronning Maud Land, Antarctica, caused changes to the rock appearance, mineralogy, and chemistry of the rocks. The main mineralogical changes are the replacement of orthopyroxene by hornblende and biotite, ilmenite by titanite, and various changes to the feldspar composition. Geochemically, these processes resulted in general gains of Si, mostly of Al, and marginally of K and Na, but losses of Fe, Mg, Ti, Ca, and P. The alteration is dated by U-Pb of newly formed titanite at 485.5 ± 1.4 Ma whereas zircon in both the altered and unaltered syenite to quartz-monzonite indicates a primary crystallization age of 520.2 ± 1.0 Ma. The Rb-Sr and Sm-Nd data for two sets of gneiss samples yield ages of 517 ± 6 Ma and 536 ± 23 Ma. The initial Sr and Nd isotopic ratios suggest derivation of the gneisses from a relatively juvenile source but with a very strong metasomatic effect that introduced radiogenic Sr into the system. The granitoid data indicate instead a derivation from the Mid-Proterozoic crust, probably with additions of mantle components.

Author Contributions: Conceptualization, A.K.E., F.C., and I.C.K.; fieldwork, A.K.E., and S.E.; formal analysis, A.K.E., F.C., I.C.K., and H.T.; investigation, data curation, A.K.E., F.C., I.C.K., H.T., and S.E.; writing—original draft preparation, A.K.E., and F.C.; writing—review and editing, A.K.E., F.C., I.C.K., H.T., and S.E. All authors have read and agreed to the published version of the manuscript.

Funding: The samples for this study were collected during the Norwegian Antarctic Research Expedition 1996/1997, funded and arranged by the Norwegian Research Council and Norwegian Polar Institute. This work was supported by the Geological Survey of Norway.

Data Availability Statement: The data presented in this study are available in the paper.

Acknowledgments: We thank Håkon Austrheim and Øyvind Sunde for fruitful discussions. We are thankful to the two anonymous reviewers whose constructive criticism improved the paper.

Conflicts of Interest: The authors declare no conflicts of interest.

References

1. FitzGerald, J.D.; Stünitz, H. Deformation of granitoids at low metamorphic grade. I: Reactions and grain size reduction. *Tectonophysics* **1993**, *221*, 269–297. [CrossRef]
2. Putnis, A. Fluid-Mineral Interactions: Controlling Coupled Mechanisms of Reaction, Mass Transfer and Deformation. *J. Petrol.* **2021**, *62*, 1–27. [CrossRef]
3. Wayte, G.J.; Worden, R.H.; Rubie, D.C.; Droop, G.T.R. A TEM study of disequilibrium plagioclase breakdown at high pressure: The role of infiltrating fluid. *Contrib. Mineral. Petrol.* **1989**, *101*, 426–437. [CrossRef]
4. Austrheim, H. Eclogitization of lower crustal granulites by fluid migration through shear zones. *Earth Planet. Sci. Lett.* **1987**, *81*, 221–232. [CrossRef]
5. Pennacchioni, G. Progressive eclogitization under fluid-present conditions of pre-Alpine mafic granulites in the Austroalpine Mt Emilius Klippe (Italian Western Alps). *J. Struct. Geol.* **1996**, *14*, 1059–1077. [CrossRef]
6. Kleine, B.I.; Skelton, A.D.L.; Huet, B.; Pitcairn, I.K. Preservation of Blueschist-facies Minerals along a Shear Zone by Coupled Metasomatism and Fast-flowing CO_2-bearing Fluids. *J. Petrol.* **2014**, *55*, 1905–1939. [CrossRef]
7. Weisheit, A.; Bons, P.D.; Elburg, M.A. Long-lived crustal-scale fluid flow: The hydrothermal mega-breccia of Hidden Valley, Mt. Painter Inlier. *South Aust. Int. J. Earth Sci.* **2013**, *102*, 1219–1236. [CrossRef]
8. Kamenetsky, V.S.; Naumov, V.B.; Davidson, P.; van Achterbergh, E.; Ryan, C.G. Immiscibility between silicate magmas and aqueous fluids; a melt inclusion pursuit into the magmatic- hydrothermal transition in the Omsukchan Granite (NE Russia). *Chem. Geol.* **2004**, *210*, 73–90. [CrossRef]
9. Putnis, A. Mineral replacement reactions: From macroscopic observations to microscopic mechanisms. *Mineral. Mag.* **2002**, *66*, 689–708. [CrossRef]
10. Ettner, D.C.; Bjørlykke, A.; Andersen, T. Fluid evolution and Au-Cu genesis along a shear zone: A regional fluid inclusion study of shear zone-hosted alteration and gold and copper mineralization in the Kautokeino greenstone belt, Finnmark, Norway. *J. Geochem. Explor.* **1993**, *49*, 233–267. [CrossRef]
11. Oliver, N.H.S.; Cleverley, J.S.; Mark, G.; Pollard, P.J.; Bin, F.; Marshall, L.J.; Rubenach, M.J.; Williams, P.J.; Baker, T. Modelling the role of sodic alteration in the genesis of iron oxide-copper-gold deposits, eastern Mount Isa block, Australia. *Econ. Geol.* **2004**, *99*, 1145–1176. [CrossRef]
12. Liebscher, A. Experimental studies in model fluid systems. In *Fluid-Fluid Interactions*; Reviews in Mineralogy and Geochemistry; Liebscher, A., Heinrich, C.A., Eds.; Walter de Gruyter GmbH & Co KG: Berlin, Germany, 2007; Volume 65, pp. 15–48.
13. Mark, G.; Foster, D.R.W. Magmatic albite-actinolite-apatite-rich rocks from the Cloncurry district, Northwest Queensland, Australia. *Lithos* **2000**, *51*, 223–245. [CrossRef]
14. Markl, G.; Henjes-Kunst, F. Magmatic conditions of formation and autometasomatism of post-kinematic charnockites in Central Dronning Maud Land, East Antarctica. *Geol. Jahrb.* **2004**, *B96*, 139–188.
15. Bucher, K.; Frost, B.R. Fluid transfer in high-grade metamorphic terrains intruded by anorogenic granites: The Thor range, Antarctica. *J. Petrol.* **2005**, *47*, 567–593. [CrossRef]
16. Engvik, A.K.; Kalthoff, J.; Bertram, A.; Stöckhert, B.; Austrheim, H.; Elvevold, S. Magma-driven hydraulic fracturing and infiltration of fluids into the damaged host rock an example from Dronning Maud Land, Antarctica. *J. Struct. Geol.* **2005**, *27*, 839–854. [CrossRef]
17. Engvik, A.K.; Stöckhert, B. The inclusion record of fluid evolution crack healing and trapping from a heterogenous system during rapid cooling of pegmatitic veins (Dronning Maud Land; Antarctica). *Geofluids* **2007**, *7*, 171–185. [CrossRef]
18. Engvik, L.; Stöckhert, B.; Engvik, A.K. Fluid infiltration, heat transport, and healing of microcracks in the damage zone of magmatic veins: Numerical modeling. *J. Geophys. Res.* **2009**, *114*, B05203. [CrossRef]
19. Stern, R.J. Arc assembly and continental collision in the Neoproterozoic East African Orogen: Implications for the consolidation of Gondwanaland. *Annu. Rev. Earth Planet. Sci.* **1994**, *22*, 319–351. [CrossRef]
20. Jacobs, J.; Thomas, R.J. Himalayan-type indenter-escape tectonics model for the southern part of the late Neoproterozoic-early Paleozoic East-African-Antarctic Orogen. *Geology* **2004**, *32*, 721–724. [CrossRef]
21. Paulsson, O.; Austrheim, H. A geochronological and geochemical study of rocks from the Gjelsvikfjella, Dronning Maud Land, Antarctica—implications for Mesoproterozoic correlations and assembly of Gondwana. *Precambrian Res.* **2003**, *125*, 113–138. [CrossRef]
22. Jacobs, J.; Bauer, W.; Fanning, C.M. New age constraints for Grenville-age metamorphism in western central Dronning Maud Land (East Antarctica), and implications for the paleogeography of Kalahari in Rodinia. *Int. J. Earth Sci.* **2003**, *92*, 301–315. [CrossRef]
23. Baba, S.; Horie, K.; Hokada, T.; Owada, M.; Adachi, T.; Shiraishi, K. Multiple Collisions in the East African-Antarctic Orogen: Constraints from Timing of Metamorphism in the Filchnerfjella and Hochlinfjellet Terranes in Central Dronning Maud Land. *J. Geol.* **2015**, *123*, 55–78. [CrossRef]

24. Jacobs, J.; Bingen, B.; Thomas, R.J.; Bauer, W.; Wingate, M.T.D.; Feitio, P. Early Paleoproterozoic orogenic collapse and voluminous late tectonic magmatism in Dronning Maud Land and Mozambique: Insight into the partially delaminated orogenic root of the East African-Antarctic Orogen? In *Geodynamic Evolution of East Antarctica: A Key to the East-West Gondwana Connection*; Satish-Kumar, M., Motoyoshi, Y., Osanai, Y., Hiroi, Y., Shiraishi, K., Eds.; Geological Society London Special Publication: London, UK, 2008; Volume 308, pp. 69–90.
25. Jacobs, J.; Bauer, W.; Fanning, C.M. Late Neoproterozoic/Early Palaeozoic events in central Dronning Maud Land and significance for the southern extension of the East African Orogen into East Antarctica. *Precambrian Res.* **2003**, *126*, 27–53. [CrossRef]
26. Elvevold, S.; Engvik, A.K.; Abu-Alam, T.S.; Myhre, P.I.; Corfu, F. Prolonged high-grade metamorphism of supracrustal gneisses from Mühlig-Hofmannfjella, central Dronning Maud Land (East Antarctica). *Precambrian Res.* **2020**, *339*, 105618. [CrossRef]
27. Bisnath, A.; Frimmel, H.E. Metamorphic evolution of the Maud Belt: P-T-t path for high-grade gneisses in Gjelsvikfjella Dronning Maud Land, East Antarctica. *J. Afr. Earth Sci.* **2005**, *43*, 505–524. [CrossRef]
28. Elvevold, S.; Engvik, A.K. Pan-African decompressional P-T path recorded by granulites from central Dronning Maud Land, Antarctica. *Mineral. Petrol.* **2013**, *107*, 651–664. [CrossRef]
29. Jacobs, J.; Klemb, R.; Fanning, C.M.; Bauer, W.; Colombo, F. Extensional collapse of the late Neoproterozoic-Early Paleozoic East African-Antarctic Orogen in central Dronning Maud Land, East Antarctica. In *Proterozoic East Gondwana: Supercontinent Assembly and Breakup*; Yoshida, M., Windley, B.F., Dasgupta, S., Eds.; Geological Society London Special Publication: London, UK, 2003; Volume 206, pp. 271–288.
30. Engvik, A.K.; Elvevold, S. Pan-African extension and near-isothermal exhumation of a granulite facies terrain, Dronning Maud Land, Antarctica. *Geol. Mag.* **2004**, *141*, 1–12. [CrossRef]
31. Hendriks, B.W.H.; Engvik, A.K.; Elvevold, S. $^{40}Ar/^{39}Ar$ record of late Pan-African exhumation of granulite facies terrain, central Dronning Maud Land, East Antarctica. *Mineral. Petrol.* **2013**, *107*, 665–677. [CrossRef]
32. Jennings, E.S.; Marschall, H.R.; Hawkesworth, C.J.; Storey, C.D. Characterization of magma from inclusions in zircon: Apatite and biotite work well, feldspar less so. *Geology* **2011**, *39*, 863–866. [CrossRef]
33. Krogh, T.E. A low-contamination method for hydrothermal decomposition of zircon and extraction of U and Pb for isotopic age determinations. *Geochim. Et Cosmochim. Acta* **1973**, *37*, 485–494. [CrossRef]
34. Mattinson, J.M. Zircon U-Pb chemical abrasion ("CA-TIMS") method: Combined annealing and multi-step partial dissolution analysis for improved precision and accuracy of zircon ages. *Chem. Geol.* **2005**, *220*, 47–66. [CrossRef]
35. Krogh, T.E. Improved accuracy of U-Pb zircon ages by the creation of more concordant systems using an air abrasion technique. *Geochim. Et Cosmochim. Acta* **1982**, *46*, 637–649. [CrossRef]
36. Corfu, F. U-Pb Age, Setting and Tectonic Significance of the Anorthosite-Mangerite-Charnockite-Granite Suite, Lofoten-Vesterålen, Norway. *J. Petrol.* **2004**, *45*, 1799–1819. [CrossRef]
37. Jaffey, A.H.; Flynn, K.F.; Glendenin, L.E.; Bentley, W.C.; Essling, A.M. Precision measurement of half-lives and specific activities of U-235 and U-238. *Phys. Rev.* **1971**, *C4*, 1889.
38. Ludwig, K.R. *Isoplot 4.1: A geochronological toolkit for Microsoft Excel*; Berkeley Geochronology Center Special Publications: Berkeley, NC, USA, 2009; Volume 4.
39. Whitney, D.L.; Evans, B.W. Abbreviations of names of rock-forming minerals. *Am. Mineral.* **2010**, *95*, 185–187. [CrossRef]
40. Stacey, J.S.; Kramers, J.D. Approximation of terrestrial lead isotope evolution by a two-stage model. *Earth Planet. Sci. Lett.* **1975**, *26*, 207–221. [CrossRef]
41. Sharp, Z.D. A laser-based microanalytical method for the in-situ determination of oxygen isotope ratios of silicates and oxides. *Geochim. Et Cosmochim. Acta* **1990**, *54*, 1353–1357. [CrossRef]
42. Rumble, D.; Hoering, T.C. Analysis of oxygen and sulfur isotope ratios in oxide and sulfide minerals by spot heating with a carbon dioxide laser in a fluorine atmosphere. *Acc. Chem. Res.* **1994**, *27*, 237–241. [CrossRef]
43. Kasemann, S.; Meixner, A.; Rocholl, A.; Vennemann, T.; Schmitt, A.; Wiedenbeck, M. Boron and oxygen isotope composition of certified reference materials NIST SRM 610/612, and reference materials JB-2G and JR-2G. *Geostand. Newsl.* **2001**, *25*, 405–416. [CrossRef]
44. Vennemann, T.W.; O'Neil, J.R. A simple and inexpensive method of hydrogen isotope and water analyses of minerals and rocks based on zinc reagent. *Chem. Geol.* **1993**, *103*, 227–234. [CrossRef]
45. Rehkamper, M.; Hofmann, A.W. Recycled ocean crust and sediment in Indian Ocean MORB. *Earth Planet. Sci. Lett.* **1997**, *147*, 93–106. [CrossRef]
46. Bouvier, A.; Vervoort, J.D.; Patchett, P.J. The Lu-Hf and Sm–Nd isotopic composition of CHUR: Constraints from unequilibrated chondrites and implications for the bulk composition of terrestrial planets. *Earth Planet. Sci. Lett.* **2008**, *273*, 48–57. [CrossRef]
47. Peucat, J.J.; Vidal, P.; Bernard-Griffith, J.; Condie, K.C. Sr, Nd, and Pb Isotopic Systematics in the Archean Low- to High-Grade Transition Zone of Southern India: Syn-Accretion vs. Post-Accretion Granulites. *J. Geol.* **1989**, *97*, 537–549. [CrossRef]
48. Engvik, A.K.; Corfu, F.; Kleinhanns, I.K.; Elvevold, S. Banded Charnockite: The Result of Crustal Magma Generation, Piecemeal Emplacement, and Fluid-Driven Mineral Replacement in High-Grade Crust (Central Dronning Maud Land, Antarctica). *J. Geol.* **2021**, *129*, 371–390. [CrossRef]
49. Engvik, A.K.; Elvevold, S. Late Pan-African fluid infiltration in the Mühlig-Hofmann and Filchnerfjella of central Dronning Maud Land, Antarctica. In *Antarctica: Contributions to Global Earth Sciences*; Damaske, D., Kleinschmidt, G., Miller, H., Tessensohn, F., Eds.; Springer: Berlin/Heidelberg, Germany; New York, NY, USA, 2006; pp. 55–62.

50. Huang, W.L.; Wyllie, P.J. Phase relationships of S-type granite with H_2O to 35 kbar: Muscovite from Harney Peak. South Dakota. *J. Geophys. Res.* **1981**, *86*, 10515–10529. [CrossRef]
51. Sheppard, S.M.F. Characterization and Isotopic Variations in Natural Waters. In *Stable Isotopes in High Temperature Geological Processes*; Reviews in Mineralogy; Valley, J.W., Taylor, H.P., O'Neil, J.R., Eds.; Walter de Gruyter GmbH & Co KG: Berlin, Germany, 1986; Volume 16, pp. 165–184.
52. Gregory, R.T.; Criss, R.E. Isotopic Exchange in Open and Closed Systems. In *Stable Isotopes in High Temperature Geological Processes*; Reviews in Mineralogy; Valley, J.W., Taylor, H.P., O'Neil, J.R., Eds.; Walter de Gruyter GmbH & Co KG: Berlin, Germany, 1986; Volume 16, pp. 91–128.
53. Taylor, H.P.; Sheppard, S.M.F. Igneous rocks: I. Processes of Isotopic Fractionation and Isotopic Systematic. In *Stable Isotopes in High Temperature Geological Processes*; Reviews in Mineralogy; Valley, J.W., Taylor, H.P., O'Neil, J.R., Eds.; Walter de Gruyter GmbH & Co KG: Berlin, Germany, 1986; Volume 16, pp. 91–128.
54. Clog, M.; Aubaud, C.; Cartigny, P.; Dosso, L. The hydrogen isotopic composition and water content of southern Pacific MORB: A reassessment of the D/H ratio of the depleted mantle reservoir. *Earth Planet. Sci. Lett.* **2013**, *381*, 156–165. [CrossRef]
55. Mysen, B. Hydrogen isotope fractionation between coexisting hydrous melt and silicate-saturated aqueous fluid: An experimental study in situ at high pressure and temperature. *Am. Mineral.* **2013**, *98*, 376–386. [CrossRef]
56. Dalou, C.; Le Losq, C.; Mysen, B.O. In situ study of the fractionation of hydrogen isotopes between aluminosilicate melts and coexisting aqueous fluids at high pressure and high temperature—Implications for the δD in magmatic processes. *Earth Planet. Sci. Lett.* **2015**, *426*, 158–166. [CrossRef]
57. Wang, Y.B.; Cody, G.D.; Cody, S.X.; Foustoukos, D.I.; Mysen, B.O. Very large intramolecular D-H oartitioning in hydrated silicate melts synthesized at upper mantle pressures and temperatures. *Am. Mineral.* **2015**, *100*, 1182–1189. [CrossRef]

Disclaimer/Publisher's Note: The statements, opinions and data contained in all publications are solely those of the individual author(s) and contributor(s) and not of MDPI and/or the editor(s). MDPI and/or the editor(s) disclaim responsibility for any injury to people or property resulting from any ideas, methods, instructions or products referred to in the content.

Article

Discrimination of Muscovitisation Processes Using a Modified Quartz–Feldspar Diagram: Application to Beauvoir Greisens

Michel Cathelineau * and Zia Steven Kahou

Université de Lorraine, CNRS, GeoRessources, F-54000 Nancy, France; zia-steven.kahou@univ-lorraine.fr
* Correspondence: michel.cathelineau@univ-lorraine.fr

Abstract: Alteration in greisen-type granites develops through the progressive replacement of feldspars by potassic micas. Under the name 'greisen', quartz–muscovite assemblages display differences and include a variety of facies with variable relative proportions of quartz and muscovite. In principle, feldspar conversion to muscovite is written usually considering constant aluminium, and should result in a modal proportion of six quartz plus one muscovite. In Beauvoir greisens, which result from albite-rich granite, the relative proportion of quartz–muscovite is in favour of muscovite. Such a balance results from a reaction that implies imputs of potassium and aluminium, thus different from the classic one. The Q'-F' diagram provides a graphical solution for discriminating between reaction paths. A representative series of greisen data from the literature is compared in this diagram: Beauvoir B1 unit, Cligga Head, Cinovec, Panasqueira, Zhengchong, and Hoggar.

Keywords: greisen; muscovite; quartz; differentiated granites; chemical–mineralogical diagram

Citation: Cathelineau, M.; Kahou, Z.S. Discrimination of Muscovitisation Processes Using a Modified Quartz–Feldspar Diagram: Application to Beauvoir Greisens. *Minerals* **2024**, *14*, 746. https://doi.org/10.3390/min14080746

Academic Editors: Ignez de Pinho Guimarães and Jefferson Valdemiro De Lima

Received: 3 July 2024
Revised: 19 July 2024
Accepted: 22 July 2024
Published: 25 July 2024

Copyright: © 2024 by the authors. Licensee MDPI, Basel, Switzerland. This article is an open access article distributed under the terms and conditions of the Creative Commons Attribution (CC BY) license (https://creativecommons.org/licenses/by/4.0/).

1. Introduction

Rocks derived from granites containing only quartz and micas are often called 'greisen' without specifying the relative abundance of quartz and micas. The origin of the word greisen dates back to the Middle Ages (Saxonian miners), and greisens are nowadays considered as a rock composed mainly of quartz and micas, often zinnwaldite, but not systematically. These minerals are accompanied by other minor mineral phases such as topaz, tourmaline, and fluorite [1–4]. The word greisen is used for quartz–muscovite alterations affecting granitic rocks, most often leucocratic differentiated granites, rich in albite, derived from a peraluminous or metaluminous magmatic suite [4–8].

$$3NaAlSi_3O_8 + K^+ + 2H^+ \leftrightarrow K(AlSi_3)(Al_2)O_{10}(OH)_2 + 3\,Na^+ + 6SiO_2$$
$$3\ Albite + K^+ + 2H^+ \leftrightarrow Muscovite + 3\,Na^+ + 6\ Quartz \tag{1}$$

The most famous greisens are those from Cligga Head, which meet the strict definition of greisens composed of quartz and mica [8,9]. The most striking feature of the Cligga Head granite is the sharp drop in sodium content, which implies almost total conversion of albite to muscovite. In principle, the reactions that form the greisens at Cligga Head follow the typical hydrothermal reaction where a feldspar (potassic or sodic) is replaced by muscovite in a reaction written considering aluminium as constant. For albite-rich granite, the reaction would be written as follows:

There are many examples of peraluminous granites with the replacement of mainly albite or several feldspars (microcline and albite) by micas. Located in the north of the French Massif Central, the albite–lepidolite–topaz Beauvoir granite represents a fascinating case of mica development. During the 'hydrothermal greisen' phase (as defined by [10] and later by [11,12]), albite, which predominates in unaltered granite, K-feldspar, which displays a variable content in the Beauvoir intrusion, and a part of lepidolite are replaced by muscovite, which is considered to be hydrothermal. The 'greisens' defined at Beauvoir contain mainly muscovite-type micas [12,13], which add to or replace the abundant fraction of phyllosilicates of magmatic origin, which occurs in the form of lepidolite at the Beauvoir B1 unit.

The albite–lepidolite–topaz Beauvoir granite represents an ideal evolved granite to study the evolution of micas and Li behaviour from the magmatic stage to the hydrothermal stage. It is located in the Sioule series of the northern French Massif Central [14]. At the Beauvoir B1 unit, lepidolite constitutes the main magmatic phyllosilicate, and its proportion is close to ~20 wt.%. During the hydrothermal greisen phase, feldspars (mostly albite) and lepidolite are replaced by hydrothermal muscovite [10,12]. Cuney et al. [10] consider that greisen from Beauvoir are "classical quartz-muscovite fissure type" and limited to the top of the Beauvoir B1 unit. Fonteilles [15] devotes a few lines to the greisen and considers that it contains muscovites I and II and sericites but does not provide a detailed description. Historical studies on the drill hole GPF-1 did not address the evolution of the relative quantities of quartz and muscovite to monitor the alteration. Moreover, only eight whole-rock analyses, mostly on fresh granites, are available on the first 200 m of the GPF drill hole, including only one sample with significant alteration [10,16]. However, these two studies carried out on the drill hole GPF-1 did not address the type of evolution of the relative quantities of quartz and muscovite as an indicator of the relative mobility of Si and Al in the vicinity of greisens.

However, an examination of the geochemical data for the greisenised Beauvoir granite showed some notable geochemical differences from other greisens. This prompted a graphical representation of the main mineral transformations to distinguish the different types of reactions and chemical balances involved in these alterations. The Q'-F' diagram [17], which breaks down the three main mineral poles of granites in the well-known quartz feldspar triangle (quartz, albite, and microcline), was used, taking into account the geochemical data for total rock as well as the crystallochemical data for the main granite-forming minerals and greisens. Significant differences emerged between the greisens from several localities prompting us also to consider data from the literature on several other well-known greisens such as those from Cligga Head [9], Cinovec (Germany-Czech Republic [6,7,18,19], Panasqueira (Portugal [5,20]), and others such as a greisen described in China in the region within the Shengdong batholith [21] and Hoggar [22].

2. Materials and Methods

2.1. Samples

Fresh granites and greisens were sampled for whole-rock analyses from three drill cores by Imerys in 2022 in the Beauvoir quarry (location map in Figure 1).

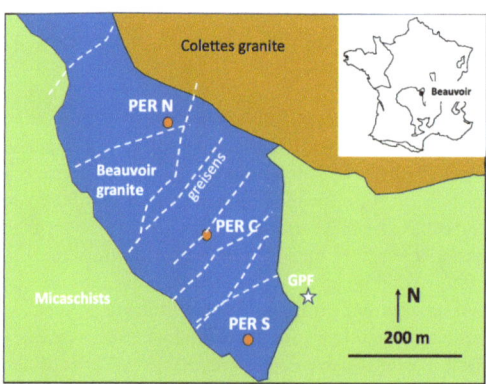

Figure 1. Location of the Beauvoir granite in the French Massif Central (inset) and location of the three studied drill holes (orange circles) from the Imerys 2022 campaign (PER N, C and S), and of the GPF deep drill hole.

2.2. Bulk Rock Analyses

Unaltered to altered granites were analysed in whole rock for significant elements. Whole-rock major and trace element analyses were performed on field samples at the Service d'Analyse des Roches et des Minéraux (SARM), Centre de Recherches Pétrographiques et Géochimiques (CRPG), Nancy, France. Major elements were analysed by inductively coupled plasma optical emission spectrometry on a Thermo-Fischer ICap 6500 173 instrument (Thermo Fisher Scientific, Waltham, MA, USA). Trace elements, including rare earth elements (REEs), were determined by inductively coupled plasma mass spectrometry (Thermo-Elemental X7, ESI, Omaha, NE, USA). Detailed analytical procedures are given in [23].

Bulk rock data on greisens from the literature for comparison are the following:

- Panasqueira: Two distinct granites are affected by greisens (partially greisenised G1, granite inferred as the main biotite granite) and pseudo-greisens affecting the G4 Rare Metal Granite from the cupola [6]. Two data sets are available: those from [6,20], which are mainly concerned with the cupola.
- Cinovec greisens occur as steep or flat zones of intensive metasomatic greisenisation along tens to hundreds of meters [7,24]. Feldspars are replaced by quartz, zinnwaldite, topaz, and fluorite. Cinovec granite geochemistry has been covered in numerous papers. However, the number of studies of greisens or greisenised granites is low. The authors of [6,7,18,25] have provided a series of whole-rock analyses on granites undergoing greisenisation. In greisen, zinnwaldite is re-equilibrated and then muscovitised [8].
- Cligga Head: Greisens conserve their granite texture and consist of quartz, protolithionite (inherited from the granite), muscovite, tourmaline, and topaz [8]. Hall [9] considers that quartz represents 52% (modal), muscovite 40%, tourmaline 5%, and topaz 0.8% of the greisen composition. The albite is entirely replaced by muscovite. Muscovite is already more abundant than protolithionite in the granite, but its abundance increases in the greisen, where protolithionite disappears [10]. Muscovites from greisen contain lithium around 2500 ppm Li_2O. A few whole-rock analyses are available in [10].
- In China, in the Zhengchong granite, several greisens were described by Liu et al. [21]. The greisen I is mainly composed of quartz (50%–55%), zinnwaldite (25%–30%), topaz (5%–10%), and fluorite (~5%). Greisen II exhibits a porphyritic-like texture where matrix and phenocrysts are observed. The matrix is mainly composed of quartz (45%–55%), zinnwaldite (15%–25%), topaz (10%–15%), and fluorite (~5%).
- In Hoggar, several greisens from the area Tamanrasset in the Pan-African Hoggar appear to result from two processes: the formation of quartz–topaz and quartz-rich greisens, followed by a quartz dissolution and its replacement by Li-rich micas, thus forming mica-rich greisens [22].

2.3. Method for Graphically Processing Geochemical Data: The Q'-F' Diagram

The chemical–mineralogical approach is based mainly on processing the analytical data for the major elements in a diagram that makes it easy to recognise the mineralogical significance of the chemical variations. It is based quantitatively on the chemical composition of the major elements analysed in bulk. It uses parametric processing of the data on the main chemical elements to reflect the actual mineralogy quickly and unambiguously, thus using 'chemo-mineralogical' diagrams in the same way as in [17,26,27].

The diagram (Figure 2) breaks down the locations of quartz, potassium feldspar, and plagioclase in a triangle close to the quartz–albite–potassium feldspar triangle [28]. Using the nomenclature of [28,29] for plutonic rocks, Debon and Lefort [27] have superimposed a new classification grid on this diagram. This diagram uses the two coordinates $Q' = Si/3 - (K + Na + 2 Ca/3)$ and $F' = K - (Na + Ca)$ expressed in gram-atoms $\times 10^3$ of each element in 100 g of rock or mineral; the values used for the diagram are thus proportional to the molar contents of each element (see Table S1 in the Supplementary Materials for the detailed calculation). Q' is proportional to the quantity by weight of quartz in common granitoid rocks, as it corresponds to the silicon not linked to feldspars and muscovite. The Q'-F' diagram (Q' = Si/3 − (K + Na) versus F' (K-Na)), modified by [17], eliminates variations in calcium content. This simplification is acceptable for hyperdifferentiated granites where most of the calcium is carried by apatite, as plagioclase is very close to albite and does not contribute to the calcium content. The Beauvoir granite, but also most of the other albitic peraluminous magmas, such as those of the Argemela intrusion [30], Segura [31,32] and references therein, which are similar to LCT pegmatites, have a calcium concentration monitored by apatite.

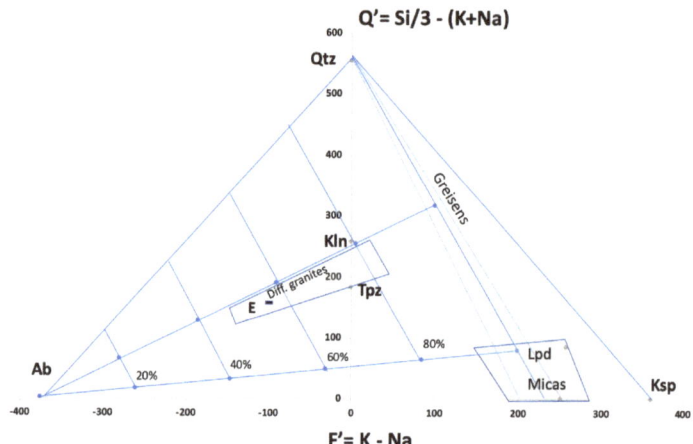

Figure 2. Principle of using the Q'-F' diagram with the location of the main rock-forming minerals and the main alteration trend. Qtz: quartz; Kln: kaolinite; Toz: topaz; Lpd: lepidolite; Ab: albite; Ksp: K-feldspar; E (square): eutectic of the granite system by [33]. Diff. granites: differentiated granite field.

This diagram is well suited to the classification of felsic plutonic rocks. Still, it has already been used to represent alteration, particularly the dissolution of quartz associated with the crystallisation of potassic minerals [17]. The diagram helps discriminate between rocks composed of quartz and muscovite issued from the alteration in felsic rocks, particularly greisen. Note that the equidimensional aspect of the triangle is not respected when expanding the F' axis to facilitate the reading of the greisen trends (see the following diagrams).

The main advantage of these chemical–mineralogical diagrams is that molar abundances are calculated, thus eliminating the effect of masses and allowing reactions between minerals to be represented as vectors whose slopes reflect the stoichiometry of reactions and structural formulae. Therefore, an albite–feldspar-K transformation is represented by a horizontal vector, and thus the transformation of albite into muscovite can be expressed—taking into account the assumptions concerning the constant element of the reaction—by vectors with well-defined and logical slopes.

3. Results

3.1. Mineralogical Analysis of Beauvoir Granite and Greisen

In the transition from fresh Beauvoir granite to greisen, the main mineral change is the replacement of albite with fine-grained muscovite. Thus, lepidolite is only of magmatic origin in the Beauvoir granite, and the newly formed micas, during later alteration, are muscovites. When present, the K-feldspar is also replaced in the most altered sample. However, the lepidolite is also replaced partially by muscovite. The shape and content of other minerals remain unchanged.

Composite and elemental maps using micro-X-ray fluorescence allowed us to determine quantitatively the relative mineral proportions [34] In the fresh sample, the mineralogy consists of albite (~45%), quartz (~25%), lepidolite (~20%), orthoclase (<10%), topaz and phosphates (<5%), and Sn-Nb-Ta oxides (<1%). In the altered facies, the feldspars tend to be replaced by hydrothermal muscovite and hydrothermal quartz. As a consequence, muscovite (~40%) and quartz (~40%) modal proportions increase while feldspars (<10%) and lepidolite (<~10%) proportions decrease.

The composition of lepidolite is relatively constant in fresh granite, and muscovite also has a well-defined composition, characterised by a very low Li content [12,13]. The analyses of the two types of K-micas were also reported in the Q'-F' diagram, where they form two very distinct clusters, both of which are aligned on the same quartz–mica line (Figure 3).

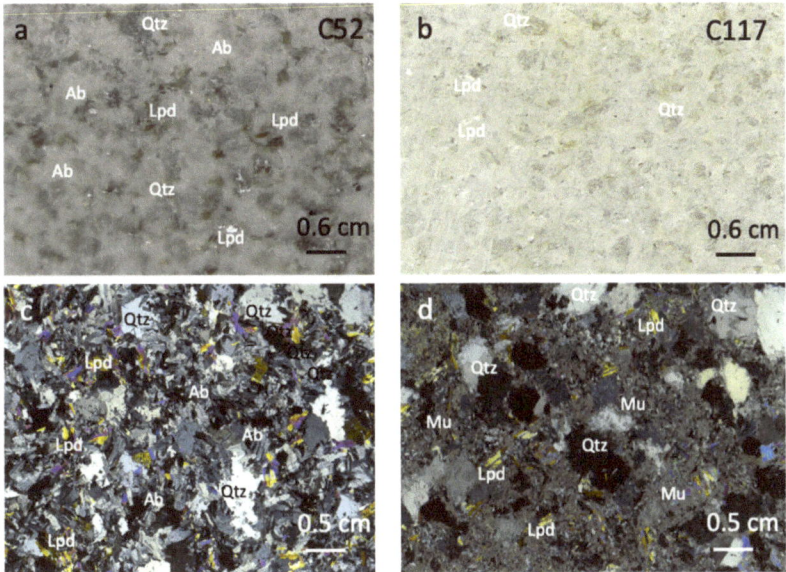

Figure 3. Petrography of the Beauvoir granite and related greisen from two samples collected from a drill hole realised in the central part of the Beauvoir quarry by Imerys (see Table 1 for whole-rock analyses). (**a,c**) fresh granite at 52 m depth, macroscopic view and corresponding thin section under crossed Nichols: coloured laths are lepidolites (Lpd), albite (Ab) is visible as elongated thin laths, and quartz (Qtz); (**b,d**) greisen at 117 m depth and corresponding thin section under crossed Nichols: albite is entirely replaced by fine-grained muscovite (Mu) and a part of the lepidolite is still visible as well as magmatic quartz. Fine-grained quartz is associated with fine-grained muscovite.

Table 1. Whole-rock analyses of representative samples of the Beauvoir granite (fresh, altered (muscovite-rich), and greisens). All numbers in %, except the two last lines, correspond to the calculated parameters Q' and F'. Sample labels correspond to two distinct drill holes: C: centre; S: south; and numbers correspond to the depth in meters. Samples are presented classified at increasing values of the parameter F'.

Sample	C123	C52	S113	C30	S58	C112	C39	C43	C42	S73	C117
	Fresh Granites					Altered Granites				Greisens	
SiO_2	66.68	68.24	68.96	68.45	68.08	69.50	67.94	65.49	65.35	71.38	72.28
TiO_2	0.00	0.00	0.00	0.00	0.00	0.00	0.00	0.00	0.00	0.00	0.00
Al_2O_3	19.35	17.28	18.19	18.13	17.31	17.69	18.02	18.12	17.54	18.22	17.88

Table 1. Cont.

Sample	C123	C52	S113	C30	S58	C112	C39	C43	C42	S73	C117
	Fresh Granites					Altered Granites				Greisens	
Fe_2O_3	0.06	0,13	0.08	0.05	0.05	0.10	0.07	0.17	0.25	0.10	0.07
FeO	bdl	0.80	0.30	0.08	0.15	0.14	0.05	0.19	0.18	0.10	0.12
MnO	0.02	0.09	0.04	0.04	0.02	0.07	0.05	0.04	0.06	0.05	0.06
MgO	0.00	0.00	0.05	0.00	0.00	0.00	0.00	0.06	0.12	0.03	0.04
CaO	0.33	0.75	0.30	0.28	0.54	0.68	0.69	1.40	2.07	0.62	0.10
Na_2O	10.36	5.95	5.84	5.80	5.26	4.70	4.46	4.25	2.04	0.09	0.11
K_2O	0.90	1.95	2.70	2.90	3.60	3.33	3.23	3.28	4.00	5.40	5.89
P_2O_5	0.50	2.45	0.35	1.14	1.21	1.26	1.09	1.29	1.62	0.51	0.10
L.O.I.	1.22	2.85	2.39	2.45	2.53	2.50	3.26	3.84	5.67	3.27	2.97
F	0.44	2.18	1.95	2.39	1.67	1.92	2.50	2.68	2.03	1.00	0.87
Total	99.43	100.56	99.22	99.32	98.76	99.99	98.86	98.14	98.93	99.78	99.53
F'	−315.0	−150.5	−131.2	−125.4	−93.5	−81.0	−75.3	−67.7	19.1	111.8	121.6
Q'	16.6	145.3	136.7	131.1	131.4	163.2	164.4	156.5	211.7	278.4	272.4

3.2. Diagram Q'-F' Applied to Granites and Greisens

3.2.1. Beauvoir Alteration Suite

Bulk-rock analyses of the Beauvoir granites and greisens are provided in Table 1. From unaltered albite-rich granites to profoundly altered and muscovite-rich granites, the main evolution of rock-forming major chemical elements is a progressive decrease in Na, correlatively to an increase in K. The unaltered granites and greisens of Beauvoir plotted in the Q'-F' diagram, covering all the facies of progressive alteration identified in petrographic studies, so-called reference series, are distributed according to a trend line from the albite-rich granites to a point situated on the line joining quartz to lepidolite and muscovite. The trend is covered, and the most altered point is relatively well defined and confined to a relative proportion of 45% and 55% of quartz and micas, respectively.

The trend line for the representative series of samples (red dots in Figure 4) is as follows:

$$Q' = 0.579 \, F' + 205.6 \qquad (R^2 = 0.97) \qquad (2)$$

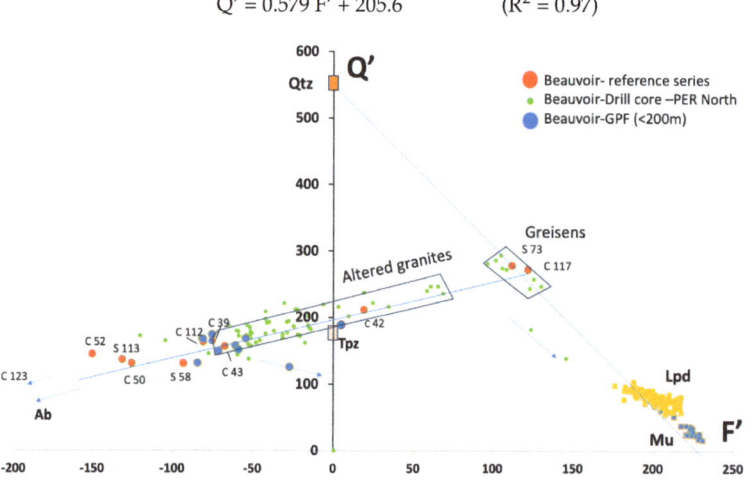

Figure 4. The Q'-F' diagram applied to the Beauvoir granite to greisen series: in red, the reference samples are characterised by petrography and quantitative mineralogy; labels of red data points refer to the samples from Table 1 (drill holes PER C and S (centre and south), numbers: depth in meter); and in green, the data for drill cores from the northern area of the quarry (PER North). The GPF data, in blue circles, are from [16]. Lpd: lepidolite population from Beauvoir after [35]). Ab: albite' Tpz: topaz; Qtz: quartz; Lpd: lepidolite; and Mu: muscovite. Lepidolite and muscovite analyses from [34]. The blue arrows indicate enrichments in micas.

The northern zone of the Beauvoir quarry is the most affected by greisens. Using the analyses carried out on 4 m samples from the north of the borehole from the 2022 campaign (Imerys analyses), all the points converge towards the same point as the analyses of the reference series (Figure 4). Two exceptions concern a muscovite-rich structure enriched for which a displacement of the data points is observed in the direction of muscovite.

3.2.2. Panasqueira

At Panasqueira, most of the available data concern the greisens developed at the expense of the granite G4 (RMG) outcropping at the apex (cupola) of the main granite. A significant part of the greisens is distributed along the quartz–muscovite line, towards the intersection of the Beauvoir trend with the quartz–lepidolite + muscovite trend of Beauvoir, where three data points display similar compositions to the Beauvoir greisen. The Q'-F' diagram, therefore, discriminates the greisenisation trend at Panasqueira from that of Beauvoir. In addition, a series of data points for greisens join the intersection with the quartz–muscovite line and the greisen from Beauvoir (orange arrow 'b' in Figure 5).

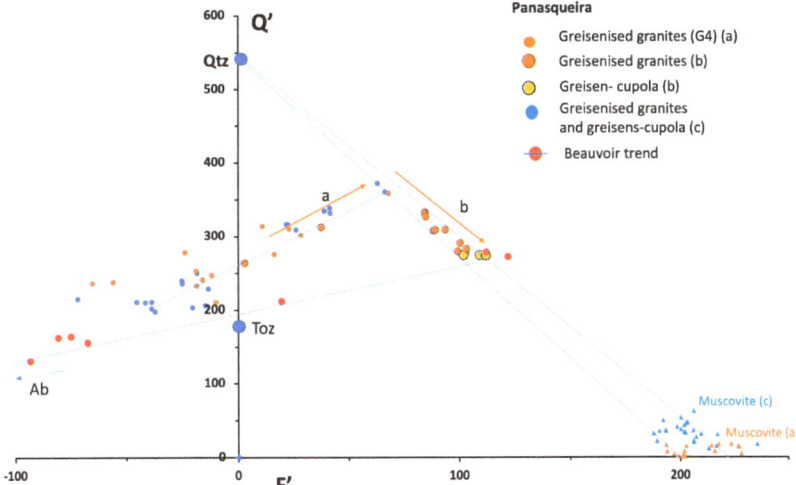

Figure 5. The Q'-F' diagram applied to the Panasqueira granite to greisen series: in red, the reference samples characterised in petrography and quantitative mineralogy from Beauvoir; in orange and yellow (data noted a and b), respectively, from [6,36] compiled in Marignac et al. [6], and in blue (data noted c from [20]). Trend I corresponds to quartz–mica development, and trend II corresponds to quartz loss and further mica enrichment. Micas are indicated by triangles using the same colours as whole rock from the same two references (micas from greisenised G4 granite (noted a) [6]; micas from cupola greisen [20] (noted c)).

3.2.3. Other Examples

Cligga Head and Cinovec

Cligga Head altered granites, as a result of incipient greisenisation, follow a trend subparallel to that of Beauvoir granites. Greisens, however, do not plot precisely in the continuity of this trend but are displaced in the direction of quartz from the intersection of this trend with the quartz–muscovite line (Figure 6).

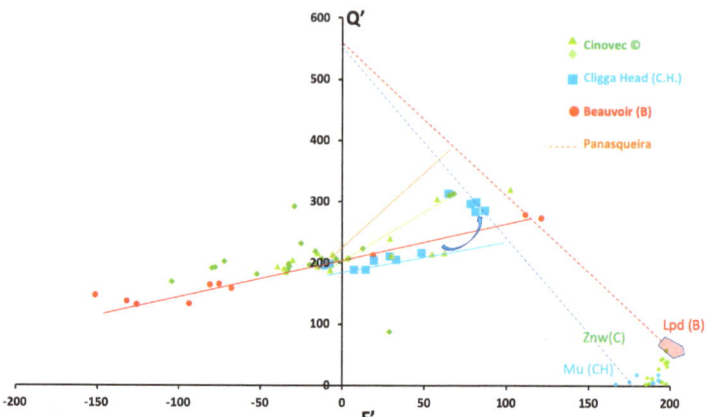

Figure 6. The Q'-F' diagram applied to the Cligga Head (CH, data from [9]) and Cinovec (C, data from [6,7,25]) granite to greisen series. The trend for Beauvoir whole-rock analyses is red (in pink, Beauvoir (noted B) lepidolite (Lpd)). Muscovite data are sourced from the same literature references as for the whole-rock analyses (Znw: zinnwaldite from Cinovec [8]; Mu: muscovites from Cligga Head [9]).

For Cinovec, the number of analyses of intermediate facies between granites and the corresponding greisens is low. Four data points are distributed around the trend obtained for Cligga Head and Beauvoir. Considering the three data points on greisens, they display similar features to those of Cligga Head, e.g., a displacement towards quartz from the intersection between the alteration trend and the quartz–zinnwaldite line (Figure 6).

The Zhengchong and Hoggar Greisens

In the Zongsheng granite, greisens are rich in topaz but plot in an intermediate domain, similar to that described for Cligga Head greisens. Data points are somewhat dispersed and do not follow clear evolutions (Figure 7).

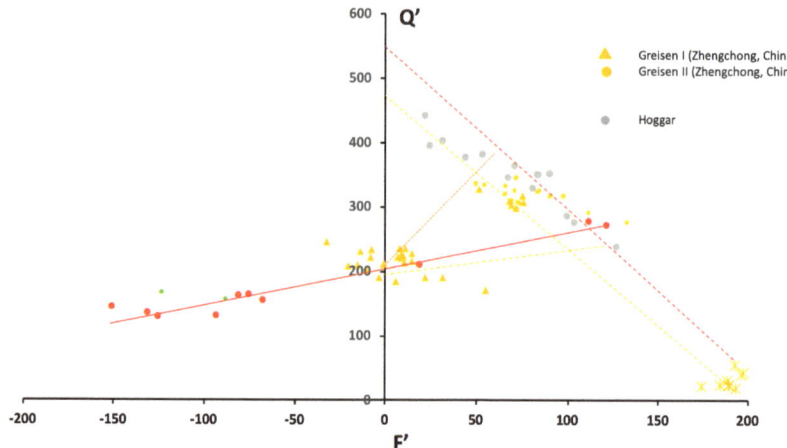

Figure 7. The Q'-F' diagram applied to the Zengchong (China, data from [21] in yellow) and Hoggar (in grey, data from [22]) granite to greisen series. Yellow stars correspond to mica analyses from [21] for Zenchong greisens. In red are the Beauvoir data points and their reference line for comparison.

Most of the samples from Hoggar greisens are characterised by a richness in quartz and a trend that follows the quartz–muscovite line in the direction of quartz. There is no clear evidence of the impact of quartz dissolution on bulk chemistry as most data points remain within the field described in all other mentioned examples, except for one data point slightly enriched in micas (Figure 7).

4. Discussion

4.1. Main Trend for the Beauvoir Granite

The primary trend corresponds to a trend from an albite-rich pole on the one hand to another pole characterised by enrichment in quartz and lepidolite. The first trend may be considered magmatic in origin and may correspond to different amounts of albite respective to lepidolite due to probable magmatic segregation among minerals.

In continuity with the magmatic trend, the alteration trend with a similar slope characterises the altered facies, where muscovite replaces mostly albite and lepidolite in the most extreme degrees of alteration. The above values are around $Q' = 150$, corresponding to $F' = -80$, and the points continue to line up on the same axis. Still, petrography indicates that points correspond to a significant albite replacement by a quartz–muscovite mixture. The quartz–muscovite amount increases to the greisen, but the relative amount of quartz versus phyllosilicates (lepidolite + muscovite) remains the same. The intersection of this trend with the quartz (lepidolite–muscovite) line corresponds to a rock constituted of around 40% quartz and 60% phyllosilicates. This trend does not correspond to a feldspar replacement trend at constant Al as would be expected in a greisen, which is a reaction generally written at constant Al for K-feldspar muscovitisation, as aluminium is typically not considered mobile at the hydrothermal stage:

$$3\ KAlSi_3O_8 + 2H^+ \leftrightarrow K(AlSi_3)(Al_2)O_{10}(OH)_2 + 2K^+ + 6SiO_2 \quad (3)$$

$$K-feldspar + 2H^+ \leftrightarrow Muscovite + 2K^+ + 6SiO_2$$

Here, the trend may be explained from the point of view of the bulk mass balance by roughly replacing albite with quartz and mica in similar amounts. Such relative amounts of quartz and K-micas pose problems concerning the mass balance of aluminium, which, in that case, cannot be considered constant.

The slope found for Beauvoir samples is relatively close to the theoretical slope defined by the following equation:

$$\begin{aligned}2\ NaAlSi_3O_8 + K^+ + Al^{3+} + 2\ H_2O \\ \leftrightarrow K(AlSi_3)(Al_2)O_{10}(OH)_2 + 3\ SiO_2 + 2\ H^+ + 2Na^+\end{aligned} \quad (4)$$

$$2\ Albite + K^+ + Al^{3+} + 2H_2O \leftrightarrow Muscovite + 3\ Quartz + 2\ H^+ + 2Na^+$$

which is an equal combination of two other equations written either at constant Si or, more traditionally, at constant Al (the classical equation proposed for greisens),

$$3\ NaAlSi_3O_8 + K^+ + 2H^+ \leftrightarrow K(AlSi_3)(Al_2)O_{10}(OH)_2 + 6SiO_2 + 3Na^+ \quad (5)$$

$$3\ Albite + K^+ + 2H^+ \leftrightarrow Muscovite + 6\ Quartz + 3\ Na^+$$

(constant Al, greisen equation),

$$NaAlSi_3O_8 + K^+ + 2Al^{3+} + 4H_2O \leftrightarrow K(AlSi_3)(Al_2)O_{10}(OH)_2 + 6H^+ + Na^+ \quad (6)$$

$$Albite + K^+ + 2Al^{3+} + 4H_2O \leftrightarrow Muscovite + 6H^+ + Na^+$$

(constant Si).

The resulting trend is noted as 'A' in Figure 8.

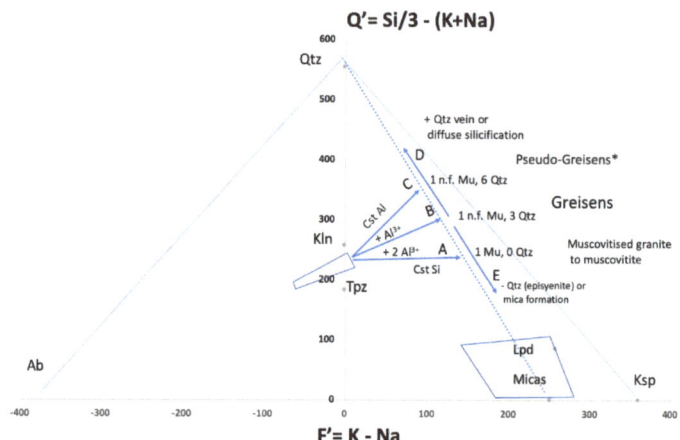

Figure 8. The Q'-F' diagram with the main alteration trends depending on the mobility of aluminium with three vectors: A: at constant Si (Cst Si); B: with the addition of one Al^{3+}, consisting of a combination of A (constant Si) and C (at constant Al); and additional processes such as quartz precipitation (D) or quartz dissolution and replacement by micas (E). *: Pseudo-greisens as defined by [6,22] as silicified granite, then dequartzified partially with quartz replacement by muscovite. n.f. Mu: newly formed muscovite; Qtz: quartz; and Lpd: lepidolite.

4.2. Differences among the Other Greisen Examples

The trends of progressive greisenisation are relatively complete at Beauvoir between fresh and altered granites. They differ significantly from those obtained for Panasqueira granites, for which the trend also almost continues towards the quartz–mica line with all intermediate compositions. It has, however, a slope distinct from that of Beauvoir. It falls onto the line of the conversion of feldspars to muscovite following Equations (3) and (4), thus with a mass balance produced by a constant aluminium reaction (trend 'B' in Figure 8).

The trends obtained at Cinovec and Cligga Head are very similar to those of Beauvoir for the first increment of greisenisation (trend 'A'). Greisens, however, are slightly displaced towards a composition more enriched in quartz and intermediate than the intersection between the greisen trend and the quartz–muscovite line, but without reaching the greisen line produced by a reaction at constant Al.

The trend obtained from Hoggar data is unclear and does not offer intermediate terms. The greisen trend reveals variable quartz contents covering greisens enriched in quartz–topaz and a muscovite greisen that are not very different from the Beauvoir greisen in composition. Both are not significantly enriched in muscovite.

The main differences between all the rocks, so-called 'greisens', concern the nature of the inherited undissolved minerals from the granite, the nature of the newly formed minerals, and the behaviour of silicon. Phyllosilicates differ, but most are muscovites, except for Cinovec, where the zinnwaldite is considered recrystallised, and muscovite late and not linked to greisenisation [7]. At Beauvoir, topaz is already present in the fresh granite, and no further newly formed topaz is identified in the greisen facies. On the contrary, topaz is linked to the greisen stage at Cligga Head and is accompanied by tourmaline and fluorite [9].

4.3. Aluminium Mobility and Muscovite Development

Many experimental works have emphasised aluminium mobility at temperatures higher than 400 °C. The richness in F, Li, and sometimes B could explain specific complexation and Al mobility in peraluminous granites ([35,37] and references therein) and porphyry–granite systems [38]. If Equation (3) represents the Beauvoir trend, it requires a contribution of potassium and aluminium from the fluid to replace the albite with muscovite. It generates two times less quartz, explaining the location of the greisens closer to the muscovite pole than the greisens from Panasqueira.

It is, however, rather difficult to identify the causes of the different trends as detailed fluid chemistry is not available. In most cases, fluids as fluid inclusions are most aqueous with low-density volatile components. The presence of high amounts of fluorine is an essential factor to consider. Burt [39] proposed two main greisenisation routes: (a) systems at high temperature (>500 °C) and relatively low pressure, where hydrothermal brines demix, thus separating a vapour phase rich in volatiles (HF and HCl) that are highly aggressive and responsible for the mica formation, and (b) greisens formed at lower temperatures (250 to 500 °C) but higher pressure without boiling. For Tagirov and Schott [40], in the case of a fluid initially containing 0.01 m of fluorine at 450 °C-1 kbar, a high mobility of Al occurs due to the formation of mixed Na-Al-O-H-F complexes. Then, during greisenisation, if the topaz precipitates, Al passes into the form of $Al(OH)^-$ and $NaAl(OH)^0$ to the detriment of the hydroxide fluoride species. The dissolution capacity of the fluid consequently decreases, and the Al mobility falls. Finally, Tagirov and Schott [41] suggest that Al-Si and Al-OH species are responsible for transporting Al in acid solutions circulating in lower-temperature hydrothermal veins, which could be the case for the Beauvoir greisens that formed along fractures. In all cases, aluminium is mobile, and its transport is facilitated by fluorine.

In the specific case of Beauvoir, the acidic fluids have been channelised within a network of fractures. These fluids, probably issued from the unmixing from the magmatic melts as suggested by [42] and confirmed by [40], were in significant disequilibrium with the granite mineral assemblage due to their richness in fluorine and pH. Albite was primarily dissolved and replaced by newly formed muscovite. The preferential dissolution of albite and its replacement by K-micas is also the main trend at Cligga Head and Cinovec. The main difference at Beauvoir is the relatively small amounts of K-feldspar in the parent granite. The amount of potassium released by the muscovitisation of K-feldspars is insufficient to account for the need for potassium. Another deficit concerns aluminium. The albite dissolution may have provided a part of the local aluminium, but another contribution is probably the transported aluminium in the solution. The origin of the aluminium contribution can be either minerals from the granite itself along the fractures or the surrounding schists, which are known to have been deeply affected by hydrothermal reaction with recrystallisation of the original metamorphic micas [10–12,43].

An overall summary of the major changes occurring among the main rock-forming minerals is proposed in Figure 9, where processes identified at Beauvoir are compared to the other considered greisens. A distinction is made between early greisens directly related to the magmatic fluid release and the greisens formed along fractures during later stages. The latter are dominated mainly by K-micas, which are close to the muscovite end-member.

4.4. Silicification in the Granite Mass or Quartz Dissolution and Quartz Vein Formation

The location along the quartz–muscovite line depends mainly on the relative abundance of newly formed quartz and muscovite during greisenisation but also on later processes such as fracture reactivation and formation of quartz vein infillings, or conversely of quartz dissolution and eventual conversion of quartz to muscovite through this process. This is the hypothesis from [5] for Panasqueira and [22] for Hoggar greisens. The suggested mechanism is a substitution of feldspar by quartz (silicification), followed by quartz dissolution (episyenite) and mica crystallisation. Such silicification processes have also been reported by [41] for porphyry–greisen alteration. In addition, at Beauvoir, fluids from the greisen stage occured at the end of the process, and quartz precipitation as veinlets, which contribute to vein-type silicification, is probably linked to cooling. Veins are highly abundant in the northern zone.

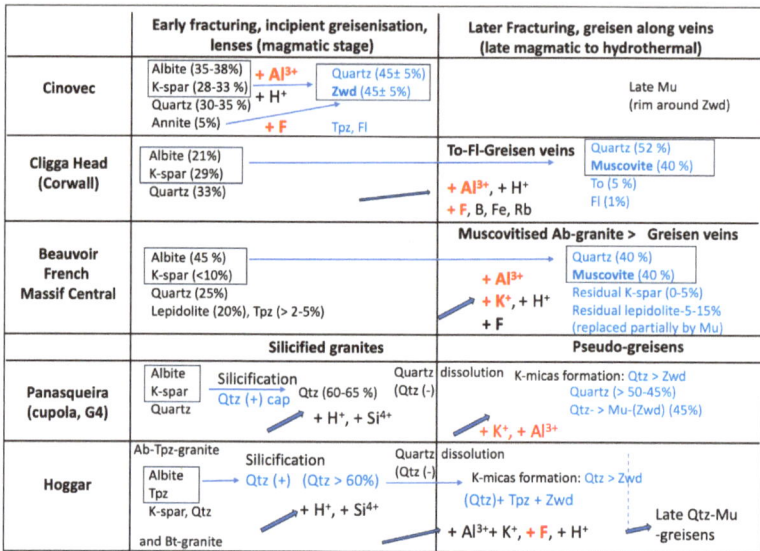

Figure 9. Summary of the main mineral reactions and transformations related to the examined greisen types. The main element supplies needed for the transformations are put forward. Zwd: zinnwaldite; Mu: muscovite; Mu(Zwd) muscovite–zinnwaldite series; Tpz: topaz; Qtz: quartz; and Fl: fluorite. Data have been simplified and summarised from the literature, except Beauvoir (Cinovec [6,7], Cligga Head [9], Panasqueira [5,20], and Hoggar [22]).

5. Conclusions

Major chemical element concentrations obtained by whole-rock chemical analyses are rich in informations and are indicative of the mass balance occurring during greisenisation. These chemical elements provide helpful information on the main trends of water–rock interactions. Aluminium appears mobile, although it is generally considered immobile in most mass balance calculations. The main trends obtained on several greisens, including Beauvoir granite, indicate that the chemical reaction usually written at constant aluminium cannot explain the slope of many trends in the quartz–feldspar–mica system. Aluminium has a specific mobility, which can be explained by acidic fluids containing fluorine.

Mineralogical specificities of each greisen testify to a distinct succession of mineral dissolution and mineral saturation depending on the fluid chemistry and temperature evolution.

In the specific case of Beauvoir, a network of fractures channelled fluids that were in significant disequilibrium with the mineral assemblage of the granite, particularly albite, and, owing to the aluminium transported, formed a substantial quantity of muscovite. The origin of the aluminium and potassium necessary for these reactions can be either minerals from the granite itself along the fractures or the surrounding schists, which are known to have been deeply affected by hydrothermal reaction with recrystallisation of the original metamorphic micas.

Supplementary Materials: The following supporting information can be downloaded at https://www.mdpi.com/article/10.3390/min14080746/s1. Table S1: Bulk-rock analyses of granites and greisens with Q'-F' parameter calculations.

Author Contributions: M.C.: sampling, data acquisition and interpretation, conceptualisation, writing (original draft preparation, review and editing), and funding acquisition; Z.S.K.: sampling, analytical data, interpretation, writing (review and editing), and funding acquisition. All authors have read and agreed to the published version of the manuscript.

Funding: This research was funded by Labex Ressources21 under the reference ANR-10-LABX-21-RESSOURCES21, supported by the Agence Nationale de la Recherche through the national programme "Investissements d'avenir and by Imerys through the collaboration programme between UL and Imerys".

Data Availability Statement: Data are partly provided in this paper. Data from Imerys are not publicly available due to ownership by Imerys.

Acknowledgments: O. Rouer and A. Lecomte (GeoRessources) are warmly acknowledged for their help in acquiring the EMPA and SEM-EDS data. M-C Boiron is acknowledged for her help in reading drafts and proofs. Imerys is warmly acknowledged for providing geochemical data and samples for the present study, particularly G. Jean, P. Fullenwarth, and B. Barré, for their fruitful discussions. Two anonymous reviewers are thanked for their constructive remarks.

Conflicts of Interest: The authors declare no conflicts of interest.

References

1. Černý, P.; Masau, M.; Goad, B.E.; Ferreira, K. The Greer Lake leucogranite, Manitoba, and the origin of lepidolite-subtype granitic pegmatites. *Lithos* **2005**, *80*, 305–321. [CrossRef]
2. Černý, P.; Blevin, P.L.; Cuney, M.; London, D. Granite-related ore deposits. *Econ. Geol.* **2005**, *100*, 37–370.
3. Shcherba, G.N. Greisens. *Int. Geol. Rev.* **1970**, *12*, 114–150. [CrossRef]
4. Štemprok, M. Greisenization (a review). *Geol. Rundsch. Stuttg.* **1987**, *76*, 169–175. [CrossRef]
5. Marignac, C.; Cuney, M.; Cathelineau, M.; Lecomte, A.; Carocci, E.; Pinto, F. The Panasqueira rare metal granite suites and their involvement in the genesis of the world-class Panasqueira W-Sn-Cu vein deposit: A petrographic, mineralogical, and geochemical study. *Minerals* **2020**, *10*, 562. [CrossRef]
6. Breiter, K.; Ďurišová, J.; Hrstka, T.; Korbelová, Z.; Vaňková, M.H.; Galiova, M.V.; Kanicky, V.; Rambousek, P.; Knesl, I.; Dobes, P.; et al. Assessment of magmatic vs. metasomatic processes in rare-metal granites: A case study of the Cínovec/Zinnwald Sn–W–Li deposit, Central Europe. *Lithos* **2017**, *292–293*, 198–217. [CrossRef]
7. Breiter, K.; Hložková, M.; Korbelová, Z.; Galiová, M.V. Diversity of lithium mica compositions in mineralised granite-greisen system: Cínovec Li-Sn-W deposit, Erzgebirge. *Ore Geol. Rev.* **2019**, *106*, 12–27. [CrossRef]
8. Charoy, B. Greisenisation, minéralisation et fluides associés à Cligga Head, Cornwall (Sud-ouest de l'Angleterre). *Bull. Mineral.* **1979**, *102*, 633–641. [CrossRef]
9. Hall, A. Greisenisation in granite of Cligga Head, Cornwall. *Proc. Geol. Assoc.* **1971**, *82*, 209–230. [CrossRef]
10. Cuney, M.; Marignac, C.; Weisbrod, A. The Beauvoir topaz-lepidolite albite granite (Massif Central, France); the disseminated magmatic Sn–Li–Ta–Nb–Be mineralisation. *Econ. Geol.* **1992**, *87*, 1766–1794. [CrossRef]
11. Monnier, L.; Salvi, S.; Jourdan, V.; Sall, S.; Bailly, L.; Melleton, J.; Béziat, D. Contrasting fluid behavior during two styles of greisen alteration leading to distinct wolframite mineralisations: The Echassières district (Massif Central, France). *Ore Geol. Rev.* **2020**, *124*, 103648. [CrossRef]
12. Monnier, L.; Salvi, S.; Melleton, J.; Lach, P.; Pochon, A.; Bailly, L.; Béziat, D.; Parseval, P.D. Mica trace-element signatures: Highlighting superimposed W-Sn mineralisations and fluid sources. *Chem. Geol.* **2022**, *600*, 120866. [CrossRef]
13. Monier, G.; Charoy, B.; Cuney, M.; Ohnenstetter, D.; Robert, J.L. Évolution spatiale et temporelle de la composition des micas du granite albitique à topaze-lepidolite de Beauvoir. *Geol. Fr.* **1987**, *2–3*, 179–188.
14. Do Couto, D.; Faure, M.; Augier, R.; Cocherie, A.; Rossi, P.; Li, X.-H.; Lin, W. Monazite U–Th–Pb EPMA and zircon U–Pb SIMS chronological constraints on the tectonic, metamorphic, and thermal events in the inner part of the Variscan orogen, example from the Sioule series, French Massif Central. *Int. J. Earth Sci.* **2016**, *105*, 557–579. [CrossRef]
15. Fonteilles, M. La composition chimique des micas lithinifères (et autres minéraux) des granites d'Echassières comme image de leur évolution magmatique. *Geol. Fr.* **1987**, *2–3*, 149–178.
16. Rossi, P.; Autran, A.; Azencott, C.; Burnol, L.; Cuney, M.; Johan, V.; Kosakevitch, A.; Ohnenstetter, D.; Monier, G.; Piantone, P.; et al. Logs pétrographique et géochimique du granite de Beauvoir dans le sondage "échassieres I" Minéralogie et géochimie comparées. *Géol. Fr.* **1987**, *2–3*, 111–135.
17. Cathelineau, M. The hydrothermal alkali metasomatism effects on granitic rocks: Quartz dissolution and related subsolidus changes. *J. Petrol.* **1986**, *27*, 945–965. [CrossRef]
18. Štemprok, M. Drill hole CS-1 penetrating the Cínovec/Zinnwald granite cupola (Czech Republic): An A-type granite with important hydrothermal mineralisation. *J. Geosci.* **2016**, *61*, 395–423. [CrossRef]
19. Hreus, S.; Výravský, J.; Cempírek, J.; Breiter, K.; Galiová, M.V.; Krátký, O.; Šešulka, V.; Skoda, R. Scandium distribution in the world-class Li-Sn-W Cínovec greisen-type deposit: Result of a complex magmatic to hydrothermal evolution, implications for scandium valorisation. *Ore Geol. Rev.* **2021**, *139*, 104433. [CrossRef]
20. Launay, G.; Sizaret, S.; Lach, P.; Melleton, J.; Gloaguen, E.; Poujol, M. Genetic relationship between greisenization and Sn-W mineralisations in vein and greisen deposits: Insights from the Panasqueira deposit (Portugal). *Bull. Soc. Geol. Fr.* **2021**, *192*, 2. [CrossRef]

21. Liu, C.; Xiao, W.; Zhang, L.; Belousova, E.; Rushmer, T.; Xie, F. Formation of Li-Rb-Cs greisen-type deposit in Zhengchong, Jiuyishan district, South China: Constraints from whole-rock and mineral geochemistry. *Geochemistry* **2021**, *81*, 125796. [CrossRef]
22. Bouguebrine, J.; Bouabsa, L.; Marignac, C. Greisen and pseudo-greisen in the Tamanrasset area (Central Hoggar, Algeria): Petrography, geochemistry and insight on the fluid origin from mica chemistry. *J. Afr. Earth Sci.* **2023**, *202*, 104898. [CrossRef]
23. Carignan, J.; Hild, P.; Mevelle, G.; Morel, J.; Yeghicheyan, D. Routine analyses of trace elements in geological samples using flow injection and low pressure on line liquid chromatography coupled to ICP-MS: A study of geochemical reference material BR, DR-N, EB-N, AN-G and GH. *Geostand. Newsl.* **2001**, *25*, 187–198. [CrossRef]
24. Johan, Z.; Strnad, L.; Johan, V. Evolution of the Cínovec (Zinnwald) granite cupola, Czech Republic: Composition of feldspars and micas, a clue to the origin of W, Sn mineralisation. *Can. Mineral.* **2012**, *50*, 1131–1148. [CrossRef]
25. Stemprok, M.; Sulcek, Z. Geochemical Profile through an Ore-Bearing Lithium Granite. *Econ. Geol.* **1969**, *64*, 392–404. [CrossRef]
26. La Roche, H. Sur l'usage du concept d'association minérale dans l'étude chimique des roches: Modèles chimiques, statistiques, représentations graphiques, classification chimico minéralogique. *C.R. Acad. Sci. Paris* **1966**, *262*, 1665–1668.
27. Debon, F.; Le Fort, P.A. Cationic classification of common plutonic rocks and their magmatic associations: Principles, method, applications. *Bull. Minéralogie* **1988**, *111*, 493–510. [CrossRef]
28. Streckeisen, A.L. Classification and nomenclature of plutonic rocks. *Geol. Rundsch.* **1974**, *63*, 773–786. [CrossRef]
29. Streckeisen, A.L. Classification of the common igneous rocks by means of their chemical composition. *Neues Jahrb Miner. Monatsh* **1976**, *1*, 1–15.
30. Michaud, J.A.S.; Pichavant, M. Magmatic fractionation and the magmatic-hydrothermal transition in rare metal granites: Evidence from Argemela (central Portugal). *Geochim. Cosmochim. Acta* **2020**, *289*, 130–157. [CrossRef]
31. Antunes, I.M.; Neiva, A.M.; Ramos, J.M.; Silva, P.B.; Silva, M.M.; Corfu, F. Petrogenetic links between lepidolite-subtype aplite-pegmatite, aplite veins and associated granites at Segura (Central Portugal). *Geochemistry* **2013**, *73*, 323–341. [CrossRef]
32. Cathelineau, M.; Boiron, M.C.; Leconte, A.; Martins, I.; Dias da Silva, I.; Mateus, A. Lithium, phosphorus, fluorine-rich intrusions and the phosphate sequence at Segura (Portugal): A comparison with other hyper-differentiated magmas. *Minerals* **2024**, *14*, 287. [CrossRef]
33. Tuttle, O.F.; Bowen, N.L. Origin of granite in the light of experimental studies in the system $NaAlSi_3O_8$-$KAlSi_3O_8$-SiO_2-H_2O. *Geol. Soc. Am.* **1958**, *74*, 153.
34. Kahou, S.; Cathelineau, M.; Boiron, M.-C. Quantitative mineralogy and lithium distribution in the upper part of the Beauvoir granite. In Proceedings of the EGU24-2024, Vienna, Austria, 14–19 April 2024. 1668-ECS, Orals, GMPV5.1.
35. Gresens, R.L. Composition-volume relationships of metasomatism. *Chem. Geol.* **1967**, *2*, 47–55. [CrossRef]
36. Pinto, F.M.V. ICP-MS Data of the SCB2 Drill-Hole. Beiralt Tin and Wolfram (Portugal) Data. Beiralt Tin and Wolfram (Portugal) Mining Company; *Unpublished Report*, 2016.
37. Errandonea-Martin, J.; Garate-Olave, I.; Roda-Robles, E.; Cardoso-Fernandes, J.; Lima, A.; Ribeiro, M.A.; Teodoro, A.C. Metasomatic effect of Li-bearing aplite-pegmatites on psammitic and pelitic metasediments: Geochemical constraints on critical raw material exploration at the Fregeneda–Almendra Pegmatite Field (Spain and Portugal). *Ore Geol. Rev.* **2022**, *150*, 105155. [CrossRef]
38. Lentz, D.R.; Gregoire, C. Petrology and mass-balance constraints on major-, trace-, and rare-earth-element mobility in porphyry-greisen alteration associated with the epizonal True Hill granite, southwestern New Brunswick, Canada. *J. Geochem. Explor.* **1995**, *52*, 303–331. [CrossRef]
39. Burt, D.M. Acidity-salinity diagrams–Application to greisen and porphyry deposits. *Econ. Geol.* **1981**, *76*, 832–843. [CrossRef]
40. Harlaux, M.; Mercadier, J.; Bonzi, W.M.E.; Kremer, V.; Marignac, C. Geochemical signature of magmatic-hydrothermal fluids exsolved from the Beauvoir rare-metal granite (Massif Central, France): Insights from LA-ICPMS analysis of primary fluid inclusions. *Geofluids* **2017**, *2017*, 1925817. [CrossRef]
41. Tagirov, B.; Schott, J. Aluminum speciation in crustal fluids revisited. *Geochim. Cosmochim. Acta* **2001**, *65*, 3965–3992. [CrossRef]
42. Aissa, M.; Weisbrod, A.; Marignac, C. Chemical and thermodynamic characteristics of hydrothermal fluid circulation at Echassieres. *Geol. Fr.* **1987**, *2–3*, 335–350.
43. Aubert, G. Les coupoles granitiques de Montebras et d'Echassières (Massif central français) et la genèse de leurs minéralisations en étain, lithium, tungstène et béryllium. *Éditions BRGM* **1969**, *46*, 1.

Disclaimer/Publisher's Note: The statements, opinions and data contained in all publications are solely those of the individual author(s) and contributor(s) and not of MDPI and/or the editor(s). MDPI and/or the editor(s) disclaim responsibility for any injury to people or property resulting from any ideas, methods, instructions or products referred to in the content.

Article

Mesoproterozoic (ca. 1.3 Ga) A-Type Granites on the Northern Margin of the North China Craton: Response to Break-Up of the Columbia Supercontinent

Bo Liu [1], Shengkai Jin [1,2,3,*], Guanghao Tian [1,4,*], Liyang Li [1], Yueqiang Qin [1], Zhiyuan Xie [1], Ming Ma [1] and Jiale Yin [1]

[1] Langfang Comprehensive Natural Resources Survey Center, China Geological Survey, Langfang 065000, China; liubo33564@163.com (B.L.); liliyang@mail.cgs.gov.cn (L.L.); qinyueqiang@mail.cgs.gov.cn (Y.Q.); xzhiyuan@mail.cgs.gov.cn (Z.X.); maming@mail.cgs.gov.cn (M.M.); yinjiale@mail.cgs.gov.cn (J.Y.)
[2] State Key Laboratory of Geological Processes and Mineral Resources, China University of Geosciences, Beijing 100083, China
[3] Command Center of Natural Resource Comprehensive Survey, China Geological Survey, Beijing 100055, China
[4] School of Civil and Resources Engineering, University of Science and Technology Beijing, Beijing 100083, China
* Correspondence: jinsklfzx@163.com (S.J.); d202220016@xs.ustb.edu.cn (G.T.)

Abstract: Mesoproterozoic (ca. 1.3 Ga) magmatism in the North China Craton (NCC) was dominated by mafic intrusions (dolerite sills) with lesser amounts of granitic magmatism, but our lack of knowledge of this magmatism hinders our understanding of the evolution of the NCC during this period. This study investigated porphyritic granites from the Huade–Kangbao area on the northern margin of the NCC. Zircon dating indicates the porphyritic granites were intruded during the Mesoproterozoic between 1285.4 ± 2.6 and 1278.6 ± 6.1 Ma. The granites have high silica contents (SiO_2 = 63.10–73.73 wt.%), exhibit alkali enrichment (total alkalis = 7.71–8.79 wt.%), are peraluminous, and can be classified as weakly peraluminous A2-type granites. The granites have negative Eu anomalies (δEu = 0.14–0.44), enrichments in large-ion lithophile elements (LILEs; e.g., K, Rb, Th, and U), and depletions in high-field-strength elements (HFSEs; e.g., Nb, Ta, and Ti). $\varepsilon_{Hf}(t)$ values range from −6.43 to +2.41, with t_{DM2} ages of 1905–2462 Ma, suggesting the magmas were derived by partial melting of ancient crustal material. The geochronological and geochemical data, and regional geological features, indicate the Mesoproterozoic porphyritic granites from the northern margin of the NCC formed in an intraplate tectonic setting during continental extension and rifting, which represents the response of the NCC to the break-up of the Columbia supercontinent.

Keywords: A-type granite; porphyritic granite; Mesoproterozoic; Columbia break-up; Huade

Citation: Liu, B.; Jin, S.; Tian, G.; Li, L.; Qin, Y.; Xie, Z.; Ma, M.; Yin, J. Mesoproterozoic (ca. 1.3 Ga) A-Type Granites on the Northern Margin of the North China Craton: Response to Break-Up of the Columbia Supercontinent. *Minerals* 2024, 14, 622. https://doi.org/10.3390/min14060622

Academic Editors: Ignez de Pinho Guimarães and Jefferson Valdemiro De Lima

Received: 21 May 2024
Revised: 15 June 2024
Accepted: 16 June 2024
Published: 18 June 2024

Copyright: © 2024 by the authors. Licensee MDPI, Basel, Switzerland. This article is an open access article distributed under the terms and conditions of the Creative Commons Attribution (CC BY) license (https://creativecommons.org/licenses/by/4.0/).

1. Introduction

The assembly and break-up of supercontinents are fundamental processes in Earth's history and have affected continent formation and destruction, the global climate, biological evolution, and large-scale mineralization events. Over the past two decades, this field has become an important area of research [1,2]. Geological records indicate the past occurrence of at least four supercontinents: Kenorland, Columbia, Rodinia, and Pangaea [3–5]. The Columbia supercontinent was assembled by a series of global-scale collisional events during 2.1–1.8 Ga [6,7] and was subsequently affected by rifting from the Paleoproterozoic to Mesoproterozoic (1.8–1.3 Ga), followed by break-up at 1.3–1.2 Ga [2,6]. During this period, Columbia experienced extensive rifting, mafic magmatism, and numerous tectonothermal events [2,8,9].

The North China Craton (NCC) was a key part of the Columbia supercontinent and encompassed both the Palaeoproterozoic Khondalite Belt (ca. 1.95 Ga) and the Central

Orogenic Belt (ca. 1.85 Ga) [10,11], which are thought to have formed by convergent processes that accompanied the assembly of the supercontinent. However, due to a lack of reliable age data for the Mesoproterozoic magmatism and tectonism, controversy exists regarding the role of the NCC during the rifting of Columbia [12–14]. Some studies have proposed that the NCC had rifted from Columbia by 1.6 Ga [15,16], suggesting it was not involved in the final break-up of the supercontinent at 1.35–1.20 Ga. However, this inference is inconsistent with palaeomagnetic data [17,18]. Shao et al. [12] identified three extensional events in the NCC, based on the following: (1) the intrusion of 1.8–1.7 Ga basaltic sills; (2) 1.3–1.2 Ga alkaline magmatism; and (3) 0.8–0.7 Ga basaltic sills. Zhai et al. [11,13] proposed that the late Paleoproterozoic to the Neoproterozoic in the NCC was characterized by four magmatic events: a large igneous event at 1.78 Ga; anorogenic magmatism during 1.72–1.62 Ga; a mafic dyke swarm during 1.37–1.32 Ga; and a mafic dyke swarm at 900 Ma. They proposed that the NCC was predominantly in an extensional setting during the Meso-Neoproterozoic. Xiang et al. [14] identified five prominent magmatic events during the late Paleoproterozoic to Mesoproterozoic along the northern margin of the NCC. These include mafic dyke swarms at 1.80–1.77 Ga, anorthosite–mangerite–charnockite–rapakivi granite assemblages (AMCG assemblage) at 1.72–1.67 Ga, alkali granites and mafic dyke swarms at 1.63–1.62 Ga, mafic dyke swarms and A-type granites at 1.33–1.30 Ga, and mafic dykes at 1.23 Ga. They proposed that Mesoproterozoic magmatism was episodic. More recently, the discovery of 1.33–1.30 Ga dolerite sills and coeval A-type granites in the northern NCC has indicated the craton underwent rifting that was coeval with the rifting of Columbia [19–21].

Regionally extensive dolerite sills were intruded into the Yanliao area at ca. 1.3 Ga in the eastern part of the northern margin of the NCC, while coeval magmatism in the central and western parts was relatively small in volume and comprised mainly granites and gabbros. Previous studies have focused mainly on the dolerite sills in the eastern part of the northern margin of the NCC, whereas few studies have examined the granitic magmatism in the central part of the craton [8,16,19].

2. Geological Setting

The NCC is located in eastern China and is bordered by the Xingmeng orogenic belt to the north and the Qinling–Dabie orogenic belt to the south (Figure 1). The NCC is one of the oldest cratons in the world and consists primarily of Precambrian crystalline basement that is overlain by thick sedimentary cover rocks [22,23]. The formation of the crystalline basement of the NCC involved mainly three processes: (1) the formation of multiple dispersed continental nuclei; (2) the amalgamation of micro-blocks; and (3) cratonization [24,25]. The Eastern and Western blocks collided along the Trans-North China orogen at ca. 1.85 Ga, leading to the final assembly of the NCC basement and its incorporation into Columbia. The Precambrian basement consists mainly of Neoarchaean–Palaeoproterozoic TTG (Tonalite-Trondhjemite-Granodiorite) rocks and 2.6–2.5 Ga granulites, amphibolites, and mafic–ultramafic intrusions [18,26]. The NCC was located on the edge of the Columbia supercontinent and adjacent to the Indian Craton [3]. During the middle to late Proterozoic, the NCC underwent large-scale extension. During this period, a radial swarm of mafic dykes formed during 1.78–1.68 Ga, and there were multiple periods of rifting, including the formation of the Yanliao, Xiong'er, and Zhaertei–Baiyunebo–Huade rifts in the NCC. The NCC experienced a period of magmatic quiescence at 1.6–1.4 Ga. The magmatism in the NCC at 1.4–1.3 Ga is mainly represented by dolerite sills and granites. The dolerite sills occur primarily in the Yanliao region, over an area that is >600 km long and 200 km wide, constituting a mafic large igneous province. The 1.3 Ga granites in the NCC are located mainly in the Shangdu area, Inner Mongolia [2].

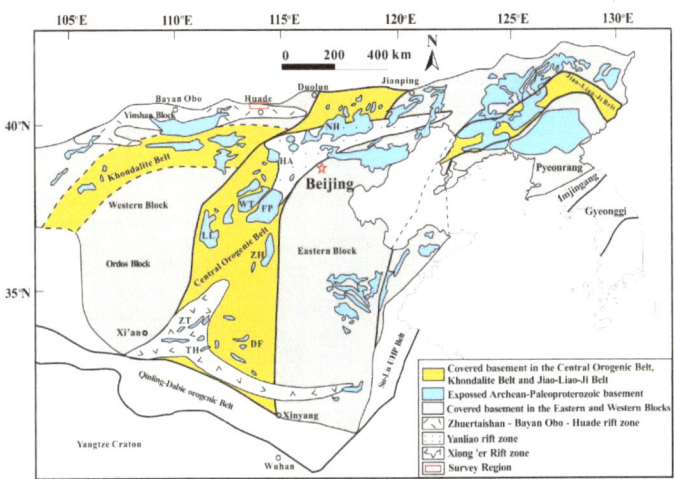

Figure 1. Schematic tectonic map of the NCC (modified after Zhao et al., 2001 [24]). Dengfeng (DF), Fuping (FP), Hengshan (HS), Huaian (HA), Lüliang (LL), Northern Hebei (NH), Taihua (TH), Wutai (WT), Zanghuang (ZH), and Zhongtiao (ZT).

The present study area is located at the northern margin of the NCC. Sedimentary rocks that crop out in the north of this area include the Mesoproterozoic Baoyintu Group, and lower Permian Sanmianjing and Elitu formations. Sedimentary rocks that crop out in the south of this area include the Mesoproterozoic Huade Group, with smaller exposures of the Middle Jurassic Tuchengzi Formation and Lower Cretaceous Zhangjiakou Formation. The Mesoproterozoic intrusions are located mainly in the northern area and trend NE–SW. Some of the intrusions have been mylonitized due to the activity of ENE–WSW-trending ductile shear zones (Figure 2).

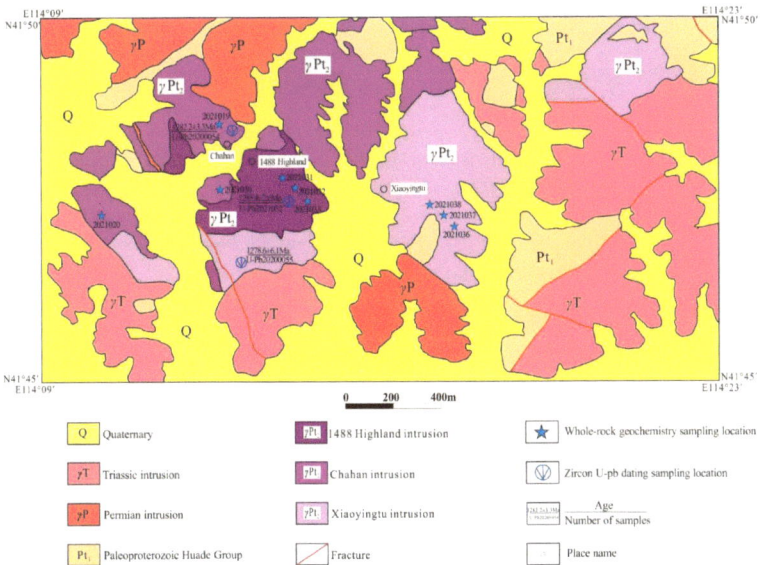

Figure 2. Simplified geologic map of the Huade area, Inner Mongolia.

3. Field Outcrops, and Petrology, and Mineralogy

Granitic samples were collected from the Huade area in Inner Mongolia and the Kangbao area in Hebei Province. Field investigations were conducted on the Xiaoyingtu, Chahan, and 1488 Highland intrusions in the study areas. These intrusions are stocks, with irregular elliptical shapes and spheroidal weathering (Figure 3a–e). Elliptical diorite xenoliths occur within the intrusions (Figure 3d).

Figure 3. Representative field and photomicrographs of the Mesoproterozoic intrusions. (**a**) field characteristics of Xiaoyingtu intrusion; (**b**) pluton intruded into the strata of the Paleoproterozoic era Huade Group; (**c**) characteristics of medium-fine grained porphyritic biotite monzogranite granite hand specimen characteristics; (**d**) gray-black diorite inclusion; (**e**) medium-fine grained porphyritic biotite monzogranite; (**f**) porphyritic biotite syenogranite; (**g**) biotite syenogranite (orthogonal polarization); (**h**) medium-fine grained porphyritic biotite syenogranite (orthogonal polarization); (**i**) fine-grained porphyritic biotite monzogranite (orthogonal polarization).

3.1. Xiaoyingtu Pluton

The Xiaoyingtu pluton consists mainly of porphyritic granodiorite and is located in the central–eastern part of the study area over an exposed area of 18.30 km^2 (Figure 3b). The granodiorite is grey-yellow in color with a porphyritic texture and massive structure and consists mainly of plagioclase (50 vol.%), K-feldspar (25 vol.%), quartz (20 vol.%), and biotite and muscovite (5 vol.%). The samples are plotted in the granodiorite and quartz monzonite fields on the granite total alkalis–SiO$_2$ (TAS) diagram. Plagioclase phenocrysts range in size from 0.5 × 0.8 to 1 × 2 cm and are euhedral–subhedral, and they exhibit occasional multiple twinning. The quartz in the matrix is granular with a grain size of 2–4 mm. Plagioclase is euhedral–subhedral with grain sizes of 1 × 3 to 2 × 5 mm, and the distribution of biotite defines banding in the granodiorite (Figure 3f).

3.2. Chahan Intrusion

The Chahan intrusion crops out over an area of 20.47 km^2 in the central part of the study area and consists of fine–medium-grained porphyritic biotite syenogranite and medium–coarse-grained porphyritic biotite syenogranite. The biotite syenogranite is light grey-brown in color with a porphyritic texture and massive structure. The phenocrysts are K-feldspar, which are subhedral and have grain sizes of 0.4×1 to 0.5×2.5 cm and comprise ~10 vol.% of the sample. The matrix consists predominantly of K-feldspar (50 vol.%), plagioclase (15 vol.%), quartz (20 vol.%), and biotite (5 vol.%), with grain sizes of 2–4 mm. K-feldspar is subhedral and primarily microcline. Plagioclase is subhedral–euhedral and zoned, while biotite exhibits localized chloritization (Figure 3g–h).

3.2.1. Highland Intrusion

The 1488 Highland intrusion is located in the central part of the study area and is a stock that has an outcrop area of 7.07 km^2. The intrusion consists mainly of fine-grained porphyritic biotite monzogranite and itself has been intruded by the Chahan intrusion. The sample is yellow-brown in color and has a porphyritic texture and massive structure. It consists of K-feldspar (35 vol.%), plagioclase (40 vol.%), quartz (20 vol.%), biotite (5 vol.%), and minor muscovite and garnet. The phenocrysts are primarily subhedral microcline and minor amounts of plagioclase, which have grain sizes of 5–10 mm. Plagioclase occurs as subhedral crystals with grain sizes of 2–5 mm that occasionally exhibit zoning. Biotite has a weak shape-preferred orientation and a grain size of 0.2–2.0 mm (Figure 3i).

4. Sampling and Methodology

Whole-rock geochemical and zircon U–Pb dating and Hf isotope analyses were carried out on samples of the Xiaoyingtu, Chahancun, and 1488 Highland intrusions.

4.1. Whole-Rock Geochemistry

Whole-rock geochemical analysis was undertaken at the Regional Geological and Mineral Survey Institute, Hebei, China, using national standards and relevant industry analytical techniques. The analyzed samples were silicate rock samples, along with standards and blanks. Approximately 50 mg of representative sample material was weighed and placed in the inner container of a closed sample digestion vessel. Subsequently, 1 mL of HF and 0.5 mL of HNO_3 were added to the sample. The vessel was sealed and placed in an oven at a temperature of 190 °C for 24 h. After cooling, the inner container was removed and heated to 200 °C on a hotplate to evaporate the acids. Finally, 0.5 mL of HNO_3 was added and evaporated. This step was repeated and then an extra 5 mL of HNO_3 was added and the container was sealed and held at 130 °C for 3 h. After cooling, the container was opened, and the contents were transferred to a cleaned plastic bottle and diluted with 50 mL of water. The mixture was shaken thoroughly before determining the trace element contents by inductively coupled plasma–mass spectrometry (ICP–MS).

The major elements were determined by X-ray fluorescence spectrometry. Approximately 0.8 g of sample was weighed into a 25 mL porcelain crucible, along with an 8 g mixture of anhydrous lithium tetraborate and lithium fluoride. The mixture was then transferred into a Pt–Au crucible and melted and cooled into a fused glass disc. For the Fe analyses, ~0.3 g of sample was weighed and placed in a Pt crucible. The sample was wetted with water, and 5 mL of HF and 10 mL of H_2SO_4 were added. The crucible was covered, placed in a pre-heated electric furnace, heated and boiled for 10 min, removed from the furnace, and immediately placed in 200 mL of water containing 25 mL of saturated boric acid. For the loss-on-ignition analyses, 15 mL of thiophosphoric acid and two drops of 1% sodium diphenylamine sulfonate solution were added. Approximately 1.0 g of the sample was weighed into a porcelain crucible. The crucible was placed in a muffle furnace and heated at 980 °C for 2 h. After heating, the crucible was removed, transferred to a desiccator, allowed to cool to room temperature for 30 min, and then weighed. The ignition process was then repeated for an additional 30 min and the crucible weighed again until a constant weight was achieved.

4.2. Zircon U–Pb Ages and Lu–Hf Isotopic Compositions

The zircon grains were separated and analyzed at the Hebei Regional Geological and Mineral Survey Institute. Subsequently, the zircons were subjected to cathodoluminescence (CL) imaging, U–Pb dating, and in situ Lu–Hf isotopic analysis at the Tianjin Geological Survey Center, China Geological Survey, Tianjin, China. The U–Pb dating was undertaken with a laser ablation (LA) system coupled to an ICP–MS (Agilent 7900) instrument (Agilent Technologies, Santa Clara, CA, USA). A RESOlution LR (ASI) LA system comprising a 193 nm ArF excimer laser was used for the analyses. Ages and concordia diagrams were obtained with Isoplot 3.0 [27]. Hafnium isotopes were determined with an LA–multiple-collector–ICP–MS (LA–MC–ICP–MS) system. The LA system was a 193 nm ArF excimer laser (model UP193-FX; ESI Company, Santa Rosa, CA, USA). The MC–ICP–MS was a Neptune Plus (a product of Thermo Fisher Scientific, which is headquartered in Waltham, MA, USA). The Hf isotope data were processed using ICPMSDataCal 9.2 software. The laser beam diameter was 50 μm, the laser energy density was 3.5 J/cm^2, and the laser frequency was 8 Hz. The ablation time was 40 s, and the ablated material was transported into the mass spectrometer using He gas.

5. Results

5.1. Zircon U–Pb Ages

Three granitic samples were selected for zircon U–Pb dating: U–Pb2020054 (porphyritic biotite syenite; 41°47′09″N, 114°15′22″E), U–Pb2020055 (porphyritic biotite syenite; 41°46′30″N, 114°13′59″E), and U–Pb2021032 (porphyritic biotite monzonitic granite; 41°48′02″N, 114°14′01″). The results are listed in Appendix A Table A1. Rock samples with minimal alteration were collected. The zircon crystals are euhedral, short columnar in shape, 90 × 70 to 120 × 300 μm in size, and they contain few inclusions and fractures (Figure 4).

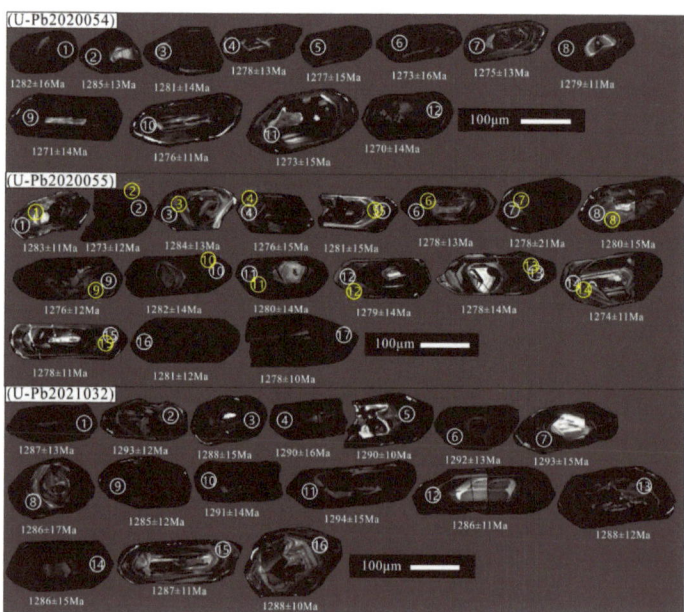

Figure 4. Cathode luminescence images of representative zircons from the Mesoproterozoic granites in the study area.

The CL images reveal oscillatory zoning, indicative of a magmatic origin. The U and Th contents are 248.93–769.10 and 17.91–308.87 ppm, respectively, with Th/U = 0.07–0.51, which are also consistent with a magmatic origin [28]. The U–Pb ages exhibit limited variations (Figure 5). The U–Pb ages for sample U–Pb2020054 range between 1270 ± 14 and 1285 ± 13 Ma, with a weighted mean age of 1282.2 ± 3.3 Ma (n = 12; MSWD = 0.008). The U–Pb ages of sample U–Pb2020055 vary between 1273 ± 12 and 1284 ± 13 Ma, with a weighted mean age of 1278.6 ± 6.1 Ma (n = 17; MSWD = 0.061). The U–Pb ages of sample U–Pb2021032 range between 1285 ± 12 and 1294 ± 15 Ma, with a weighted mean age of 1285.4 ± 2.6 Ma (n = 16; MSWD = 0.100). The zircon U–Pb ages of all samples are the same within error, suggesting the porphyritic granites are Mesoproterozoic in age.

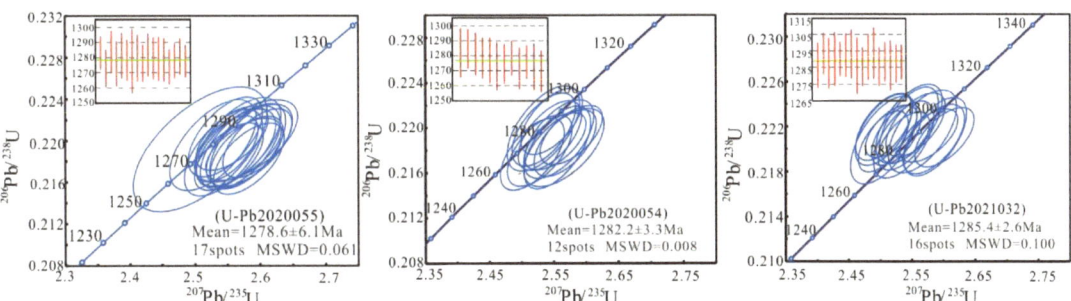

Figure 5. LA-ICP-MS zircon U-Pb concordia diagrams of the Mesoproterozoic granite samples in the study area.

5.2. Major Elements

The SiO_2 contents of the analyzed samples range from 63.10 to 73.73 wt.%. The total alkali ($K_2O + Na_2O$) contents are 6.28–8.79 wt.% (Figure 6a). The results are listed in Table A2. The Al_2O_3 contents are 12.04–17.64 wt.%, with an average of 14.54 wt.%. The A/CNK ratios are >1.0 (1.07–1.90; average = 1.22). Sample 2021037 has A/CNK > 1.90, potentially due to late-stage metamorphism, which led to the development of metamorphic minerals such as andalusite and sillimanite. The A/NK values range from 1.24 to 2.33, with an average of 1.54. All samples are plotted in the peraluminous field in an A/CNK versus A/NK diagram (Figure 6b). In a SiO_2–FeOT/(FeOT + MgO) diagram (Figure 6c), most samples are plotted in the magnesian granite field. In a SiO_2–K_2O diagram (Figure 6d), the samples are plotted in the shoshonite field.

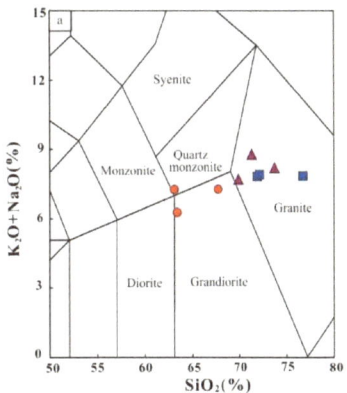
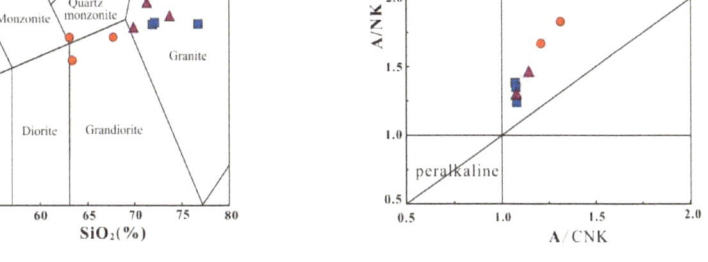

Figure 6. *Cont.*

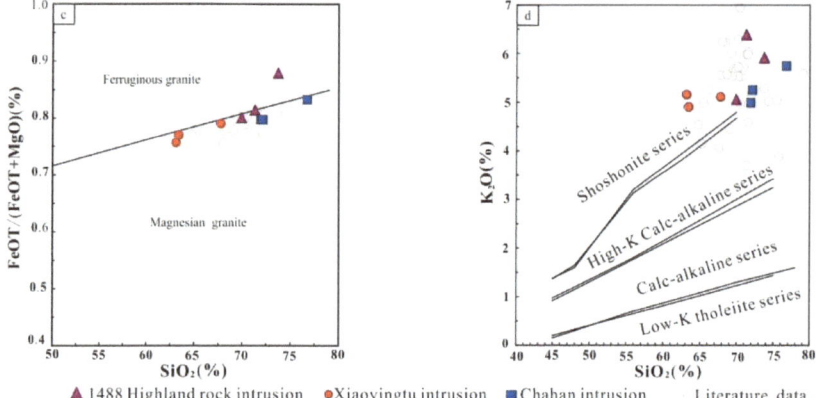

Figure 6. TAS (total-alkali-SiO$_2$) diagram (**a**), after Middlemost,1994 [29]), A/CNK-A/NK diagram (**b**), after Maniar and Piccoli, 1989 [30]), SiO$_2$-FeO/(FeO + MgO) diagram (**c**), after Frost et al., 2001 [31]), SiO$_2$-K$_2$O covariant diagram (**d**), after Peccerillo and Taylor, 1976 [32]). Literature data from Zhang Shuanhong, 2014 [33]; Meng Baohang, 2016 [34]; Phase Vibration Group, 2020 [14].

5.3. Trace Elements

The total rare earth element ($\sum$REE) contents of the analyzed samples vary between 168.85 and 690.06 ppm, with an average of 374.64 ppm and significant variations between samples. The samples exhibit enrichments in light REE and depletions in heavy REE (Figure 7a), with ratios of light to heavy REE of 7.46–24.0. (La/Yb)$_N$ ratios vary from 5.37 to 61.49, with an average of 22.87. The chondrite-normalized REE patterns exhibit negative Eu anomalies (δEu = 0.14–0.54), indicative of plagioclase fractionation. In a primitive-mantle-normalized multi-element diagram (Figure 7b), the samples are enriched in large-ion lithophile elements (LILEs; e.g., K and U) and exhibit small negative Zr and large negative Ti anomalies. In general, the trace element patterns of the studied granites are similar, indicating the granites are cogenetic.

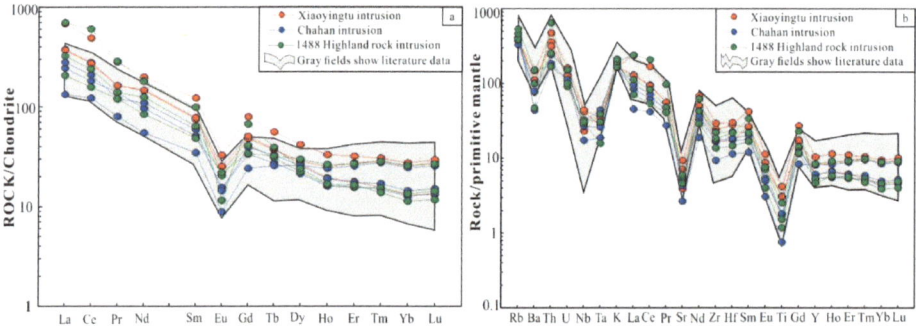

Figure 7. Chondrite-normalized REE patterns and (**a**) a primitive mantle-normalized multi-element diagram (**b**) of the Mesoproterozoic granites in the study area; standard reference values of chondrites and primitive mantle are from Sun and McDonough, 1989 [35]. Literature data from Zhang Shuanhong, 2014 [33]; Meng Baohang, 2016 [34]; Phase Vibration Group, 2020 [14].

5.4. Zircon Hf Isotopes

Hafnium isotopic compositions were determined for zircons in one granite sample (2021055). The results are listed in Table 3. The geochemical properties of Hf are similar to those of Zr, and Hf readily substitutes for Zr in the zircon crystal lattice, which results in low zircon Lu/Hf ratios. As such, the ^{176}Hf/^{177}Hf ratio changes little over time due to the in situ decay of ^{176}Lu. Therefore, magmatic zircon effectively retains the initial ^{176}Hf/^{177}Hf ratio of the melt it crystallizes from [36]. The zircon ^{176}Lu/^{177}Hf ratios (Figure 8b) range between 0.0004 and 0.0017, and ^{176}Hf/^{177}Hf ratios vary between 0.281807 and 0.282080. ^{176}Yb/^{177}Hf ratios are 0.0155–0.0671. Zircon $f_{Lu/Hf}$ and $\varepsilon_{Hf}(t)$ values and T_{DM1} and T_{DM2} ages vary from −0.98 to −0.95, −6.43 to +2.41, 1676 to 2022 Ma, and 1905 to 2462 Ma (Figure 8a), respectively.

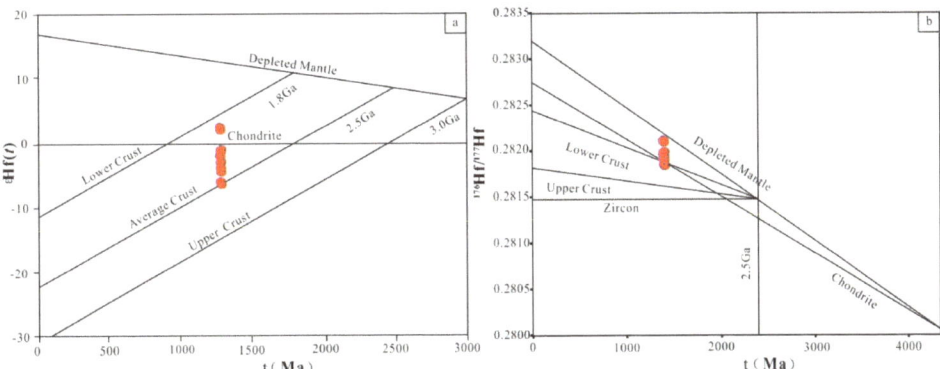

Figure 8. Correlation between Hf isotopic compositions and crystallization ages of porphyritic biotite monzogranite.

6. Discussion

6.1. Granite Petrogenesis

The primary minerals in the Mesoproterozoic granites are alkali feldspar, plagioclase, quartz, and biotite, with K-feldspar megacrysts present in most samples. Samples are enriched in Ga (19.50–28.46 ppm), with 1000 Ga/Al values of 2.64–3.10. Most samples are plotted in the A-type granite field on 10,000 Ga/Al–Ce, 10,000 Ga/Al–Zr, and 10,000 Ga/Al–K$_2$O/MgO diagrams (Figure 9a–c; [37]). Similarly, on a Zr + Nb + Ce + Y–TFeO/MgO diagram (Figure 9d), all samples are plotted in the A-type granite field. In addition, the absence of inherited zircons in the Mesoproterozoic granites, which are typically present in S-type granites, indicates the source region of the Mesoproterozoic granites underwent high-temperature melting. This characteristic is typical of A-type granites. Most samples are plotted in the A2-type granite fields in Nb–Y–Ce and Nb–Y–3Ga ternary diagrams (Figure 9e–f).

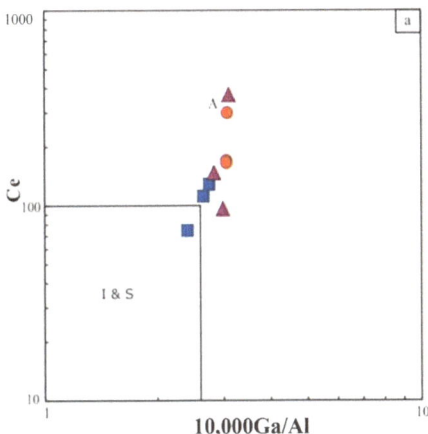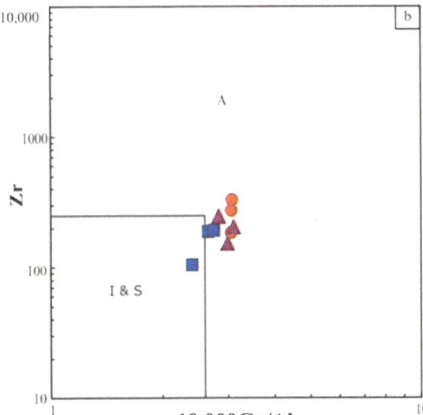

Figure 9. Cont.

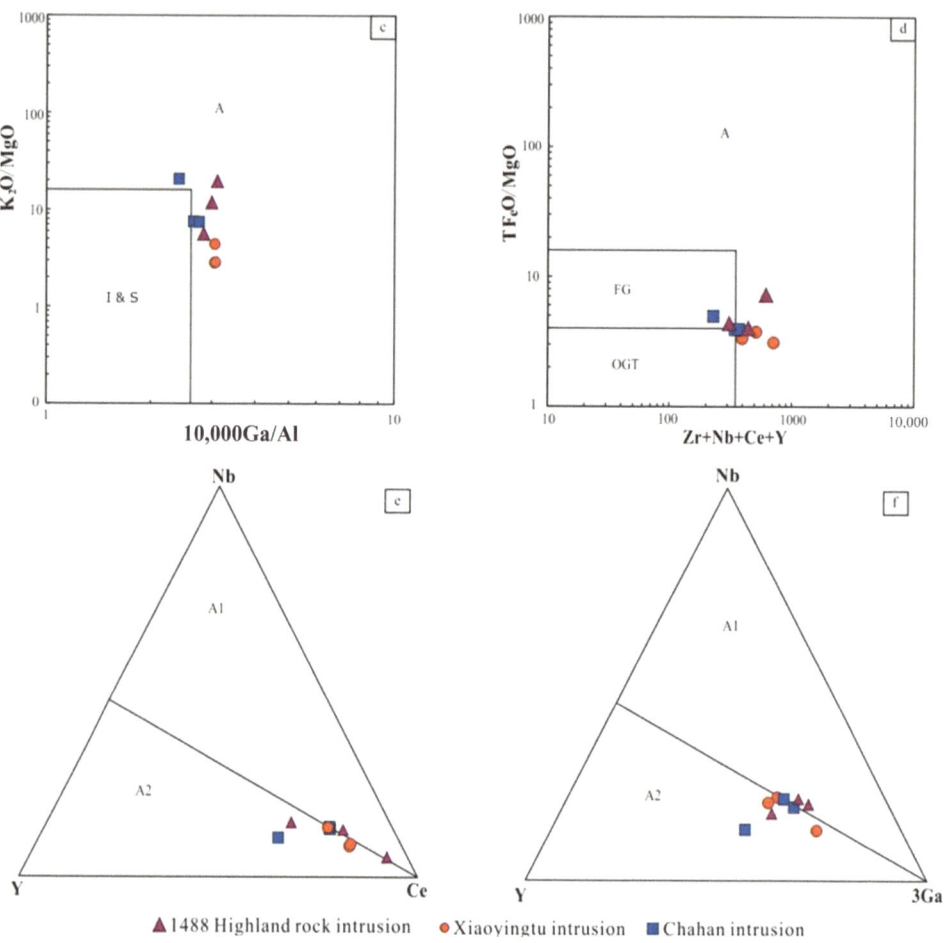

▲ 1488 Highland rock intrusion   ● Xiaoyingtu intrusion   ■ Chahan intrusion

Figure 9. Classification diagrams for the genetic types of Mesoproterozoic granites in the study area: (**a**) 10,000 Ga/Al versus Ce; (**b**) 10,000 Ga/Al versus Zr; (**c**) 10,000 Ga/Al versus K_2O/MgO; (**d**) (Zr + Nb + Ce + Y) versus TFeO/MgO; (**e**) Nb-Y-Ce; (**f**) Nb-Y-3Ga; (**a–d**) are after Whalen et al. (1987) [37], and (**e**) and (**f**) are after Eby. (1992) [38]; A-, I- and S-, A-, I- and S-type granite; FG. Differentiated I-type granite area; OGT. Undifferentiated I- and S-type granite area; A1. Non-orogenic granite; A2. Post-orogenic granite.

6.2. Magma Sources and Evolution

There are eight main hypotheses regarding the origins of A-type granites [39]: (1) differentiation of alkaline magmas derived from the mantle, resulting in residual A-type granitic melts [38,40]; (2) extreme differentiation of tholeiitic magma derived from the mantle or low-degree partial melting of tholeiitic rocks [41–43]; (3) interaction of alkali magmas derived from the mantle with crustal materials, leading to the formation of a syenitic magma source area, which is further differentiated or mixed with crustal materials [44,45]; (4) partial melting of F-rich granulite residue after extraction of I-type granitic magma from the lower crust [37]; (5) direct melts of igneous crustal rocks (i.e., tonalite and granodiorite) [46]; (6) melting of crustal rocks due to magmatic underplating [47,48]; (7) partial melting of lower crustal rocks due to the addition of volatiles from the mantle [49]; and (8) mixing between crust- and mantle-derived magmas [50].

The absence of coeval (ultra)mafic rocks precludes the involvement of mantle-derived (alkali) basalt magma. The granite samples have high SiO_2 and $Na_2O + K_2O$ contents, and low TiO_2, total Fe_2O_3, and MgO contents. A/CNK ratios indicate the granites are strongly peraluminous. The depletions in Nb, Ta, Sr, and Ti and enrichments in K, Rb, Th, and U in the granites suggest their parental melt(s) were derived from the crust [51]. $\varepsilon_{Hf}(t)$ values of −1.08 to −6.43 (except one analysis of 2.41) are indicative of a crustal source. The zircon T_{DM1} ages of 1676–2022 Ma (average = 1864 Ma) correspond to a period of uplift and rifting of NCC basement, and anorogenic magmatism. The zircon T_{DM2} ages of 1905–2462 Ma are consistent with the timing of development of the Palaeoproterozoic orogenic belt in the NCC [52]. Harker diagrams (Figure 10) exhibit negative linear correlations for TiO_2, MgO, and Al_2O_3. Furthermore, the trace element patterns all exhibit depletions in Ba, Nb, Ta, Sr, Ti, and Eu, indicating the sources of the granites were similar and that the generated magmas underwent significant fractional crystallization. In conclusion, the Mesoproterozoic granites were generated by the melting of crustal rocks due to magmatic underplating, and the parental magmas underwent significant crystal fractionation.

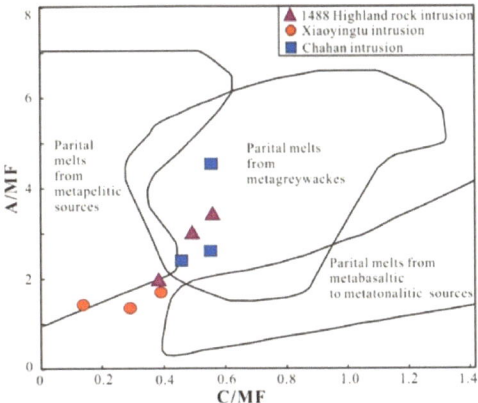

Figure 10. C/MF-A/MF diagram of Mesoproterozoic granite in the study area (Alther et al., 2000 [52]).

6.3. Tectonic Setting

The ca. 1.30 Ga magmatism along the northern margin of the NCC includes dolerite dykes and sills in the eastern part of the Yanliao rift, the Shangdu–Huade–Kangbao granite, and the Baiyun Obo carbonatite. Dolerite sills (1.33–1.30 Ga) associated with continental rifting have intruded the Shimaling, Tieling, Wumoushan, and Gaoqizhuang formations in the northern NCC [5,9,14]. In addition, 1.3 Ga carbonatite and mafic dykes occur in the Baiyun Obo area. Previous studies of carbonatites in the Baiyun Obo area, middle Mesoproterozoic granites, and large-scale dolerite sills from the northern margin of the NCC indicate that rifting persisted until the middle Mesoproterozoic [53,54].

The studied Mesoproterozoic granites are peraluminous A-type granites, and most studies consider that A-type granites form in extensional tectonic settings, such as those associated with post-orogenic extension, continental margins, and intraplate rifts. The Mesoproterozoic A-type granites have REE characteristics indicative of a crustal origin, similar to intraplate granites formed by the melting of continental crust. Eby [39] categorized A-type granites into two sub-types, A1 and A2. The A1 sub-type is thought to be derived by the differentiation of ocean island basalt magma, while the A2 sub-type may be derived by the interaction between mantle-derived magmas and continental crust. Granite samples from the study area plot in the A2 field in Nb–Y–Ce and Nb–Y–3Ga diagrams, indicating their formation in a stable extensional setting during the late stages of an orogeny. Pearce [55] used Yb + Ta–Rb, Nb + Y–Rb, Yb–Ta, and Y–Nb diagrams to identify the tectonic settings of granitic magmas. Most of the studied granites plot in the post-collisional to intraplate fields in these diagrams (Figure 11a–d). Therefore, the Mesoproterozoic granites were formed in an intraplate tectonic setting during continental extension and rifting.

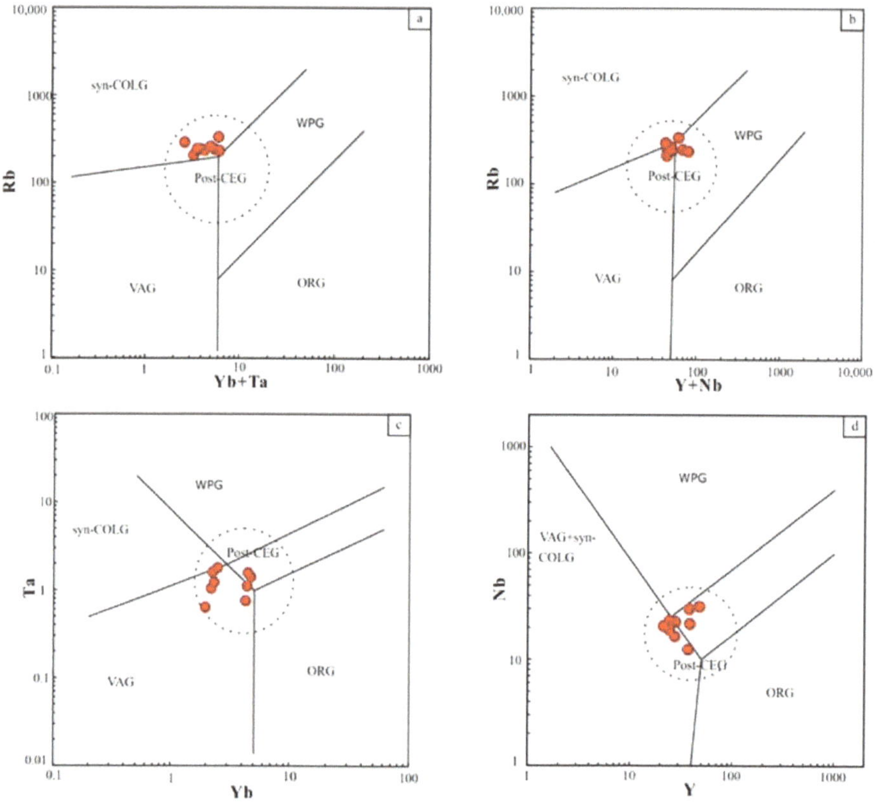

Figure 11. Diagrams of the tectonic environment of trace elements of Mesoproterozoic granite in the study area (after Pearce et al., 1984 [55]). WPG-within plate granite; ORG- ocean ridge granite; VAG-volcanic arc granite; syn-COLG-syn-collision granite; Post-CEG-post collision granite.

6.4. Relationship to the Break-Up of Columbia

The ca. 1.3 Ga magmatism was global in extent, suggesting the extensional/rifting event during this period was a global phenomenon. Mafic dyke swarms ranging in age from 1.3 to 1.2 Ga have been identified in the Mackenzie Mountains in Canada, the São Francisco craton in southeastern Brazil, and Australia [16,56,57]. In addition to mafic magmatism during the middle Mesoproterozoic along the northern margin of the NCC, granites were formed at this time in the study area and surrounding regions due to continental extension and rifting [8,18–20,53]. These late Mesoproterozoic dolerites, granites, and carbonatites form a bimodal magmatic assemblage in the NCC associated with the late-stage break-up of Columbia. Palaeomagnetic data for the period 1.77–1.50 Ga have shown that the palaeomagnetic poles of the NCC closely resemble those of North America (Laurentia) and Siberia [58]. However, during 1.35–1.20 Ga, the palaeomagnetic poles of North China exhibited a ~90° rotation relative to Laurentia (North America; [58]). Evans et al. [59] reconstructed the initial fragmentation of the Mesoproterozoic Columbia supercontinent based on tectonic, stratigraphic, and palaeomagnetic data for Siberia, Laurentia, and Baltica (Figure 12). Their findings indicate that the southern and eastern margins of Siberia were directly adjacent to the Urals margin of Baltica. Furthermore, they proposed the NCC may have fully separated from Columbia during the middle Mesoproterozoic (1.50–1.27 Ga).

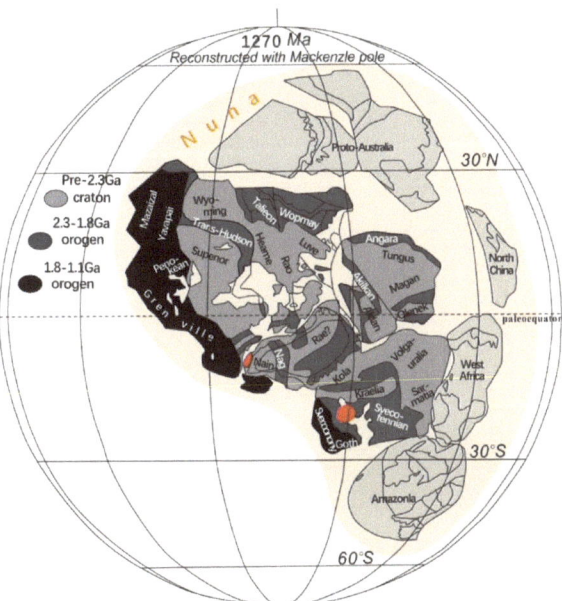

Figure 12. Reconstruction of the initial Mesoproterozoic fragmentation of the Columbia supercontinent (Modified by Evans, 2011 [59]).

In summary, it can be inferred that the Mesoproterozoic dolerite dykes along the northern margin of the NCC represent magmas derived from the mantle during continental rifting, whereas the granites represent the melting of the ancient crust. Both types of magmatism were related to the break-up of Columbia. The initial fragmentation of Columbia in different cratons may have varied in time. However, the ca. 1.30 Ga magmatic event in the study area likely corresponds to the last break-up phase of Columbia and highlights the response of the NCC to this final break-up event.

7. Conclusions

1. (Zircon U–Pb ages of porphyritic granites in the central part of the northern margin of the NCC range from 1285.4 ± 2.6 to 1278.6 ± 6.1 Ma; these are the same within error.
2. The Mesoproterozoic granites are weakly peraluminous A2-type granites. Trace element data suggest they were formed by partial melting of ancient crustal materials and may have formed in an intraplate tectonic setting affected by continental extension and rifting.
3. The Mesoproterozoic granites were formed in an extensional rift setting that also generated widespread coeval doleritic dykes in the NCC. These igneous rocks are also coeval with the 1.3–1.2 Ga extensional events that occurred worldwide. Therefore, the extensional setting at the northern margin of the NCC reflects a response to the late break-up stage of the Columbia supercontinent.

Author Contributions: Investigation, L.L. and Z.X.; Resources, M.M.; Data curation, S.J.; Writing—original draft, B.L.; Writing—review & editing, G.T.; Visualization, J.Y.; Project administration, Y.Q. All authors have read and agreed to the published version of the manuscript.

Funding: This research was funded by the Geological Survey Project of China Geological Survey, grant number DD20230251 and DD20208003.

Data Availability Statement: Data for this research are included in this paper.

Conflicts of Interest: The authors declare no conflict of interest.

Appendix A

Table A1. LA-ICP-MS zircon U-Pb data of the volcanic rocks from the Mesoproterozoic granite-like zircon.

The Number of the Measurement Point	Elemental Content (10⁻⁶)			Th/U	Isotope Ratio and 1σ								Isotope Age Values and 1σ						Confidence
	Pb	^{232}Th	^{238}U		^{207}Pb/^{206}Pb	1σ	^{207}Pb/^{235}U	1σ	^{206}Pb/^{238}U	1σ	^{208}Pb/^{232}Th	1σ	^{208}Pb/^{232}Th	1σ	^{207}Pb/^{235}U	1σ	^{206}Pb/^{238}U	1σ	
U-Pb20200054.1	90	56	364	0.15	0.0831	0.0011	2.5340	0.0462	0.2200	0.0031	0.0831	0.0011	0.0640	0.0014	1282	13	1282	16	99%
U-Pb20200054.2	124	48	489	0.10	0.0833	0.0011	2.5467	0.0392	0.2206	0.0024	0.0833	0.0011	0.0630	0.0013	1285	11	1285	13	99%
U-Pb20200054.3	121	46	489	0.09	0.0831	0.0010	2.5345	0.0404	0.2199	0.0027	0.0831	0.0010	0.0648	0.0014	1282	12	1281	14	99%
U-Pb20200054.4	108	138	405	0.34	0.0836	0.0010	2.5440	0.0404	0.2193	0.0025	0.0836	0.0010	0.0676	0.0013	1285	12	1278	13	99%
U-Pb20200054.5	117	102	455	0.22	0.0843	0.0012	2.5576	0.0446	0.2190	0.0029	0.0843	0.0012	0.0684	0.0020	1289	13	1277	15	99%
U-Pb20200054.6	121	120	469	0.26	0.0844	0.0010	2.5553	0.0425	0.2183	0.0030	0.0844	0.0010	0.0634	0.0015	1288	12	1273	16	98%
U-Pb20200054.7	139	48	549	0.09	0.0858	0.0011	2.6046	0.0382	0.2186	0.0021	0.0858	0.0011	0.0673	0.0018	1302	11	1274	11	97%
U-Pb20200054.8	183	46	702	0.07	0.0856	0.0013	2.6003	0.0404	0.2186	0.0022	0.0856	0.0013	0.0767	0.0029	1301	11	1274	12	97%
U-Pb20200054.9	95	51	373	0.14	0.0845	0.0011	2.5676	0.0396	0.2187	0.0024	0.0845	0.0011	0.0652	0.0015	1291	11	1275	13	98%
U-Pb20200054.10	162	309	593	0.52	0.0846	0.0012	2.5771	0.0398	0.2194	0.0022	0.0846	0.0012	0.0570	0.0011	1294	11	1279	11	98%
U-Pb20200054.11	86	66	336	0.20	0.0864	0.0013	2.6158	0.0444	0.2181	0.0023	0.0864	0.0013	0.0586	0.0012	1305	13	1272	12	97%
U-Pb20200054.12	121	56	498	0.11	0.0836	0.0011	2.5281	0.0412	0.2180	0.0026	0.0836	0.0011	0.0624	0.0014	1280	12	1271	14	99%
U-Pb20200054.13	95	60	376	0.16	0.0840	0.0011	2.5523	0.0391	0.2188	0.0021	0.0840	0.0011	0.0651	0.0014	1287	11	1276	11	99%
U-Pb20200054.14	125	45	493	0.09	0.0846	0.0013	2.5656	0.0512	0.2183	0.0028	0.0846	0.0013	0.0743	0.0035	1291	15	1273	15	98%
U-Pb20200054.15	187	297	681	0.44	0.0834	0.0011	2.5166	0.0408	0.2177	0.0026	0.0834	0.0011	0.0627	0.0012	1277	12	1270	14	99%
U-Pb20200055.1	129	42	511	0.08	0.0838	0.0010	2.5570	0.0365	0.2202	0.0020	0.0834	0.0012	1288	23	1288	10	1283	11	99%
U-Pb20200055.2	144	182	541	0.34	0.0855	0.0009	2.5828	0.0357	0.2183	0.0023	0.0662	0.0010	1328	21	1296	10	1273	12	98%
U-Pb20200055.3	84	30	328	0.09	0.0836	0.0009	2.5466	0.0376	0.2204	0.0025	0.0604	0.0015	1283	22	1285	11	1284	13	99%
U-Pb20200055.4	79	32	324	0.10	0.0841	0.0009	2.5439	0.0397	0.2189	0.0028	0.0652	0.0014	1294	19	1285	11	1276	15	99%
U-Pb20200055.5	84	36	344	0.11	0.0845	0.0010	2.5665	0.0432	0.2199	0.0028	0.0679	0.0014	1306	23	1291	12	1281	15	99%
U-Pb20200055.6	99	39	400	0.10	0.0845	0.0009	2.5631	0.0401	0.2193	0.0025	0.0661	0.0014	1306	21	1290	11	1278	13	99%
U-Pb20200055.7	68	22	270	0.08	0.0827	0.0016	2.5077	0.0680	0.2192	0.0040	0.0618	0.0021	1265	38	1274	20	1278	21	99%
U-Pb20200055.8	62	18	249	0.07	0.0835	0.0011	2.5278	0.0438	0.2196	0.0029	0.0654	0.0019	1281	24	1280	13	1280	15	99%
U-Pb20200055.9	141	47	545	0.09	0.0854	0.0009	2.5794	0.0369	0.2189	0.0023	0.0658	0.0013	1324	20	1295	11	1276	12	98%

Table A1. Cont.

The Number of the Measurement Point	Elemental Content (10^{-6})			Th/U	Isotope Ratio and 1σ								Isotope Age Values and 1σ						Confidence
	Pb	^{232}Th	^{238}U		^{207}Pb/^{206}Pb	1σ	^{207}Pb/^{235}U	1σ	^{206}Pb/^{238}U	1σ	^{208}Pb/^{232}Th	1σ	^{208}Pb/^{232}Th	1σ	^{207}Pb/^{235}U	1σ	^{206}Pb/^{238}U	1σ	
U-Pb2020055.10	94	29	356	0.08	0.0849	0.0015	2.5782	0.0522	0.2200	0.0026	0.0812	0.0036	1322	34	1294	15	1282	14	99%
U-Pb2020055.11	91	30	358	0.08	0.0854	0.0013	2.5834	0.0435	0.2196	0.0026	0.0850	0.0052	1326	29	1296	12	1280	14	98%
U-Pb2020055.12	104	38	423	0.09	0.0853	0.0009	2.5807	0.0386	0.2195	0.0027	0.0670	0.0014	1324	21	1295	11	1279	14	98%
U-Pb2020055.13	105	35	435	0.08	0.0848	0.0009	2.5694	0.0405	0.2193	0.0027	0.0697	0.0015	1311	16	1292	12	1278	14	98%
U-Pb2020055.14	106	45	432	0.10	0.0846	0.0009	2.5520	0.0349	0.2185	0.0021	0.0647	0.0012	1306	20	1287	10	1274	11	98%
U-Pb2020055.15	195	115	769	0.15	0.0842	0.0008	2.5487	0.0320	0.2192	0.0021	0.0586	0.0009	1298	19	1286	9	1278	11	99%
U-Pb2020055.16	92	54	375	0.15	0.0856	0.0010	2.5956	0.0348	0.2198	0.0022	0.0657	0.0011	1329	21	1299	10	1281	12	98%
U-Pb2020055.17	60	46	250	0.19	0.0851	0.0011	2.5744	0.0383	0.2192	0.0020	0.0645	0.0013	1318	26	1293	11	1278	10	98%
U-Pb2021032.1	98	49	389	0.13	0.0845	0.0011	2.5846	0.0400	0.2209	0.0025	0.0687	0.0014	1306	25	1296	11	1287	13	99%
U-Pb2021032.2	137	279	487	0.57	0.0830	0.0010	2.5563	0.0387	0.2221	0.0023	0.0654	0.0010	1269	19	1288	11	1293	12	99%
U-Pb2021032.3	125	52	492	0.11	0.0845	0.0010	2.5871	0.0407	0.2211	0.0028	0.0667	0.0013	1306	24	1297	12	1288	15	99%
U-Pb2021032.4	142	85	564	0.15	0.0835	0.0010	2.5668	0.0444	0.2216	0.0030	0.0630	0.0014	1281	23	1291	13	1290	16	99%
U-Pb2021032.5	102	34	400	0.09	0.0820	0.0010	2.5176	0.0335	0.2216	0.0019	0.0672	0.0016	1256	24	1277	10	1290	10	98%
U-Pb2021032.6	72	65	273	0.24	0.0814	0.0010	2.5063	0.0371	0.2220	0.0024	0.0636	0.0011	1232	23	1274	11	1292	13	98%
U-Pb2021032.7	135	51	549	0.09	0.0827	0.0009	2.5456	0.0385	0.2221	0.0028	0.0679	0.0015	1261	21	1285	11	1293	15	99%
U-Pb2021032.8	124	42	491	0.08	0.0861	0.0011	2.6303	0.0365	0.2203	0.0019	0.0986	0.0040	1343	25	1309	10	1284	10	98%
U-Pb2021032.9	89	36	364	0.10	0.0818	0.0009	2.5056	0.0431	0.2208	0.0032	0.0640	0.0015	1240	20	1274	13	1286	17	99%
U-Pb2021032.10	109	62	438	0.14	0.0820	0.0009	2.5048	0.0324	0.2207	0.0022	0.0632	0.0011	1244	28	1273	9	1285	12	99%
U-Pb2021032.11	108	45	438	0.10	0.0815	0.0009	2.5055	0.0370	0.2217	0.0026	0.0665	0.0013	1233	22	1274	11	1291	14	98%
U-Pb2021032.12	85	59	340	0.17	0.0826	0.0011	2.5459	0.0432	0.2222	0.0029	0.0667	0.0014	1259	26	1285	12	1294	15	99%
U-Pb2021032.13	122	65	494	0.13	0.0829	0.0009	2.5374	0.0336	0.2207	0.0021	0.0658	0.0012	1266	16	1283	10	1286	11	99%
U-Pb2021032.14	92	42	380	0.11	0.0825	0.0009	2.5258	0.0324	0.2213	0.0022	0.0672	0.0013	1257	22	1279	9	1288	12	99%
U-Pb2021032.15	107	41	448	0.09	0.0829	0.0009	2.5373	0.0393	0.2208	0.0028	0.0665	0.0014	1278	22	1283	11	1286	15	99%
U-Pb2021032.16	82	138	302	0.46	0.0867	0.0014	2.6398	0.0436	0.2203	0.0028	0.0723	0.0015	1354	31	1312	12	1284	15	97%
U-Pb2021032.17	109	121	426	0.28	0.0851	0.0011	2.6017	0.0384	0.2210	0.0021	0.0626	0.0011	1318	26	1301	11	1287	11	98%
U-Pb2021032.18	73	41	299	0.14	0.0843	0.0011	2.5772	0.0376	0.2212	0.0019	0.0468	0.0012	1298	26	1294	11	1288	10	99%

Table A2. Analysis results of major elements (%), rare earth elements (10^{-6}), and trace elements (10^{-6}) of Mesoproterozoic granites in Inner Mongolia.

Sample	2021036	2021037	2021038	2021019	2021020	2021030	2021031	2021032	2021034
	(Xiaoyingtu Intrusion)			(Chahan Intrusion)			(1488. Highland Intrusion)		
SiO_2	67.72	63.39	63.10	71.90	72.13	76.71	73.73	69.91	71.30
TiO_2	0.67	0.66	0.91	0.38	0.40	0.16	0.26	0.55	0.33
Al_2O_3	15.23	17.64	16.59	13.93	13.59	12.04	13.23	14.44	14.19
Fe_2O_3	1.20	3.59	2.35	0.94	0.90	0.45	0.75	0.98	0.65
FeO	3.28	2.58	3.52	1.77	1.97	0.98	1.52	2.72	1.78
MnO	0.041	0.061	0.046	0.03	0.032	0.017	0.023	0.034	0.035
MgO	1.16	1.73	1.81	0.66	0.71	0.28	0.30	0.90	0.54
CaO	1.92	0.93	1.97	1.61	1.42	0.80	1.17	1.53	1.26
Na_2O	2.17	1.37	2.11	2.84	2.66	2.11	2.29	2.65	2.39
K_2O	5.11	4.91	5.16	4.99	5.25	5.75	5.92	5.06	6.39
P_2O_5	0.164	0.091	0.098	0.15	0.136	0.051	0.049	0.122	0.170
LOI	0.78	2.58	1.72	0.49	0.46	0.47	0.51	0.65	0.58
Total	99.82	99.83	99.76	99.87	99.87	99.93	99.91	99.84	99.83
Rb	246	246	235	209.92	241	263	295	247	338
Ba	776	779	1036	540	552	308	331	698	1056
Th	30.4	26.9	40.2	15.75	20.6	14.40	54.9	21.5	14.4
U	2.72	2.65	3.39	2.36	3.18	2.37	1.92	3.23	2.11
Nb	30.1	16.5	31.7	19.00	22.8	12.40	20.8	23.4	21.8
Ta	1.13	1.63	1.44	1.07	1.82	0.77	0.65	1.25	1.60
La	89.3	88.5	162	58.1	66.4	31.47	166	76.94	49.25
Ce	170	166	299	112	128	74.70	369	147	96.3
Pr	15.5	15.5	26.9	13.1	11.5	7.58	27.1	13.2	11.4
Sr	150	82.4	196	90.6	108	56.5	97.5	118	125
Nd	67.8	68.2	93.2	44.8	50.8	25.60	84.3	58.2	39.3
Zr	275	186	331	189	193	105	205	248	153
Hf	8.77	5.40	9.35	5.37	5.68	3.53	6.72	6.94	4.53
Sm	11.5	11.9	18.8	7.83	9.06	5.31	15.1	9.77	7.41
Eu	1.27	1.47	1.90	0.90	0.84	0.52	0.67	1.19	1.27
Gd	9.87	10.4	16.4	6.89	8.07	4.97	13.8	8.45	7.00
Tb	1.39	1.41	2.10	1.02	1.22	0.96	1.46	1.16	1.20
Dy	7.25	6.80	10.6	5.43	6.50	6.57	6.10	5.66	7.54
Y	37.9	27.3	47.4	24.2	27.7	36.90	21.2	24.4	38.3
Ho	1.42	1.10	1.87	0.91	1.08	1.37	0.95	0.93	1.49
Er	4.51	2.94	5.29	2.58	2.88	4.24	2.72	2.60	4.44
Tm	0.73	0.40	0.78	0.38	0.43	0.70	0.35	0.39	0.72
Yb	4.32	2.22	4.66	2.16	2.45	4.20	1.93	2.29	4.39
Lu	0.70	0.34	0.75	0.35	0.38	0.65	0.30	0.36	0.68

Table 3. Zircon in situ Hf isotope analysis results of medium-grained porphyritic biotite monzogranite.

Periods	Age (Ma)	$^{176}Yb/^{177}Hf$	2σ	$^{176}Lu/^{177}Hf$	2σ	$^{176}Hf/^{177}Hf$	2σ	$\varepsilon_{Hf}(t)$	$\varepsilon_{Hf}(0)$	T_{DM1}	T_{DM2}	$f_{Lu/Hf}$
						U-Pb20200055						
−1	1283	0.0310	0.0006	0.0008	0.0000	0.281956	0.000027	−1.08	−28.86	1812	2130	−0.98
−2	1273	0.0608	0.0010	0.0016	0.0000	0.282080	0.000031	2.41	−24.47	1676	1905	−0.95
−3	1284	0.0324	0.0004	0.0009	0.0000	0.281807	0.000032	−6.43	−34.13	2022	2462	−0.97
−4	1276	0.0594	0.0006	0.0016	0.0000	0.282065	0.000028	1.94	−25.00	1698	1936	−0.95
−5	1281	0.0354	0.0003	0.0010	0.0000	0.281906	0.000028	−3.07	−30.63	1891	2252	−0.97
−6	1278	0.0275	0.0003	0.0008	0.0000	0.281819	0.000023	−6.05	−33.70	2001	2434	−0.98
−7	1278	0.0671	0.0045	0.0017	0.0001	0.281925	0.000032	−3.07	−29.95	1900	2249	−0.95
−8	1280	0.0155	0.0002	0.0004	0.0000	0.281866	0.000026	−4.00	−32.04	1916	2309	−0.99
−9	1276	0.0303	0.0004	0.0008	0.0000	0.281945	0.000027	−1.62	−29.25	1827	2158	−0.98
−10	1282	0.0285	0.0003	0.0008	0.0000	0.281860	0.000022	−4.51	−32.25	1944	2342	−0.98
−11	1280	0.0187	0.0001	0.0006	0.0000	0.281902	0.000018	−2.90	−30.77	1877	2240	−0.98
−12	1279	0.0318	0.0004	0.0009	0.0000	0.281883	0.000020	−3.84	−31.44	1918	2298	−0.97
−13	1278	0.0317	0.0002	0.0008	0.0000	0.281946	0.000018	−1.54	−29.21	1826	2155	−0.98
−14	1274	0.0283	0.0003	0.0008	0.0000	0.281933	0.000018	−2.10	−29.67	1844	2186	−0.98
−15	1278	0.0310	0.0006	0.0008	0.0000	0.281956	0.000027	−1.20	−28.86	1812	2133	−0.98

References

1. Roberts, N.M.W. The boring billion?—Lid tectonics, continental growth and environmental change associated with the Columbia supercontinent. *Geosci. Front.* 2013, 4, 681–691. [CrossRef]
2. Zhang, S.-H.; Ernst, R.E.; Yang, Z.; Zhou, Z.; Pei, J.; Zhao, Y. Spatial distribution of 1.4–1.3 Ga LIPs and carbonatite-related REE deposits: Evidence for large-scale continental rifting in the Columbia (Nuna) supercontinent. *Earth Planet. Sci. Lett.* 2022, 597, 117815. [CrossRef]
3. Li, S.; Yu, S.; Zhao, S.; Zhang, G.; Liu, X.; Cao, H.; Xu, L.; Dai, L.; Li, T. Perspectives of Supercontinent Cycle and Global Plate Reconstruction. *Mar. Geol. Quat. Geol.* 2015, 1, 51–60.
4. Pan, G.; Lu, S.; Xiao, Q.; Zhang, K.; Yin, F.; Hao, G.; Luo, M.; Ren, F.; Yuan, S. Division of tectonic stages and tectonic evolution in China. *Earth Sci. Front.* 2016, 23, 10.
5. Li, C.; Liu, Z.; Xu, Z.; Dong, X.; Liu, J.; Cheng, Y.; Zhang, N. Mesoproterozoic (~1.3 Ga) S–type granites in Shangdu area, Inner Mongolia of the North China Craton (NCC): Implications for breakup of the NCC from the Columbia supercontinent. *Precambrian Res.* 2022, 369, 106515. [CrossRef]
6. Zhao, G.; Sun, M.; Wilde, S.A.; Li, S. A Paleo-Mesoproterozoic supercontinent: Assembly, growth and breakup. *Earth-Sci. Rev.* 2004, 67, 91–123. [CrossRef]
7. Volante, S.; Pourteau, A.; Collins, W.J.; Blereau, E.; Li, Z.; Smit, M.; Evans, N.J.; Nordsvan, A.R.; Spencer, C.J.; McDonald, B.J.; et al. Multiple *P-T-d-t* paths reveal the evolution of the final Nuna assembly in northeast Australia. *J. Metamorph. Geol.* 2020, 38, 593–627. [CrossRef]
8. Zhang, S.-H.; Zhao, Y.; Liu, Y. A precise zircon Th-Pb age of carbonatite sills from the world's largest Bayan Obo deposit: Implications for timing and genesis of REE-Nb mineralization. *Precambrian Res.* 2017, 291, 202–219. [CrossRef]
9. Zhang, S.-H.; Zhao, Y.; Li, X.-H.; Ernst, R.E.; Yang, Z.-Y. The 1.33–1.30 Ga Yanliao large igneous province in the North China Craton: Implications for reconstruction of the Nuna (Columbia) supercontinent, and specifically with the North Australian Craton. *Earth Planet. Sci. Lett.* 2017, 465, 112–125. [CrossRef]
10. Zhong, Y.; Chen, Y.L.; Zhai, M.G.; Ma, X.D. Stratigraphical correlation and lithofacies paleogeography of khondalite series in the western North China Craton. *Acta Petrol. Sin.* 2016, 32, 713–726.
11. Zhai, M.G. Ocean and Continent in Archean. *J. Palaeogeogr.* 2022, 24, 825–847. [CrossRef]
12. Shao, J.; Zhang, L.Q.; Li, D.M. Three Proterozoic extensional events in North China Craton. *Acta Petrol. Sin.* 2002, 18, 152–160.
13. Zhai, M.G. Multi-stage crustal growth and cratonization of the North China Craton. *Geosci. Front.* 2014, 5, 457–469. [CrossRef]
14. Xiang, Z.Q.; Lu, S.N.; Li, H.K.; Tian, H.; Liu, H.; Zhang, K. Mesoproterozoic magmatic events in the North China Craton. *Geol. Surv. Res.* 2020, 43, 137–152.
15. Lu, S.; Zhao, G.; Wang, H.; Hao, G. Precambrian metamorphic basement and sedimentary cover of the North China Craton: A review. *Precambrian Res.* 2007, 160, 77–93. [CrossRef]
16. Rogers, J.J.; Santosh, M. Configuration of Columbia, a Mesoproterozoic Supercontinent. *Gondwana Res.* 2002, 5, 5–22. [CrossRef]
17. Wu, H.; Zhang, S.; Li, Z.X.; Li, H.; Dong, J. New paleomagnetic results from the Yangzhuang Formation of the Jixian System, North China, and tectonic implications. *Sci. Bull.* 2005, 50, 1483–1489. [CrossRef]

18. Zhang, S.H.; Zhao, Y. The 1.33–1.30 Ga mafic large igneous province and REE-Nb metallogenic event in the northern North China Craton. *Earth Sci. Front.* **2018**, *25*, 34–50. [CrossRef]
19. Li, H.K.; Lu, S.N.; Li, H.M.; Sun, L.X.; Xiang, Z.Q.; Geng, J.Z.; Zhou, H.Y. Zircon and beddeleyite U-Pb precision dating of basic rock sills intruding Xiamaling Formation, North China. *Geol. Bull. China* **2009**, *10*, 1396–1404.
20. Zhang, H.F. Peridotite-melt interaction: A key point for the destruction of cratonic lithospheric mantle. *Chin. Sci. Bull.* **2009**, *54*, 3417–3437. [CrossRef]
21. Yang, K.-F.; Fan, H.-R.; Santosh, M.; Hu, F.-F.; Wang, K.-Y. Mesoproterozoic mafic and carbonatitic dykes from the northern margin of the North China Craton: Implications for the final breakup of Columbia supercontinent. *Tectonophysics* **2011**, *498*, 1–10. [CrossRef]
22. Şengör, A.M.C.; Natal'In, B.A.; Burtman, V.S. Evolution of the Altaid tectonic collage and Palaeozoic crustal growth in Eurasia. *Nature* **1993**, *364*, 299–307. [CrossRef]
23. Jiang, N.; Guo, J.H.; Zhai, M.G.; Zhang, S.Q. ~2.7 Ga crust growth in the North China craton. *Precambrian Res.* **2010**, *179*, 37–49. [CrossRef]
24. Zhao, G.C.; Wilde, S.A.; Cawood, P.A.; Sun, M. Archean blocks and their boundaries in the North China Craton: Lithological, geochemical, structural and P-T path constraints and tectonic evolution. *Precambrian Res.* **2001**, *107*, 45–73. [CrossRef]
25. Zhai, M.G. The maSin old lands in China and assembly of Chinese unified continent. *Sci. China Earth Sci.* **2013**, *56*, 1829–1852. [CrossRef]
26. Zhai, M.G. Tectonic evolution and metallogenesis of North China Craton. *Miner. Depos.* **2010**, *29*, 24–36.
27. Ludwig, K. User's Manual for Isoplot/Ex version 3.00-A Geochronology Toolkit for Microsoft Excel. *Berkeley Geochronological Cent. Spec. Publ.* **2003**, *4*, 1–70.
28. Hoskin, P.W.O.; Schaltegger, U. The Composition of Zircon and Igneous and Metamorphic Petrogenesis. *Rev. Miner. Geochem.* **2003**, *53*, 27–62. [CrossRef]
29. Middlemost, A.K. Naming materials in the magma/igneous rock system. *Earth-Sci. Rev.* **1994**, *37*, 215–224. [CrossRef]
30. Maniar, P.D.; Piccoli, P.M. Tectonic discrimination of granitoids. *Geol. Soc. Am. Bull.* **1989**, *101*, 635–643. [CrossRef]
31. Frost, B.R.; Barnes, C.G.; Collins, W.J.; Arculus, R.J.; Ellis, D.J.; Frost, C.D. A Geochemical Classification for Granitic Rocks. *J. Petrol.* **2001**, *42*, 2033–2048. [CrossRef]
32. Peccerillo, A.; Barberio, M.R.; Yirgu, G.; Ayalew, D.; Barbieri, M.; Wu, T.W. Relationship between mafic and peralkaline felsic magmatism in continental rift settings: A petrological, geochemical and isotopic study of the Gedemsa Volcano, Central Ethiopian Rift. *J. Petrol.* **2003**, *44*, 2003–2032. [CrossRef]
33. Zhang, S.H. *Mesoproterozoic Magmatism and Metallogenic Setting in Northern North China Craton*; Institute of Geomechanics, Chinese Academy of Geological Sciences: Beijing, China, 2014; pp. 137–187.
34. Meng, B.H. *Petrogenesis of the Lujiaying Megaporphyric Granite and Its Tectonic Implications within Kangbao County, North Hebei Province*; Chengdu University of Technology: Chengdu, China, 2016; pp. 48–62.
35. Sun, S.S.; McDonough, W.F. Chemical and isotopic systematics of oceanic basalts: Implications for mantle composition and processes. In *Magmatism in the Ocean Basin*; Geological Society Special Publication: London, UK, 1989; Volume 42, pp. 313–345.
36. Wu, F.Y.; Li, X.H.; Zheng, Y.F.; Gao, S. Lu-Hf isotopic systematics and their applications in petrology. *Acta Petrol. Sin.* **2007**, *23*, 185–220.
37. Whalen, J.B.; Currie, K.L.; Chappell, B.W. A-type granites: Geochemical characteristics, discrimination and petrogenesis. *Contrib. Miner. Petrol.* **1987**, *95*, 407–419. [CrossRef]
38. Eby, G.N. Chemical subdivision of the A-type granitoids: Petrogenesis and tectonic implications. *Gelology* **1992**, *20*, 641–644. [CrossRef]
39. Jia, X.H.; Wang, Q.; Tang, G.J. A-type Granites:Research Progress and Implications. *Geotecton. Metallog.* **2009**, *3*, 465–480. [CrossRef]
40. Eby, G.N. The A-type granitoids: A review of their occurrence and chemical characteristics and speculations on their petrogenesis. *Lithos* **1990**, *26*, 115–134. [CrossRef]
41. Frost, C.D.; Frost, B.R. Reduced rapakivi-type granites: The tholeiite connection. *Geology* **1997**, *25*, 647–650. [CrossRef]
42. Frost, C.D.; Frost, B.R.; Chamberlain, K.R.; Edwards, B. Petrogenesis of the 1.43 Ga Sherman batholith, SE Wyoming, USA: A reduced, rapakivi-type anorogenic granite. *J. Petrol.* **1999**, *40*, 1771–1802. [CrossRef]
43. Zhao, Z.H.; Wang, Z.G.; Zhou, T.R.; Masuda, A. Study on Petrogenesis of alkali-rich intrusive rocks of Ulungur, Xinjiang. *Geochimica* **1996**, *25*, 205–220. [CrossRef]
44. Litvinovsky, B.A.; Steele, I.M.; Wickham, S.M. Silicic Magma Formation in Overthickened Crust: Melting of Charnockite and Leucogranite at 15, 20 and 25 kbar. *J. Petrol.* **2000**, *41*, 717–737. [CrossRef]
45. Litvinovsky, B.A.; Jahn, B.-M.; Zanvilevich, A.; Saunders, A.; Poulain, S.; Kuzmin, D.; Reichow, M.; Titov, A. Petrogenesis of syenite–granite suites from the Bryansky Complex (Transbaikalia, Russia): Implications for the origin of A-type granitoid magmas. *Chem. Geol.* **2022**, *189*, 105–133. [CrossRef]
46. Creaser, R.A.; Price, R.C.; Wormald, R.J. A-type granites revisited: Assessment of a residual-source model. *Geology* **1991**, *19*, 163–166. [CrossRef]
47. Wu, F.-Y.; Sun, D.-Y.; Li, H.; Jahn, B.-M.; Wilde, S. A-type granites in northeastern China: Age and geochemical constraints on their petrogenesis. *Chem. Geol.* **2002**, *187*, 143–173. [CrossRef]

48. Rämö, O.T.; McLemore, V.T.; Hamilton, M.A.; Kosunen, P.J.; Heizler, M.; Haapala, I. Intermittent 1630–1220 Ma magmatism in central Mazatzal province: New geochronologic piercing points and some tectonic implications. *Geology* **2003**, *31*, 335–338. [CrossRef]
49. Pei, J.; Yang, Z.; Zhao, Y. A Mesoproterozoic paleomagnetic pole from the Yangzhuang Formation, North China and its tectonics implications. *Precambrian Res.* **2006**, *151*, 1–13. [CrossRef]
50. Yang, J.-H.; Wu, F.-Y.; Chung, S.-L.; Wilde, S.A.; Chu, M.-F. A hybrid origin for the Qianshan A-type granite, northeast China: Geochemical and Sr–Nd–Hf isotopic evidence. *Lithos* **2006**, *89*, 89–106. [CrossRef]
51. Wu, F.Y.; Li, X.H.; Yang, J.H.; Zheng, Y. Discussions on the petrogenesis of granites. *Acta Petrol. Sin.* **2007**, *23*, 1217–1238.
52. Altherr, R.; Holl, A.; Hegner, E.; Langer, C.; Kreuzer, H. High-potassium, calc-alkaline I-type plutonism in the European Variscides: Northern Vosges (France) and northern Schwarzwald (Germany). *Lithos* **2000**, *50*, 51–73. [CrossRef]
53. Sun, J.; Fang, N.; Li, S.Z.; Chen, Y.L.; Zhu, X.K. Magnesium isotopic constraints on the genesis of Bayan Obo ore deposit. *Acta Petrol. Sin.* **2012**, *28*, 2890–2902. [CrossRef]
54. Zhang, S.-H.; Zhao, Y.; Yang, Z.-Y.; He, Z.-F.; Wu, H. The 1.35Ga diabase sills from the northern North China Craton: Implications for breakup of the Columbia (Nuna) supercontinent. *Earth Planet. Sci. Lett.* **2009**, *288*, 588–600. [CrossRef]
55. Pearce, J.A.; Harris, N.B.W.; Tindle, A.G. Trace element discrimination diagrams for the tectonic interpretation of granitic rocks. *J. Petrol.* **1984**, *25*, 956–983. [CrossRef]
56. Rogers, J.J.; Santosh, M. Supercontinents in Earth History. *Gondwana Res.* **2003**, *6*, 357–368. [CrossRef]
57. Pesonen, L.; Elming, S.; Mertanen, S.; Pisarevsky, S.; D'Agrella-Filho, M.; Meert, J.; Schmidt, P.; Abrahamsen, N.; Bylund, G. Palaeomagnetic configuration of continents during the Proterozoic. *Tectonophysics* **2003**, *375*, 289–324. [CrossRef]
58. Pei, J.L.; Yang, Z.Y.; Zhao, Y. New Mesoproterozoic paleomagnetic results in North China and its implication for the Columbia supercontinent. *Geol. Bull. China* **2005**, *24*, 496–498.
59. Evans, D.A.D.; Mitchell, R.N. Assembly and breakup of the core of Paleoproterozoic-Mesoproterozoic supercontinent Nuna. *Geology* **2011**, *39*, 443–446. [CrossRef]

Disclaimer/Publisher's Note: The statements, opinions and data contained in all publications are solely those of the individual author(s) and contributor(s) and not of MDPI and/or the editor(s). MDPI and/or the editor(s) disclaim responsibility for any injury to people or property resulting from any ideas, methods, instructions or products referred to in the content.

Article

Petrogenesis and Geodynamic Evolution of A-Type Granite Bearing Rare Metals Mineralization in Egypt: Insights from Geochemistry and Mineral Chemistry

Mohamed M. Ghoneim [1,*], Ahmed E. Abdel Gawad [1,*], Hanaa A. El-Dokouny [2], Maher Dawoud [2], Elena G. Panova [3], Mai A. El-Lithy [2] and Abdelhalim S. Mahmoud [4]

[1] Nuclear Materials Authority, El-Maadi, Cairo P.O. Box 530, Egypt
[2] Geology Department, Faculty of Science, Menofia University, Shebin El Koum 32511, Egypt; hanaaabdelnaby4@gmail.com (H.A.E.-D.); maielleithy24@gmail.com (M.A.E.-L.)
[3] Department of Geochemistry, Saint Petersburg State University, 199034 Saint Petersburg, Russia; elena-geo@list.ru
[4] Department of Geology, Fayoum University, Al-Fayoum 63514, Egypt; halim.geologist@mail.ru
* Correspondence: moh.gho@mail.ru (M.M.G.); gawadnma@gmail.com (A.E.A.G.)

Citation: Ghoneim, M.M.; Abdel Gawad, A.E.; El-Dokouny, H.A.; Dawoud, M.; Panova, E.G.; El-Lithy, M.A.; Mahmoud, A.S. Petrogenesis and Geodynamic Evolution of A-Type Granite Bearing Rare Metals Mineralization in Egypt: Insights from Geochemistry and Mineral Chemistry. *Minerals* **2024**, *14*, 583. https://doi.org/10.3390/min14060583

Academic Editor: Jaroslav Dostal

Received: 22 April 2024
Revised: 22 May 2024
Accepted: 29 May 2024
Published: 31 May 2024

Copyright: © 2024 by the authors. Licensee MDPI, Basel, Switzerland. This article is an open access article distributed under the terms and conditions of the Creative Commons Attribution (CC BY) license (https://creativecommons.org/licenses/by/4.0/).

Abstract: During the Late Precambrian, the North Eastern Desert of Egypt underwent significant crustal evolution in a tectonic environment characterized by strong extension. The Neoproterozoic alkali feldspar granite found in the Homret El Gergab area is a part of the Arabian Nubian Shield and hosts significant rare metal mineralization, including thorite, uranothorite, columbite, zircon, monazite, and xenotime, as well as pyrite, rutile, and ilmenite. The geochemical characteristics of the investigated granite reveal highly fractionated peraluminous, calc–alkaline affinity, A-type granite, and post-collision geochemical signatures, which are emplaced under an extensional regime of within-plate environments. It has elevated concentrations of Rb, Zr, Ba, Y, Nb, Th, and U. The zircon saturation temperature ranges from 753 °C to 766 °C. The formation of alkali feldspar rare metal granite was affected by extreme fractionation and fluid interactions at shallow crustal levels. The continental crust underwent extension, causing the mantle and crust to rise, stretch, and become thinner. This process allows basaltic magma from the mantle to be injected into the continental crust. Heat and volatiles were transferred from these basaltic bodies to the lower continental crust. This process enriched and partially melted the materials in the lower crust. The intrusion of basaltic magma from the mantle into the lower crust led to the formation of A-type granite.

Keywords: alkali feldspar granite; geochemistry; mineral chemistry; rare metals mineralization; Egypt

1. Introduction

The Arabian Nubian Shield (ANS) is widely distributed throughout East African Orogen (EAO) and Western Arabia, including Egypt, Sudan, Ethiopia, Eritrea, Somalia, Saudi Arabia, Yemen, Oman, Jordan, and Palestine. In contrast, the southern part of the ANS is located along the Mozambique Belt (Figure 1a). The Eastern Desert of Egypt is a part of the ANS and can be subdivided into three major structural tectonic provinces, according to Stern and Hedge [1], as follows: (1) the South Eastern Desert (SED) is the oldest and most highly deformed province, and it is distinguished by the predominance of compressional and extrusion-related structures having WNW–ESE to NW–SE trends in the western part and N–S to NE–SW trends in the eastern part; (2) the Central Eastern Desert (CED) is characterized by prominent transpressional and extensional related structures having NW–SE and WNW–ESE trends, dissected by younger shearing NE–SW trend; (3) the Northern Eastern Desert (NED) is dominated by the so-called younger granite (c. 580 Ma), with extensional structures E–W and NE–SW trends. These three provinces are separated from each other by major tectonic discontinuities: the Qena–Safaga Shear Belt between the NED and CED, and most likely, the Wadi Kharit–Wadi Hodein Shear Belt between the CED and SED. The Eastern Desert of Egypt is a part of the northern ANS [2], and it is characterized by a pronounced distribution of post- to late-collisional granites containing rare metals mineralization. This region is dated between 620 and 580 Ma [3–7].

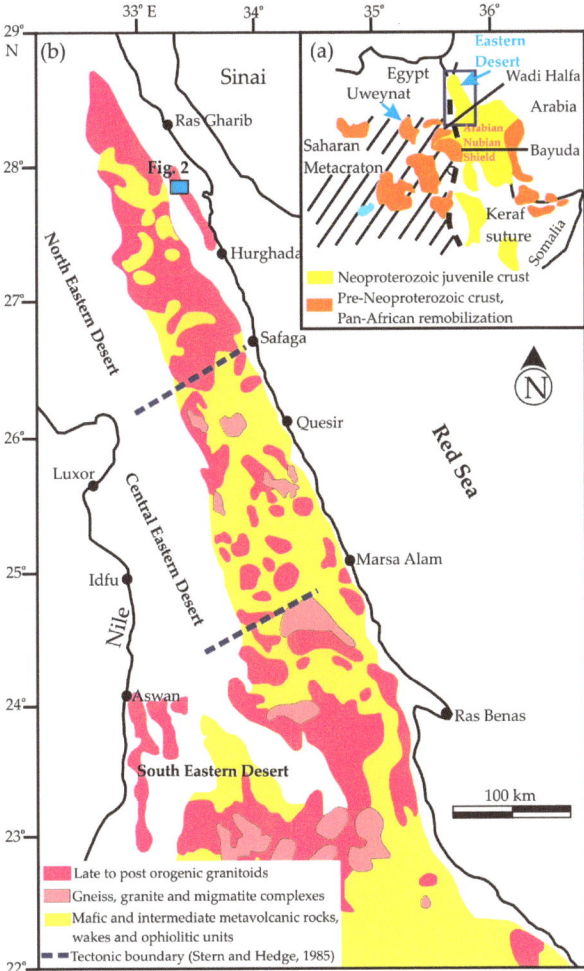

Figure 1. (**a**) Geologic map showing the Arabian Nubian Sheild (ANS). (**b**) Geological map showing the distribution of the Neoproterozoic basement rocks in the Eastern Desert, Egypt [2].

Rare metals mineralization has a markedly wide distribution in granite and associated pegmatite, mylonite, and rhyolite flow tuffs, as well as in lamprophyre and felsite dikes and quartz veins, which could have originated from magmatic and/or metasomatic processes [8–14]. Rare metal granite is widely distributed in the Eastern Desert (Figure 1b). It is highly evolved granite that is distinguished by calc–alkaline, alkaline to peralkaline affinities, and A-type granite. They are widespread in orogenic belts. These rocks are subdivided into metaluminous granite (Nb, Zr, and Y rich), peraluminous granite (Ta is the most predominant one, followed by Nb, Sn, W, Be, and Li), and metasomatized granite (Nb is the most abundant, followed by Ta, Sn, Zr, Y, U, Be, and W) [15–18].

The distribution of alkali feldspar granite in the North Eastern Desert of Egypt has several favorable implications. First, crustal growth occurred, whereas the magma was derived from the mantle, then ascended through the crust, and crystallized as granite. Second, intracontinental rifting processes, which are characterized by fragmentation and splitting of continents without the involvement of plate boundaries, likely occurred. Third, the study of alkali feldspar granite can provide insights into the petrogenesis and melting processes occurring within the ANS. Fourth, heat and fluid transfers within the lithospheric layer occurred. Finally, the distribution and characteristics of alkali feldspar granite can provide insights into the tectonic setting and regional geology of the ANS. Further research and geodynamic modeling can provide deeper insights into these implications and their significance in the broader context of the ANS.

The main objective of the present work is to study the geochemical and mineralogical features of alkali feldspar granite in the ANS to understand the petrogenesis of the granite and its tectonic evolution in the studied area. This knowledge can improve our understanding of the geological and geochemical signatures of A-type granite as a good resource for rare metals mineralization in the ANS. Therefore, we proposed a geodynamic model in order to explain the origin of the Homret El Gergab alkali feldspar granite.

2. Geologic Setting

This study area is characterized by the presence of basement rocks, which are predominantly covered by Dokhan Volcanics, alkali feldspar granite, and post-granite dikes and veins, including microgranite, basaltic–andesite dikes, and quartz veins (Figure 2). The term "Dokhan Volcanics" is used to refer to a thick sequence of multicolored stratified lava flows with their pyroclastics. These rocks are hard, massive, fine-grained, and vary in color from black to greenish gray to dark gray, and from buff, pinkish-red, to reddish-brown. The pyroclastics include ash tuffs, lapilli tuffs, and agglomerates with purple ignimbrite [19–21]. They are mainly composed of basalt, andesite, rhyolite, rhyodacite, and dacite. They were dissected by quartz veins (Figure 3a).

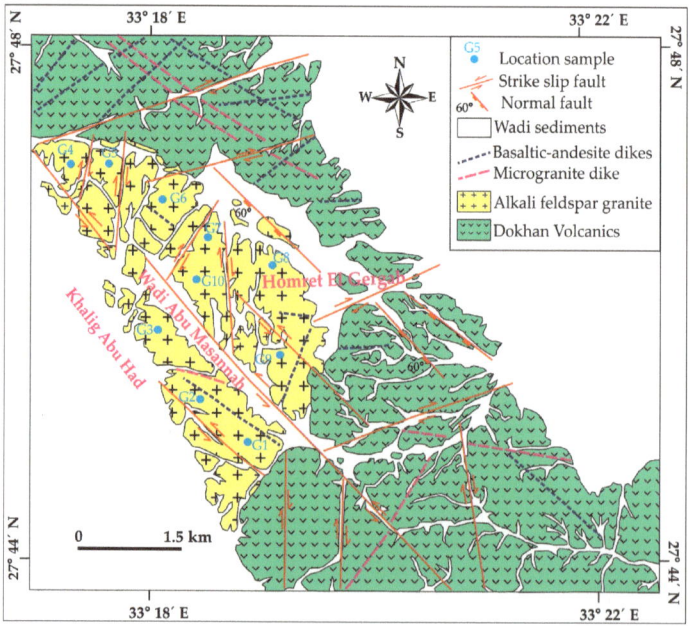

Figure 2. Geologic map of Homret El Gergab, North Eastern Desert, Egypt, modified after Abd El-Hadi [22].

Figure 3. (**a**) Quartz vein (Qz) cutting the Dokhan Volcanics (DV). (**b**) Sharp intrusive contact between the alkali feldspar granite (Gr) and Dokhan Volcanics (DV). (**c**) Exfoliation in alkali feldspar granite, (**d**,**e**). Open cut in granite for prospecting feldspars. (**f**) Distribution of alkali feldspar granite at Wadi Abu Masananah.

The Homret El Gergab granite is hard, blocky, medium- to coarse-grained, and varies in color from pink to reddish-pink and crimson. The granitic pluton occurs in the central western part of the study area and has a semicircular shape with rough terrain. It has moderately to highly elevated peaks that rise to a height of 433 m above sea level. The alkali feldspar granite intruded the Dokhan Volcanics with a markedly sharp intrusive contact (Figure 3b). Exfoliation structures are prominent features resulting from stress release, especially along the margins of the granitic pluton (Figure 3c). It is composed essentially of K-feldspar, quartz, plagioclase, and biotite. The alkali feldspar granite is cut by microgranite, basaltic, and andesite dikes and is strongly affected by faults and shear zones having E–W, NW–SE, NE–SW, and N–S trends. It shows well-defined joints and fractures. Locally, the joints and fracture planes show reddish-brown or brick-red staining, probably due to mineral solutions enriched in iron oxides. Pegmatite pockets and quartz veins are very present in the studied granite. The feldspars found in the study area are derived from alkali feldspar granite [22], specifically the Homret El Gergab pluton, which covers approximately 10 km^2 (Figure 3d–f).

3. Petrography

The alkali feldspar granite exhibits a medium- to coarse-grained hypidiomorphic texture. It is composed mainly of K-feldspar, quartz, plagioclase, and biotite. The alteration products are kaolinite, sericite, chlorite, and muscovite. Zircon, accompanied by opaque Fe-Ti oxides, are the main accessory minerals (Figure 4a–f). Perthite, a component of alkali feldspar granite, is present as well-defined crystals, in which albite lamellae are bound by/or enclosed in the perthite crystals (Figure 4a,b). The mantle is dominated by patchy and flame textures (Figure 4a). Plagioclase crystals occur as medium- to coarse-grained, subhedral to anhedral, tabular, and some show signs of erosion and are trapped within perthite. Some plagioclase crystals are displaced from their original positions and show discoloration due to the presence of iron oxides. Quartz exists as anhedral to subhedral crystals that occupy the interstitial spaces between various mineral components. In particular, it displays prominent strain patterns, including cracking and waviness, with a tendency toward undulating extinction. Biotite occurs as subhedral to anhedral crystals, and certain flakes have been partially or completely altered to chlorite and muscovite (Figure 4c,d). Zircons occur as high-relief prismatic crystals, exhibiting zonation, and are poikilitically enclosed in quartz, biotite, and opaque materials (Figure 4e,f).

Figure 4. Photomicrographs of the studied alkali feldspar granite at Homret El Gergab, North Eastern Desert of Egypt, clarifying that (**a**) flame and patchy perthites are associated with antiperthite; (**b**) perthite encloses plagioclase; (**c**) biotite is highly altered to ferrichlorite and is associated with quartz; (**d**) fine muscovite flacks are associated with quartz; (**e**) quartz encloses euhedral zircon crystal; and (**f**) zircon is enclosed in iron oxides. Abbreviations: Per, perthite; Ant, antiperthite; Plg, plagioclase; Bt, Biotite; Qz, quartz; Mus, muscovite; Kfs, K-feldspar; Zrn, zircon; Irx, iron oxide.

4. Materials and Analytical Methods

Twenty samples were collected from the host alkali feldspar granite, and then twenty polished thin sections were obtained to identify rare metals mineralization. This work was carried out using a scanning electron microscope (FEI, Eindhoven, The Netherlands, 2006), which was mounted on the base of the analytical complex Pegasus 4000 (EDAX, Mahwah, NJ, USA). Microanalyses were performed using an electron microprobe equipped with a CAMECA SX 100 (CAMECA, Courbevoie, Île-de-France, France). The diameter of the analyzed spot on the surface of the sample was less than 1 μm at the standard temperature (15 nA beam current, 15 kV acceleration potential, 10 s counting time for peak and 5 s counting time for background; sample: iron oxide; volume: 3×10^{-19} m^3; mass: 2×10^{-12} g). The electron microprobe was equipped with five automated wavelength dispersive spectrometers (WDS) and an energy dispersive spectrometer (EDS). The WDS diffraction crystals included LiF, PET, TAP, PC0, PC1, PC2, and PC3, while the spectrometer could vary from 0.22 to 0.83 and the sin-theta had a resolution of 10^{-5}. The following natural standards were used: orthoclase for Si (Kα), albite for Al (Kα), olivine for Mg (Kα), hematite for Fe (Kα), wollastonite for Ca (Kα), and monazite for Yb (Lα). The synthetic compounds used include ThO_2, UO_2, $PbCrO_4$, ZrO_2, $CePO_4$, $NdPO_4$, $SmPO_4$, GdTiGe, $DyRu_2Ge_2$, and YPO_4, which are the corresponding standards for Th (Mα), U (Mβ), Pb (Mα), Zr (Lα), Ce (Lα), Nd (Lα), Sm (Lα), Gd (Lα), Dy (Lα), and Y (Lα), respectively.

Ten granitic samples were selected and then crushed into ten mesh particles. These particles were finely ground to a size of 200 mesh. Major oxides and trace elements were analyzed by X-ray fluorescence using an ARL 9800 f. ARL X-ray spectrometer at the Central Laboratories of St. Petersburg State University, St. Petersburg, Russia. For XRF analyses, granite powder samples were prepared with Mowiol II polyvinyl alcohol and fused with tetraborate pellets. The detection limits are 0.01% for major oxides and 1–4 ppm for trace elements.

5. Geochemistry

Whole-rock chemical compositions of the major and trace element concentrations in the Homret El Gergab alkali feldspar granite are listed in Table 1. The geochemical behavior of the major element oxides in the studied granitic samples shows that the SiO_2 content ranges from 71.2 to 74.7 wt%, the Al_2O_3 content ranges from 13.7 to 15.4 wt%, the Na_2O content ranges from 3.7 to 4.7 wt%, and the K_2O content ranges from 4.9 to 6.1 wt%. On the other hand, Fe_2O_3, CaO, MgO, and TiO_2 have low concentrations, reaching 1.8, 1.2, 0.6, and 0.3 wt%, respectively (Table 1). The Harker variation diagrams (Figure 5) illustrate a marked decrease in Al_2O_3, Fe_2O_3, and Na_2O contents with increasing SiO_2, but TiO_2, K_2O, and CaO contents show a scattered distribution of the analyzed samples.

Among the analyzed trace elements, alkali feldspar granite has a high concentration of large-ion lithophile elements (LILEs) and high-field strength elements (HFSEs), such as Rb, Zr, Ba, Y, Nb, and Th, reaching 289, 202, 180, 89, 84, and 68 ppm, respectively (Table 1). The Harker variation diagrams show a marked increase in trace elements, especially Rb, Y, Nb, Ba, and U, with increasing SiO_2, but a marked decrease in Sr, Zr, and Cr (Figure 5).

Many geochemical classifications have been proposed for igneous rocks using various geochemical parameters. The normative An-Ab-Or (anorthite-albite-orthoclase) diagram by Barker [23] shows that the analyzed samples plot in the granite field (Figure 6a). SiO_2 versus ($Na_2O + K_2O$) the classification diagram by Middlemost [24] (Figure 6b) reveals the studied samples plot in the alkali feldspar granite field.

Table 1. Whole-rock major element oxides (wt%), calculated CIPW norm and trace elements (ppm) of the analyzed Alkali feldspar granite, Homret El Gergab, North Eastern Desert, Egypt.

Samples	G1	G2	G3	G4	G5	G6	G7	G8	G9	G10
SiO_2	72.9	72.3	73.4	74.7	72.6	72.9	72.4	73.5	71.9	71.2
Al_2O_3	14.4	14.8	14.5	13.7	15.4	14.9	14.9	14.6	14.8	15.2
K_2O	6.10	5.90	4.90	5.60	5.10	5.20	5.30	5.10	5.70	5.00
Na_2O	3.66	4.17	4.12	4.04	4.41	4.4	4.52	4.32	4.23	4.66
Fe_2O_3	1.35	1.78	1.44	1.21	1.25	1.33	1.39	1.10	1.52	1.61
CaO	0.89	0.47	0.91	0.52	0.78	0.71	0.8	0.76	1.05	1.16
MgO	0.29	0.20	0.40	0.07	0.19	0.23	0.26	0.16	0.14	0.58
TiO_2	0.24	0.22	0.15	0.12	0.12	0.15	0.14	0.15	0.17	0.29
MnO	0.04	0.03	0.02	0.02	0.01	0.02	0.02	0.02	0.03	0.06
P_2O_5	0.01	0.03	0.02	0.03	0.02	0.01	0.05	0.01	0.04	0.03
L.O.I.	0.20	0.10	0.20	0.10	0.10	0.20	0.20	0.30	0.20	0.10
Total	100	99.9	100	99.9	99.9	100	100	100	99.7	99.9
Q	26.0	24.3	28.2	28.6	25.5	25.6	23.9	27.0	23.2	21.7
C	0.20	0.80	0.80	0.10	1.30	0.80	0.40	0.60	0.00	0.10
Or	36.0	34.9	29.0	33.1	30.1	30.7	31.3	30.1	33.7	29.5
Ab	31.0	35.3	34.9	34.2	37.3	37.2	38.2	36.6	35.8	39.4
An	4.40	2.10	4.40	2.40	3.70	3.50	3.60	3.70	4.60	5.60
Di	0.00	0.00	0.00	0.00	0.00	0.00	0.00	0.00	0.00	0.00
Hy	0.70	0.50	1.00	0.20	0.50	0.60	0.60	0.40	0.30	1.40
Mt	0.00	0.00	0.00	0.10	0.00	0.00	0.00	0.00	0.00	0.00
Il	0.10	0.10	0.00	0.00	0.00	0.00	0.00	0.00	0.10	0.10
Hm	1.40	1.80	1.40	1.20	1.30	1.30	1.40	1.10	1.50	1.60
Tn	0.00	0.00	0.00	0.00	0.00	0.00	0.00	0.00	0.30	0.00
Ru	0.20	0.20	0.10	0.00	0.10	0.10	0.10	0.10	0.00	0.20
Ap	0.00	0.10	0.00	0.10	0.00	0.00	0.10	0.00	0.10	0.10
Trace elements (ppm)										
V	48.0	46.0	36.0	26.0	26.0	38.0	34.0	31.0	29.0	63.0
Cr	44.0	23.0	15.0	11.0	17.0	19.0	18.0	8.00	19.0	37.0
Co	26.0	15.0	10.0	19.0	9.80	66.0	12.0	26.0	7.00	39.0
Cu	3.00	10.0	18.0	8.00	11.0	75.0	6.00	28.0	16.0	7.00
Zn	18.0	28.0	21.0	14.0	32.0	11.0	14.0	11.0	18.0	18.0
As	5.00	2.00	7.00	5.00	4.00	9.00	6.00	2.00	6.00	5.00
Zr	185	202	127	112	138	142	145	142	196	139
Mo	13.0	9.00	5.00	22.0	9.00	8.00	6.00	10.0	8.00	6.00
Cd	21.0	14.0	5.00	4.00	12.0	23.0	18.0	23.0	31.0	12.0
Sn	5.00	21.0	8.00	4.00	3.00	7.60	19.0	5.90	5.00	4.70
Pb	15.0	14.0	14.0	9.00	12.0	9.00	15.0	12.0	11.0	11.0
U	25.0	23.0	27.0	21.0	26.0	29.0	27.0	23.0	25.0	22.0
Th	34.0	39.0	42.0	44.0	33.0	45.0	43.0	36.0	46.0	45.0
Rb	224	202	269	281	289	252	232	245	277	210
Sr	37.0	41.0	27.0	16.0	16.0	35.0	27.0	21.0	26.0	39.0
Ni	5.30	4.60	3.00	6.50	6.00	8.00	4.00	9.00	5.00	3.00
Nb	74.0	66.0	84.0	59.0	74.0	66.0	71.0	80.0	52.0	59.0
Ga	16.0	22.0	26.0	15.0	8.00	19.0	23.0	9.50	28.0	22.0
Y	67.0	54.0	89.0	69.0	73.0	82.0	44.0	73.0	71.0	60.0
Hf	2.30	3.10	0.90	2.10	0.80	0.70	2.10	3.50	3.90	4.40
Ba	115	120	143	180	98.0	150	144	167	84.0	139
TZr °C	795	805	767	753	775	775	773	775	794	766
Th/U	0.74	0.59	0.64	0.48	0.79	0.64	0.63	0.64	0.54	0.49
Rb/Sr	6.05	4.93	9.96	17.56	18.06	7.20	8.59	11.67	10.65	5.38

TZr °C (zircon saturation temperature).

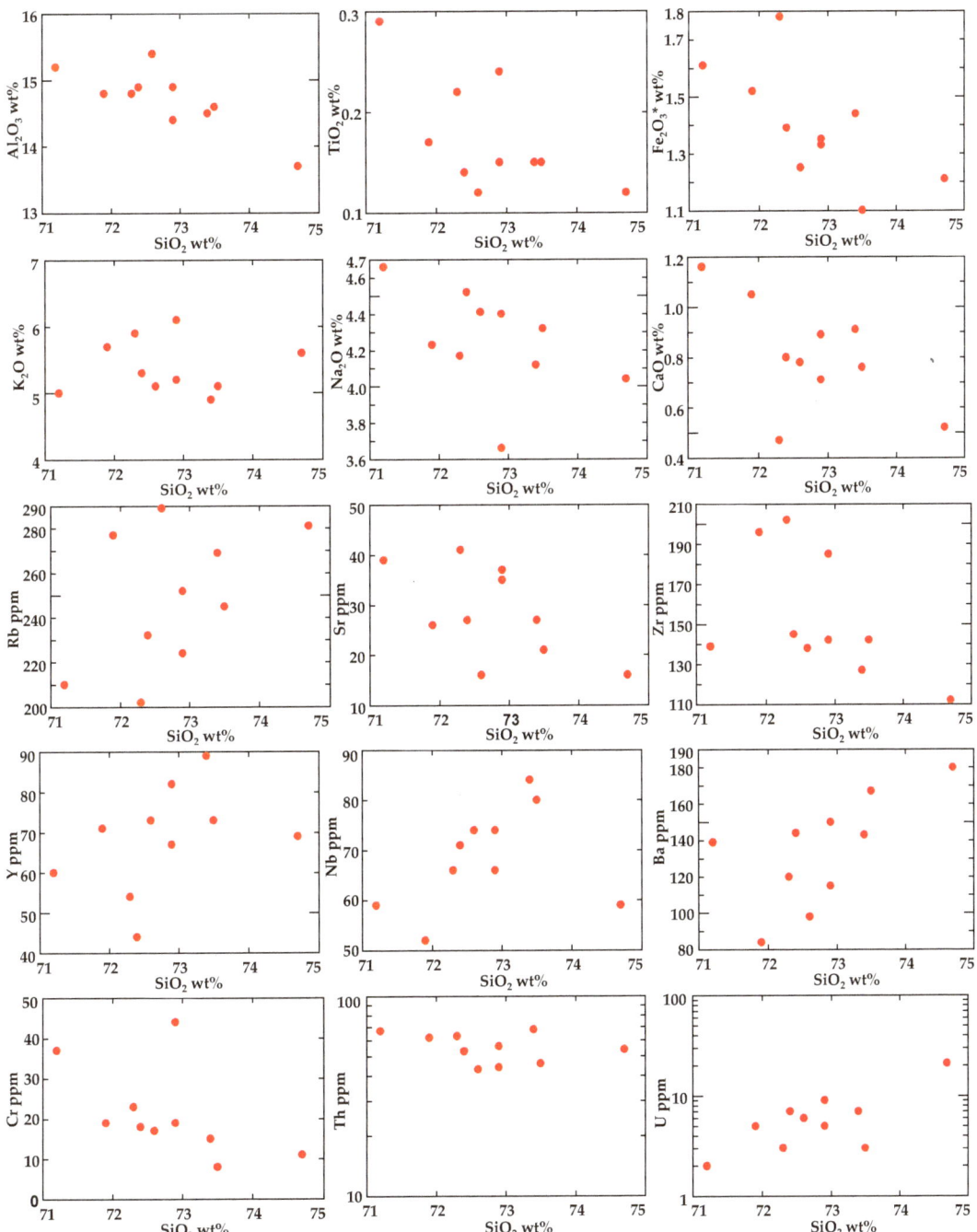

Figure 5. Harker variation diagrams illustrate the distributions of major oxides and trace elements in relation to silica in the studied granite.

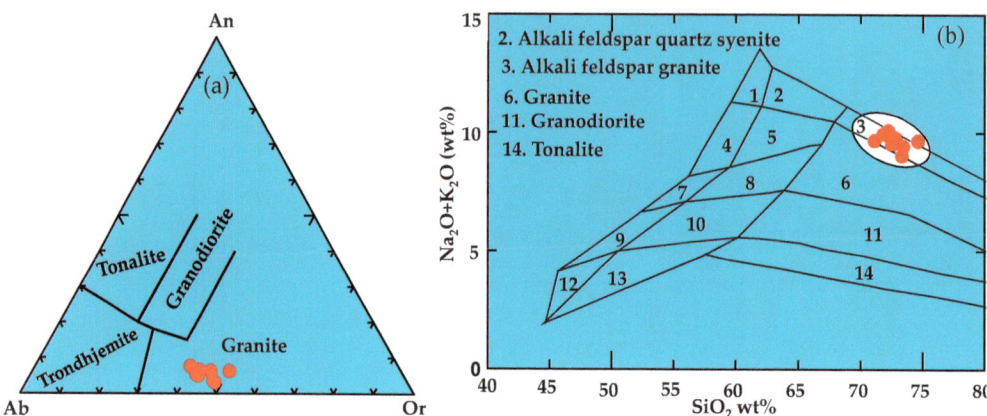

Figure 6. Geochemical discrimination diagrams present the nomenclature of the studied granite. (**a**) Ternary An–Ab–Or normative diagram by Barker [23]. (**b**) Binary diagram shows the total alkalis versus silica, according to Middlemost [24].

The concentrations of multiple elements were normalized to those of the primitive mantle values of Sun and McDonough [25], which could provide a general indication of the source and tectonic affinities of the studied granite (Figure 7). The analyzed alkali feldspar granite is strongly enriched in Rb, as a large ion lithophile element (LILE), when compared with high-field strength elements (HFSE; Nb, Zr, Y), but depleted in Ba, Sr, P, and Ti elements with marked troughs. The negative anomalies in P and Sr elements could mark fractionation of apatite and feldspars at the source or during differentiation. The differences in trace element contents in the studied granite may be a clue to the different degrees of partial melting or fractional crystallization. Ti depletion could be related to the fractionation of biotite or titanomagnetite.

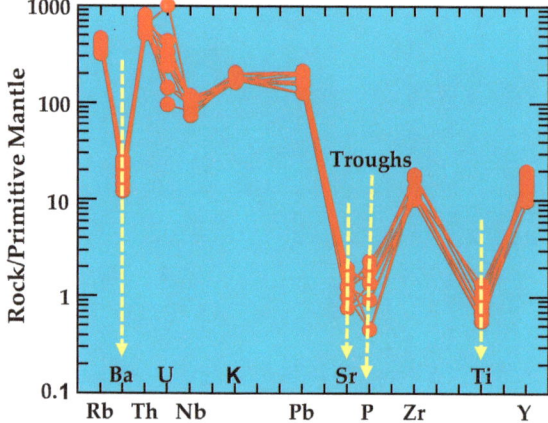

Figure 7. Normalized multi-element pattern according to Sun and McDonough [25] of the studied granite.

Figure 8 illustrates the geochemical behavior of U and Th in the studied alkali feldspar granite. The histograms show that Th ranges from 33 to 46 ppm and U ranges from 21 to 29 ppm (Table 1). The Th/U ratio of the analyzed alkali feldspar granite ranges from 0.48 to 3.8, which is lower than the upper crust value of 3.8 [26]. Thorium is an immobile element; therefore, this ratio mainly depends on the content of uranium, which is a mobile element. The analyzed granite samples have a lower Th/U ratio, which could be related to U enrichment.

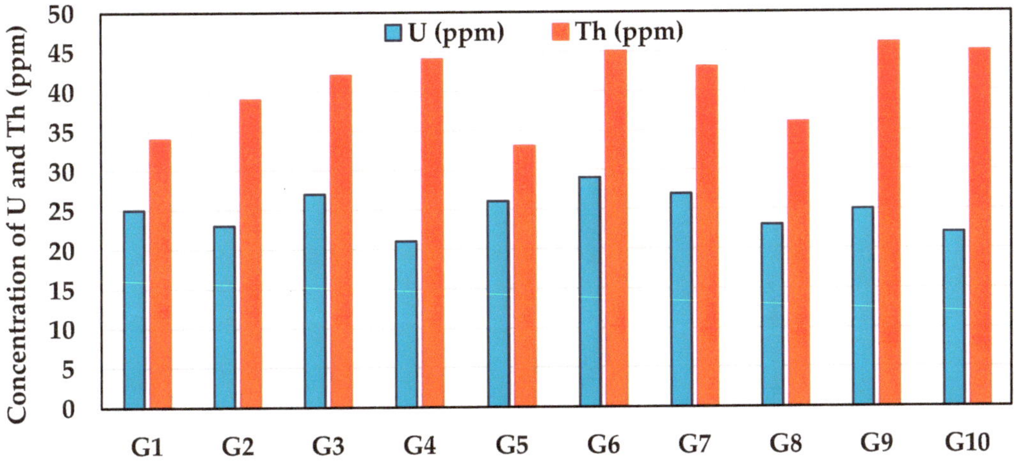

Figure 8. Histograms showing the concentrations of U and Th in the studied granitic samples from Homret El Gergab, North Eastern Desert, Egypt.

6. Mineral Chemistry

6.1. Radioactive Minerals

6.1.1. Uranothorite

It usually occurs as tiny individual crystals (<10 μm) that are anhedral and are often associated with monazite (Figure 9a). The EPMA data show that uranothorite is composed essentially of ThO_2, ranging from 53.74 to 54.24 wt%; SiO_2, from 15.66 to 16.27 wt%; and UO_2, from 13.25 to 14.21 wt%. LREEs, including La, Ce, Nd, and Sm, have been reported, and $\Sigma LREE_2O_3$ (La–Sm) ranges from 3.71 to 4.45 wt%. Fe_2O_3, CaO, and P_2O_5 are well documented in small amounts (Table 2).

Table 2. Representative EMPA of uranothorite (oxides in wt%) from Alkali feldspar granite at Homret El Gergab, North Eastern Desert, Egypt.

ThO_2	53.85	53.79	54.24	53.74
SiO_2	15.98	16.01	16.27	15.66
UO_2	13.25	14.21	14.12	13.92
La_2O_3	0.25	0.31	0.34	0.27
Ce_2O_3	1.78	1.86	1.91	1.81
Nd_2O_3	1.24	1.61	1.58	1.31
Sm_2O_3	0.44	0.67	0.53	0.46
CaO	1.53	1.78	1.12	1.45
P_2O_5	0.84	0.68	0.61	0.43
Fe_2O_3	1.49	1.57	1.56	1.61
Total	90.65	92.49	92.28	90.66
$\Sigma LREE_2O_3$	3.71	4.45	4.36	3.85
		Calculated formulae (apfu)		
Th	0.71	0.70	0.70	0.72
Si	0.92	0.91	0.93	0.92
U	0.17	0.18	0.18	0.18
La	0.01	0.01	0.01	0.01
Ce	0.04	0.04	0.04	0.04
Nd	0.03	0.03	0.03	0.03
Sm	0.01	0.01	0.01	0.01
Ca	0.09	0.11	0.07	0.09
P	0.04	0.03	0.03	0.02
Fe	0.06	0.07	0.07	0.07
Sum	2.07	2.09	2.07	2.08

Calculated chemical formula based on 4 oxygen (apfu) for uranothorite.

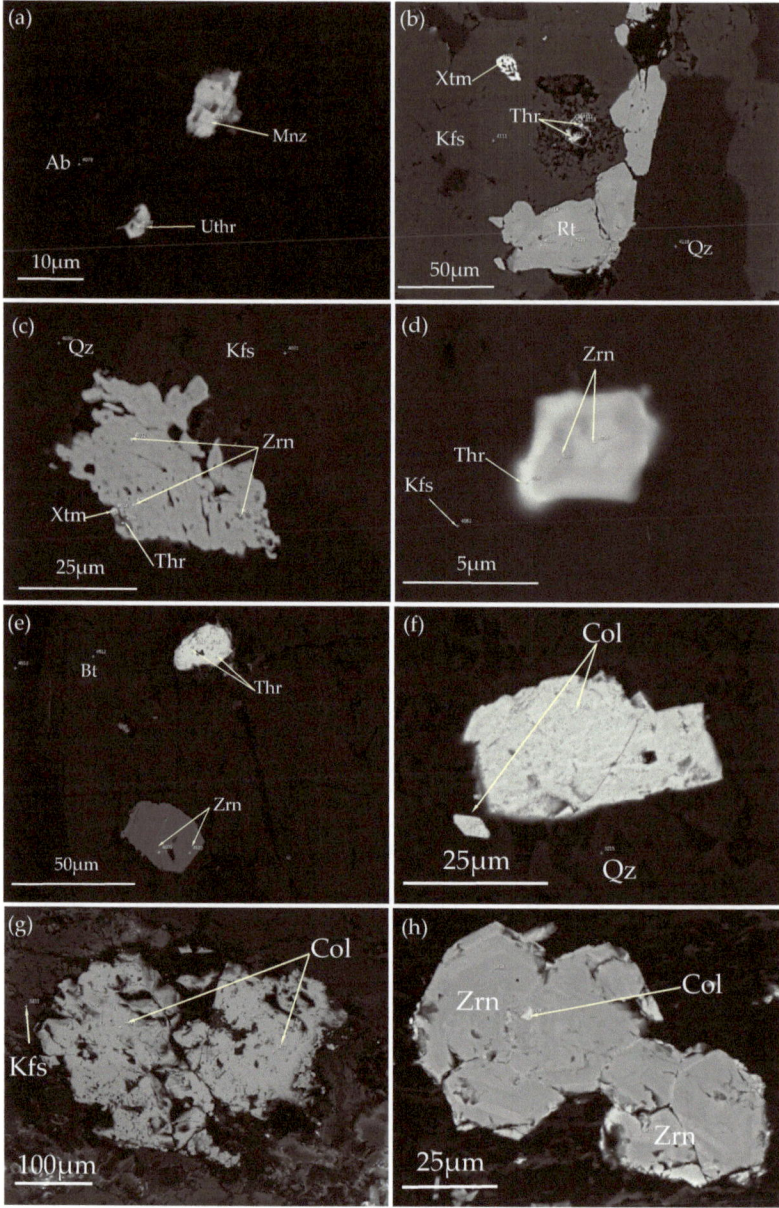

Figure 9. Back-scattered images (BSE) of radioactive minerals associated with other accessory minerals from alkali feldspar granite, Homret El Gergab, North Eastern Desert, Egypt. (**a**) Fine-grained uranothorite is associated with monazite and is enclosed in feldspar. (**b**) Fine-grained xenotime and thorite are associated with rutile. (**c**) Fine-grained xenotime and thorite occur along the rims of zircon. (**d**) Thin films of thorite occur along the periphery of zircon. (**e**) Thorite and zircon are enclosed in biotite. (**f**) Columbite crystals are enclosed in quartz. (**g**) Highly deformed columbite crystals are enclosed in K-feldspar. (**h**) Large zircon crystals enclose microinclusions of columbite. Abbreviations: uthr, uranothorite; thr, thorite; Zrn, zircon; Mnz, monazite; Xtm, xenotime; Col, columbite; Rt, rutile; Qz, quartz; Ab, Albite; Kfs, K-feldspar; Bt, Biotite.

6.1.2. Thorite

Th-silicate minerals are abundant and predominant in the investigated alkali feldspar granite in the study area. These grains occur as fine individual crystals and/or adjacent thin films (bright colors) along the peripheries of zircon crystals (Figure 9b–e). It consists of anhedral crystals that are <10 μm in length, is accompanied by zircon, xenotime, and rutile, and is always enclosed in feldspar and quartz crystals. It occurs as bright microinclusions and/or adjacent to zircon crystals (Figure 9b–e). The analyzed EPMA data (Table 3) show that thorite is essentially composed of ThO_2 ranging from 51.27 to 52.73 wt% and SiO_2 ranging from 16.17 to 18.35 wt%. UO_2 is well represented in thorite, with higher concentrations reaching up to 7.1 wt% (Table 3). Y_2O_3-bearing thorite is present and ranges from 2 to 3.2 wt%. The higher concentrations of ΣREE_2O_3 (La–Er) range from 9.35 to 14.1 wt%. Ce_2O_3 is the most abundant REE and reaches up to 4.3 wt% of the other elements. Moreover, CaO, P_2O_5, Fe_2O_3, and Al_2O_3 are well represented in the analyzed thorite at low concentrations (Table 3).

Table 3. Representative EMPA of thorite (oxides in wt%) from alkali feldspar granite at Homret El Gergab, North Eastern Desert, Egypt.

ThO_2	51.35	51.27	52.73	52.17	51.33	52.22
SiO_2	18.35	17.27	17.73	16.17	17.33	16.22
UO_2	2.94	5.66	3.16	7.1	4.73	2.93
La_2O_3	0.54	1.31	1.23	1.24	1.29	1.76
Ce_2O_3	1.98	1.32	1.51	1.92	0.96	4.30
Nd_2O_3	1.29	1.17	1.33	1.66	1.45	2.30
Sm_2O_3	1.32	1.29	1.37	1.40	1.46	1.48
Gd_2O_3	1.74	1.61	1.81	1.87	1.94	1.97
Dy_2O_3	2.10	2.15	1.99	2.00	1.94	1.97
Er_2O_3	0.54	1.19	0.73	0.66	0.31	0.32
Y_2O_3	3.20	2.51	1.84	2.59	2.89	2.01
P_2O_5	1.73	1.28	1.08	1.65	1.58	2.06
CaO	0.77	0.92	0.78	1.04	1.04	0.74
Fe_2O_3	1.69	1.76	2.02	1.39	1.34	1.22
Al_2O_3	0.91	0.87	0.76	0.69	0.85	0.94
Total	90.45	91.58	90.07	93.55	90.44	92.44
ΣREE_2O_3	9.51	10.04	9.97	10.75	9.35	14.1
Calculated formulae (apfu)						
Th	0.61	0.62	0.65	0.64	0.63	0.64
Si	0.96	0.92	0.96	0.87	0.93	0.87
U	0.03	0.07	0.04	0.09	0.06	0.03
La	0.01	0.03	0.02	0.02	0.03	0.03
Ce	0.04	0.03	0.03	0.04	0.02	0.08
Nd	0.02	0.02	0.03	0.03	0.03	0.04
Sm	0.02	0.02	0.03	0.03	0.03	0.03
Gd	0.03	0.03	0.03	0.03	0.03	0.03
Dy	0.04	0.04	0.03	0.03	0.03	0.03
Er	0.01	0.02	0.01	0.01	0.01	0.01
Y	0.09	0.07	0.05	0.07	0.08	0.06
P	0.08	0.06	0.05	0.08	0.07	0.09
Ca	0.04	0.05	0.05	0.06	0.06	0.04
Fe	0.07	0.07	0.08	0.06	0.05	0.05
Al	0.06	0.05	0.05	0.04	0.05	0.06
Sum	2.10	2.11	2.10	2.10	2.10	2.11

Calculated chemical formula based on 4 oxygen (apfu) for thorite.

6.2. Columbite

Columbite occurs as tabular, subhedral to anhedral crystals that are massive and vary in size from 100 to 400 μm (Figure 9f,g). They are commonly enclosed in K-feldspar, quartz, and plagioclase crystals, whereas others are commonly dispersed as microinclusions in zircon crystals (Figure 9f,g). The EPMA data show that columbite is composed of Nb_2O_5 at concentrations ranging from 66.49 to 74.7 wt% and Fe_2O_3 at concentrations ranging from 17.27 to 26.42 wt%, whereas MnO has a lower concentration ranging from 0.5 to 5.78 wt%. TiO_2, Ta_2O_5, and Y_2O_3 have low concentrations (Table 4). The radioactive elements uranium and thorium have low contents in the analyzed columbite, reaching 3.09 wt% for UO_2 and 1.53 wt% for ThO_2.

Table 4. Representative EMPA of columbite (oxides in wt%) from alkali feldspar granite at Homret El Gergab, North Eastern Desert, Egypt.

Nb_2O_5	70	69.21	70.09	74.7	70.95	72.8	66.49	67.76	70.84
Ta_2O_5	0.77	n.d.	n.d.	n.d.	3.44	3.44	3.03	n.d.	2.55
TiO_2	2.67	0.80	1.73	2.82	1.78	0.63	2.84	1.33	1.51
Fe_2O_3	23.56	23.39	25.35	20.93	19.44	17.27	23.91	26.42	22.6
MnO	1.58	0.57	1.01	1.46	1.25	5.78	n.d.	0.68	0.50
Y_2O_3	n.d.	2.92	n.d.	n.d.	n.d.	n.d.	2.46	1.74	1.93
ThO_2	0.44	1.53	0.64	n.d.	n.d.	n.d.	1.40	1.11	n.d.
UO_2	1.00	1.53	1.24	n.d.	3.09	n.d.	n.d.	0.98	n.d.
Total	100	100	100	100	100	100	100	100	100
				Calculated formulae (apfu)					
Nb	1.69	1.71	1.70	1.78	1.76	1.80	1.62	1.65	1.72
Ta	0.01				0.05	0.05	0.04		0.04
Ti	0.11	0.03	0.07	0.11	0.07	0.03	0.12	0.05	0.06
Fe	0.95	0.96	1.02	0.83	0.80	0.71	0.97	1.07	0.91
Mn	0.07	0.03	0.05	0.07	0.06	0.27		0.03	0.02
Y		0.08					0.07	0.05	0.06
Th	0.01	0.02	0.01				0.02	0.01	
U	0.01	0.02	0.01		0.04			0.01	
$\sum$cation	2.85	2.85	2.85	2.79	2.78	2.85	2.84	2.88	2.81
Ta/(Ta + Nb)	0.01				0.03	0.03	0.02		0.02
Mn/(Mn + Fe)	0.07	0.03	0.05	0.08	0.07	0.28		0.03	0.02

The chemical formula was calculated for columbite based on 6 oxygen atoms (apfu). n.d. not determined.

6.3. Zircon

Zircon crystals occur in various shapes, ranging from prismatic to subhedral or anhedral crystals (Figures 9 and 10). Zircon inclusions can range in size from being as small as microinclusions in K-feldspar crystals to larger single crystals that can reach up to 50 μm in size. Some of these zircon crystals appear to be heavily corroded and contain dark patches. In addition, bright xenotime and thorite can be found at the edges of the zircon crystals. The EPMA data show that the zircon crystals are composed of ZrO_2, which ranges from 58.05 and 61.1 wt%, and SiO_2, which ranges from 29.1 and 32.67 wt%. Other elements present include HfO_2, which ranges from 1.43 to 3.61 wt%, and Sc_2O_3, which from 0.26 to 0.64 wt%. Al_2O_3, CaO, MnO, and Fe_2O_3 are recorded in low concentrations (Table 5). The two radioactive elements reached 4.38 wt% for UO_2 and 3.89 wt% for ThO_2.

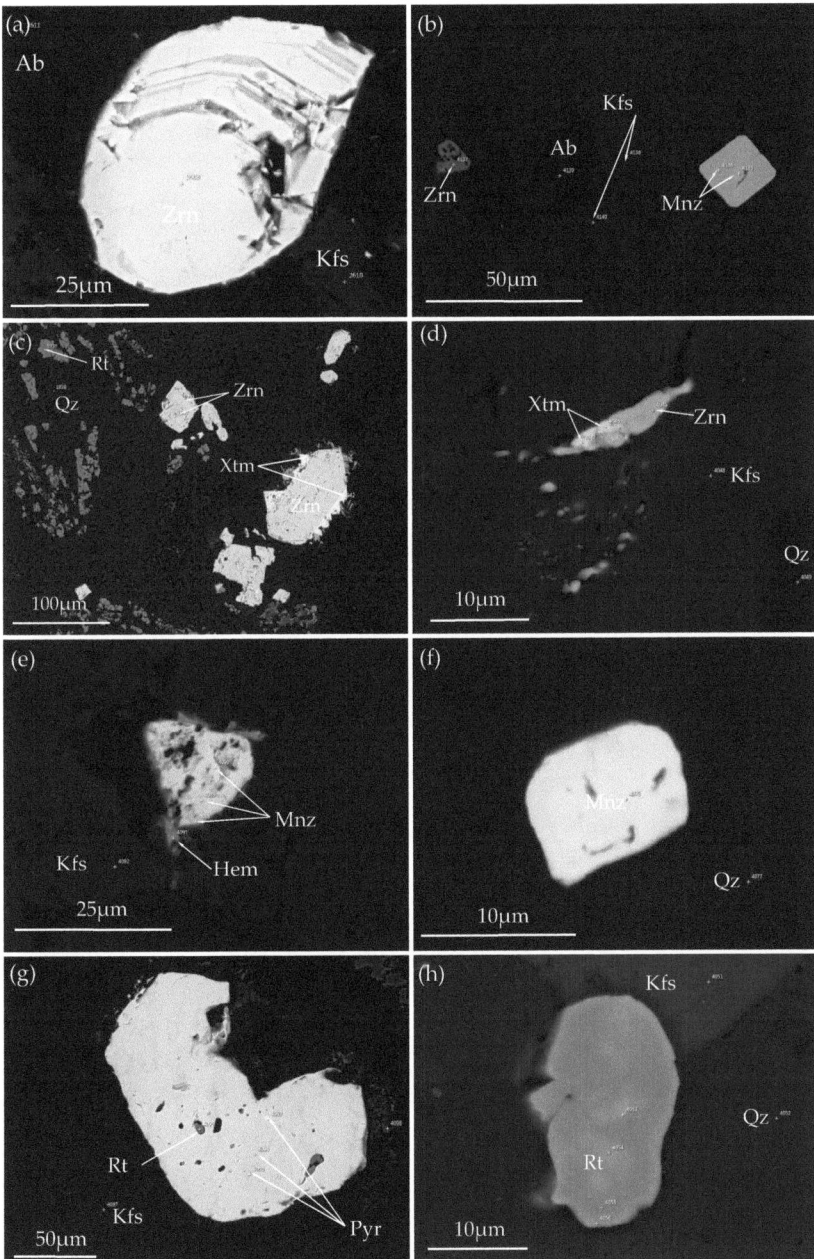

Figure 10. Back-scattered images (BSE) of Zr and REE minerals associated with other accessories from alkali feldspar granite, Homret El Gergab, North Eastern Desert, Egypt. (**a**) Zoned zircon crystals are enclosed in feldspar. (**b**) Fine-grained monazite and zircon are enclosed in feldspar. (**c**) Xenotime is overgrown along the rims of zircon. (**d**) Fine-grained xenotime is adjacent to zircon. (**e**) Hematite is along the periphery of monazite. (**f**) Quartz encloses monazite. (**g**) Pyrite encloses rutile. (**h**) Quartz encloses rutile. Abbreviations: Zrn, zircon; Mnz, monazite; Xtm, xenotime; Rt, rutile; Hem, hematite; Qz, quartz; Ab, Albite; Kfs, K-feldspar.

Table 5. Representative EMPA of zircon (oxides in wt%) from alkali feldspar granite at Homret El Gergab, North Eastern Desert, Egypt.

SiO_2	30.09	31.2	31.04	31.75	31.12	30.11	30.07	32.16	32.13	29.56	29.97	29.1	32.67	30.5	31.14
ZrO_2	58.25	60.33	59.28	59.59	61.1	58.25	58.24	59.78	58.55	58.05	58.25	58.15	59.36	61.03	59.33
HfO_2	2.84	3.61	3.57	3.25	3.53	3.02	3.31	2.91	1.90	1.43	1.51	1.52	1.76	3.60	3.57
Sc_2O_3	0.38	0.31	0.41	0.26	0.28	0.34	0.38	0.35	0.29	0.31	0.38	0.64	0.39	0.42	0.41
CaO	0.84	0.76	1.12	1.05	0.62	0.92	0.69	0.81	0.97	1.01	0.85	1.15	0.81	0.76	1.12
Fe_2O_3	1.53	0.7	0.79	1.06	0.56	1.02	1.09	0.97	0.89	1.04	2.27	1.46	1.03	0.70	0.79
Al_2O_3	0.59	0.49	0.79	0.53	0.36	0.66	0.47	0.59	0.64	0.42	0.54	0.58	0.59	0.49	0.79
MnO	0.5	0.61	0.56	0.6	0.44	0.39	0.45	0.45	0.41	0.53	0.38	0.57	0.64	0.61	0.56
ThO_2	3.89	1.32	1.15	0.97	1.12	3.29	3.35	0.91	0.85	3.31	3.41	3.57	1.25	1.21	1.05
UO_2	1.17	0.67	1.33	1.00	0.93	2.00	1.95	1.10	3.46	4.38	2.44	3.27	1.59	0.67	1.33
Total	100	100	100	100	100	100	100	100	100	100	100	100	100	100	100
						Calculated formulae (apfu)									
Si	0.96	0.98	0.98	0.99	0.98	0.96	0.96	1.00	1.00	0.96	0.98	0.94	1.01	0.96	0.98
Zr	0.91	0.92	0.91	0.91	0.94	0.91	0.91	0.91	0.89	0.92	0.90	0.92	0.89	0.94	0.91
Hf	0.03	0.03	0.03	0.03	0.03	0.03	0.03	0.03	0.02	0.01	0.01	0.01	0.02	0.03	0.03
Sc	0.01	0.01	0.01	0.01	0.01	0.01	0.01	0.01	0.01	0.01	0.01	0.01	0.01	0.01	0.01
Ca	0.03	0.03	0.04	0.04	0.02	0.03	0.02	0.03	0.03	0.04	0.00	0.04	0.03	0.03	0.04
Fe	0.04	0.02	0.02	0.02	0.01	0.02	0.03	0.02	0.02	0.03	0.05	0.04	0.02	0.02	0.02
Al	0.02	0.02	0.03	0.02	0.01	0.02	0.02	0.02	0.02	0.02	0.02	0.02	0.02	0.02	0.03
Mn	0.01	0.02	0.01	0.02	0.01	0.01	0.01	0.01	0.01	0.01	0.01	0.02	0.02	0.02	0.01
Th	0.03	0.01	0.01	0.01	0.01	0.02	0.02	0.01	0.01	0.02	0.02	0.02	0.03	0.01	0.01
U	0.01	0.00	0.01	0.01	0.01	0.01	0.01	0.01	0.02	0.03	0.02	0.02	0.01	0.00	0.01
Sum	2.04	2.03	2.04	2.04	2.03	2.04	2.03	2.03	2.04	2.04	2.04	2.05	2.04	2.03	2.04

Calculated chemical formula based on 4 oxygen (apfu) for zircon.

6.4. REE Phosphates

Phosphate minerals are recorded in alkali feldspar granite. They include monazite and xenotime, and they are considered good sources of LREEs and HREEs, respectively, as well as Y (Section 6.4.2).

6.4.1. Monazite-Ce

Monazite-Ce is a common LREE phosphate mineral that occurs as euhedral to anhedral and fine- to medium-grained (Figure 10b,e,f). From the analyzed EPMA data, monazite chemistry reveals that P_2O_5 ranges from 27.37 to 29.84 wt%, with a marked enrichment in LREEs, while the $\sum REE_2O_3$ (La_2O_3–Gd_2O_3) concentration ranges from 59.15 wt% to 65.75 wt% (Table 6). Ce_2O_3 is the most abundant LREE in the analyzed monazite grains, reaching 30.51 wt%, followed by Nd_2O_3, reaching 16.76 wt%, La_2O_3, reaching 12.93 wt%, and others (Table 6). Gd_2O_3 is the only recorded HREE in the analyzed monazite and ranges from 1.29 to 3.19 wt%. Enrichment in ThO_2 was clearly observed in the monazite grains, ranging from 4.85 to 10.41 wt% (Table 6). CaO was recorded in smaller amounts, ranging from 0.69 to 3.37 wt%. The average empirical formula for monazite is $(Ce_{0.42}Nd_{0.19}La_{0.15}Pr_{0.05}Sm_{0.06}Gd_{0.03}Th_{0.07})(P_{0.98})O_4$. Monazite has shown a predominance of Ce (0.42 apfu) over other REEs corresponding to monazite-Ce. The geochemical compositions of the monazite crystals are remarkably similar, and the chemical reactions are carried out by the following substitutions:

$$Th^{4+} + Si^{4+} \leftrightarrow REE^{3+} + P^{5+}$$

$$Th^{4+} + Ca^{2+} \leftrightarrow 2REE^{3+}$$

As recorded by Watt [27], Abd El Ghaffar [28], and Abdel Gawad [29].

Table 6. Representative EMPA of monazite (oxides in wt%) from alkali feldspar granite at Homret El Gergab, North Eastern Desert, Egypt.

P_2O_5	27.45	27.37	29.69	29.84	29.04	28.5	28.53	29.17	29.48	28.45	28.2	28.88	27.42
CaO	1.06	1.16	1.19	1.15	3.37	1.11	0.70	0.69	2.69	0.71	1.19	1.19	0.84
La_2O_3	8.03	12.29	11.9	11.99	12.93	9.69	10.71	8.39	10.92	8.78	7.00	10.17	8.89
Ce_2O_3	28.31	29.48	28.52	29.95	29.61	28.46	30.18	28.05	30.51	28.38	28.42	27.35	28.11
Pr_2O_3	3.92	4.22	3.75	2.66	4.07	4.04	3.97	2.72	2.98	3.23	3.71	3.73	3.87
Nd_2O_3	14.13	11.57	10.25	9.70	11.13	15.2	13.44	15.92	11.85	15.27	13.41	15.98	16.76
Sm_2O_3	4.26	4.24	3.31	3.64	3.06	4.58	3.79	4.96	2.74	4.59	4.95	3.99	5.04
Gd_2O_3	2.46	1.29	1.42	1.52	1.75	2.26	2.46	2.59	1.83	3.19	3.01	1.96	3.08
ThO_2	10.41	8.46	9.92	9.14	4.85	6.15	6.19	7.60	6.65	7.39	10.18	6.85	6.00
Total	100	100	99.95	99.59	99.81	100	100	100	99.65	100	100	100	100
ΣREE_2O_3	61.11	63.09	59.15	59.46	62.55	64.23	64.55	62.63	60.83	63.44	60.5	63.18	65.75
Calculated formulae (apfu)													
P	0.96	0.95	1.00	1.00	0.97	0.97	0.98	0.99	0.98	0.98	0.97	0.98	0.95
Ca	0.05	0.05	0.05	0.05	0.14	0.05	0.03	0.03	0.11	0.03	0.05	0.05	0.04
La	0.12	0.19	0.17	0.18	0.19	0.14	0.16	0.12	0.16	0.13	0.11	0.15	0.13
Ce	0.43	0.44	0.41	0.43	0.43	0.42	0.45	0.41	0.44	0.42	0.42	0.4	0.42
Pr	0.06	0.06	0.05	0.04	0.06	0.06	0.06	0.04	0.04	0.05	0.05	0.05	0.06
Nd	0.21	0.17	0.15	0.14	0.16	0.22	0.19	0.23	0.17	0.22	0.19	0.23	0.25
Sm	0.06	0.06	0.05	0.05	0.04	0.06	0.05	0.07	0.04	0.06	0.07	0.06	0.07
Gd	0.03	0.02	0.02	0.02	0.02	0.03	0.03	0.03	0.02	0.04	0.04	0.03	0.04
Th	0.10	0.08	0.09	0.08	0.04	0.06	0.06	0.07	0.06	0.07	0.09	0.06	0.06
Sum	2.01	2.02	1.99	1.99	2.05	2.01	2.01	1.99	2.03	2.00	2.01	2.01	2.02

Calculated chemical formula based on 4 oxygen (apfu) for monazite; ΣREE_2O_3 (La_2O_3–Gd_2O_3).

6.4.2. Xenotime

Xenotime is anhedral to subhedral, fine-grained, and often enclosed in K-feldspar crystals. It is predominantly accompanied by zircon, thorite, and rutile (Figures 9b,c and 10c,d). EPMA data show that the analyzed xenotime is a good source of HREEs (Table 7). The chemical data show that P_2O_5 ranges from 30.23 to 32.15 wt%, Y_2O_3 ranges from 45.22 to 48.64 wt%, and $\Sigma HREE_2O_3$ (Gd_2O_3–Lu_2O_3) ranges from 15.45 to 17.9 wt%. In addition, xenotime has higher concentrations of Dy_2O_3, Er_2O_3, and Yb_2O_3 (Table 7). It has higher concentrations of radioactive elements, especially ThO_2, which ranges from 1.55 to 5.44 wt%, whereas UO_2 reaches up to 1.51 wt%. CaO and Fe_2O_3 were presented as minor constituents.

Table 7. Representative EMPA of xenotime (oxides in wt%) from studied alkali feldspar granite at Homret El Gergab, North Eastern Desert, Egypt.

Y_2O_3	48.64	45.36	46.4	45.22	46.3
Gd_2O_3	2.69	2.74	2.26	2.49	2.39
Dy_2O_3	4.50	4.40	3.09	4.11	4.21
Ho_2O_3	1.15	0.77	0.78	1.09	0.66
Er_2O_3	3.02	2.79	2.98	3.46	3.46
Tm_2O_3	1.61	1.54	1.51	1.67	1.62
Yb_2O_3	2.16	2.13	3.11	3.33	3.33
Lu_2O_3	1.34	1.36	1.72	1.75	1.69
P_2O_5	30.35	30.23	31.33	32.15	31.65
CaO	1.33	1.36	1.41	1.38	1.36
Fe_2O_3	0.81	0.75	0.83	0.79	0.80
ThO_2	1.55	5.44	3.07	1.64	1.64
UO_2	0.93	1.13	1.51	0.94	0.94
Total	100	100	100	100	100
$\Sigma HREE_2O_3$	16.47	15.73	15.45	17.90	17.36
Calculated formulae (apfu)					
Y	0.91	0.85	0.86	0.83	0.85

Table 7. Cont.

	1	2	3	4	5
Gd	0.03	0.03	0.03	0.03	0.03
Dy	0.05	0.05	0.03	0.05	0.05
Ho	0.01	0.01	0.01	0.01	0.01
Er	0.03	0.03	0.03	0.04	0.04
Tm	0.02	0.02	0.02	0.02	0.02
Yb	0.02	0.02	0.03	0.03	0.04
Lu	0.01	0.01	0.02	0.02	0.02
P	0.90	0.91	0.92	0.94	0.93
Ca	0.05	0.05	0.05	0.05	0.05
Fe	0.02	0.02	0.02	0.02	0.02
Th	0.01	0.04	0.02	0.01	0.01
U	0.01	0.01	0.01	0.01	0.01
Sum	2.08	2.06	2.06	2.05	2.06

Calculated chemical formula based on 4 oxygen (apfu) for xenotime; ΣREE_2O_3 (Gd_2O_3–Lu_2O_3).

6.5. Other Accessory Minerals

Pyrite occurs as coarse-grained, equant plates, euhedral to subhedral, and encloses rutile microinclusions (gray to black colors) (Figure 10g). It is enclosed in feldspar. The analyzed EPMA data show that pyrite is essentially composed of Fe ranging from 46.25 to 47 wt% and S ranging from 53 to 53.75 wt%.

Rutile is the most abundant accessory mineral. It occurs as anhedral to subhedral, fine- to medium-grained microinclusions in ilmenite and is associated with thorite, zircon, monazite, xenotime, and ilmenite (Figures 9b and 10c,h). The EMPA data indicate that rutile is composed of TiO_2 in the range of 98.12 to 99.9 wt%, Fe_2O_3 from 0.61 to 3.66 wt%, and Nb_2O_5 from 3 to 4.62 wt%.

Ilmenite occurs along the periphery of rutile and is enclosed in feldspars. It is associated with monazite and zircon. The EMPA data show that ilmenite is mainly composed of TiO_2, ranging from 46.25 to 47.66 wt%; Fe_2O_3, from 51.01 to 53.2 wt%; and MnO, from 0.45 to 0.87 wt%.

7. Discussion

7.1. Magma Type and Tectonic Setting Signature

Several discrimination diagrams have been proposed and used to elucidate the magma types of igneous rocks. Maniar and Piccoli [30] used $(Al_2O_3)/(Na_2O + K_2O + CaO)$ versus $(Al_2O_3)/(Na_2O + K_2O)$ to distinguish between peraluminous, metaluminous, and peralkaline rocks. Figure 11a shows the granite under study concentrated in the peraluminous field. For the subalkaline rocks, Rickwood [31] used the K_2O versus SiO_2 discrimination diagram (Figure 11b) to differentiate between the shoshonite series, high-K calc–alkaline series, medium-K calc–alkaline series, and low-K tholeiites. The studied granite samples plot in the calc–alkaline series high-K field.

Several tectonic setting discrimination variation diagrams have been proposed for granites. Rb versus (Y + Nb) and Nb versus Y discrimination diagrams by Pearce et al. [32] are used to distinguish between volcanic arc granite (VAG), syn-collision granite (Syn-COLG), ocean ridge granite (ORG), and within-plate granite (WPG). The figures show that the study samples are plotted in the within-plate regime field (Figure 12a,b). By utilizing various diagrams, Whalen et al. [33] distinguished between two groups of granites with different tectonic settings. Granites are subdivided into (1) those generated during the evolution of fold belts (orogenic), such as I-, S-, and M-types [34–36], and (2) those associated with uplift and major strike-slip faulting (anorogenic), such as A-types [37–39]. In Figure 12c,d, the plotting of 10,000*Ga/Al versus K_2O/MgO and Nb indicates that the studied alkali feldspar granite is plotted in A-type granite.

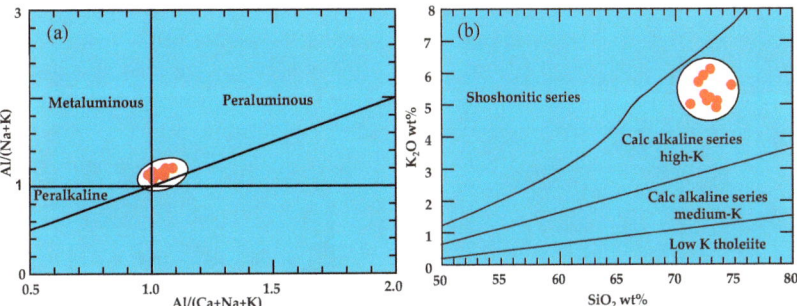

Figure 11. Magma-type discrimination diagrams of the studied granite. (**a**) Binary molar $Al_2O_3/(Na_2O + K_2O)$ versus $Al_2O_3/(CaO + Na_2O + K_2O)$ diagram of Maniar and Piccoli [30]. (**b**) Binary K_2O versus SiO_2 diagram of Rickwood [31].

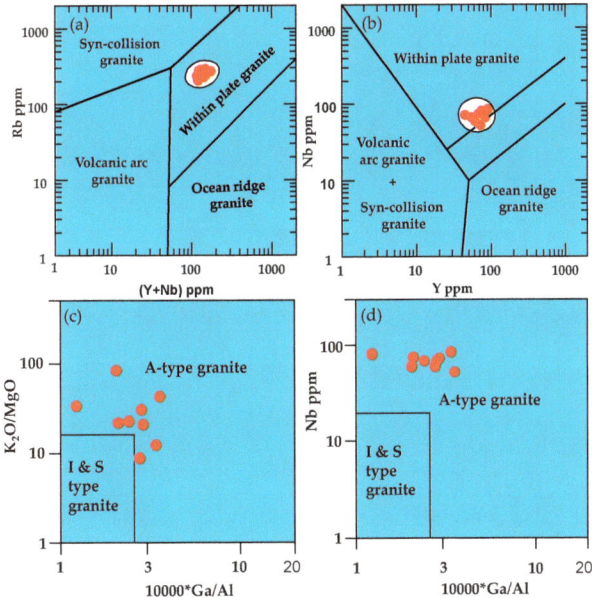

Figure 12. Tectonic setting discrimination diagrams of the studied granite. (**a**) Binary Rb versus (Y + Nb) diagram of Pearce et al. [32]. (**b**) Binary Nb versus Y diagram of Pearce et al. [32]. (**c**) Binary discrimination K_2O/MgO versus 10,000*Ga/Al diagram of Whalen et al. [33]. (**d**) Binary discrimination Nb versus 10,000*Ga/Al diagram of Whalen et al. [33].

7.2. Petrogenesis of Homret El Gergab Alkali Feldspar Granite

The relationships between Rb, Sr, and Ba are helpful in inferring the origin of granites, either from partial melting or fractional crystallization. The K/Rb ratio is considered a good petrogenetic indicator that decreases during magmatic fractionation. K_2O concentration could reflect the enrichment of Rb in igneous rock types. K_2O versus Rb discrimination diagram shows that the alkali feldspar granite samples plot in the field of the Ras ed Dome ring complex, Sudan [37]. In addition, the analyzed granite samples plotted very closely to the pegmatitic-hydrothermal trend of Shaw [40] (Figure 13a), and some samples that deviated could be post-magmatic and/or auto-metasomatic alterations [41]. The behavior of these elements in granitic systems is strongly controlled by plagioclase, K-feldspar, and mica. A binary discrimination diagram of Rb versus Sr (Figure 13b) confirms the role of K-feldspar fractionation in the studied granite. The average Ba/Rb ratio in the crust is approximately 4.4 [42]. Figure 13c shows that the studied granite samples occurred between 1 and 0.4 and are strongly enriched in Rb, which indicates the contribution of crustal material to the magmatic differentiation in the studied rock.

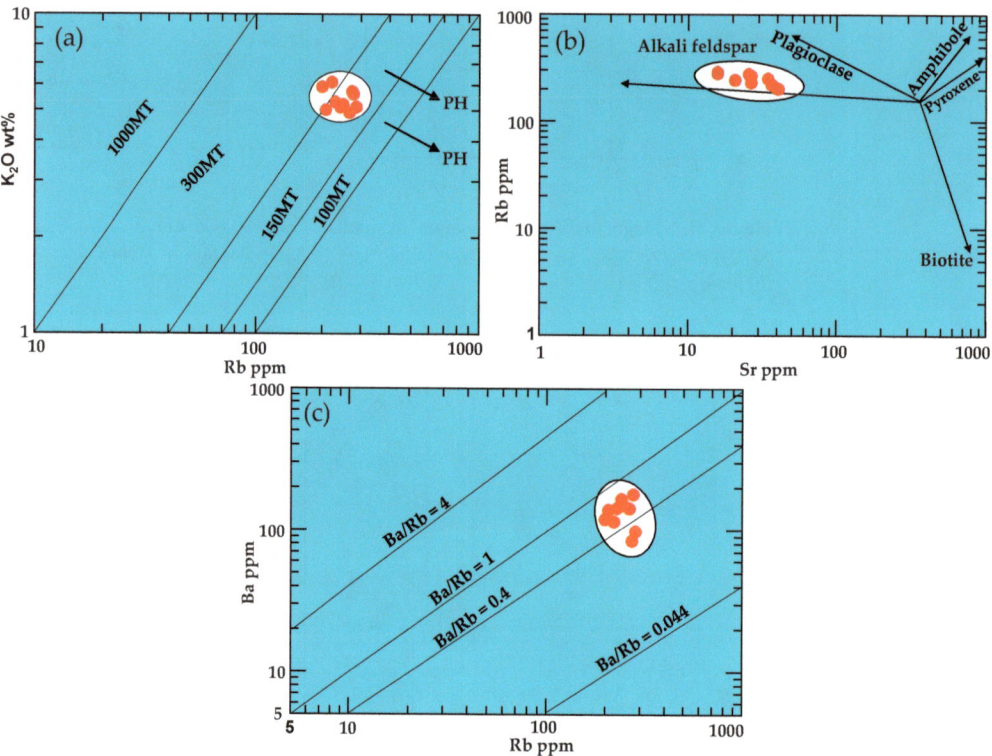

Figure 13. Petrogenesis discrimination diagrams show the distribution of the analyzed alkali feldspar granite samples. (**a**) Binary K$_2$O versus Rb diagram, where MT refers to the magmatic trend and PH refers to the pegmatitic hydrothermal trend, according to Shaw [40]. The shaded area illustrates the field of the Ras ed Dome ring complex, Sudan, according to O'Halloran [37]. (**b**) Binary Rb versus Sr diagram. (**c**) Binary Ba versus Rb diagram of Mason [42].

7.3. Zircon-Saturation Temperatures

The relationship between zircon solubility and the temperature of the melt is important for the Zr geothermometer. This approach could help determine the temperature and amount of Zr required for zircon formation in specific granite samples. Experimental research conducted by Watson and Harrison [43] on zircon saturation in various igneous rocks provides valuable data. The zircon saturation temperatures for the granite samples are listed in Table 1. The average estimated temperature for the alkali feldspar granite samples is close to 778 °C. The higher temperatures could suggest that the granite formed from a hotter magmatic source, potentially indicating the involvement of more evolved silica-rich melts. This could be crucial for understanding the most important conditions under which alkali feldspar granite has formed.

7.4. Insights into Rare Metals Mineralization

The enriched rare metal mineralization could being attributed to be crystallized from late magmatic melts. In addition, the higher concentrations of high-field strength elements (HFSEs) in the Homret El Gergab granite are considered good indicators of late-stage differentiation. Thorite can be crystallized from a melt of Th-Y-Zr-rich silicate. Abdel-Karim [44] and Abdel Gawad [29] stated that during the progressive phase of magmatic crystallization, the residual melts of SiO_2 could have forced most of the monazite to crystallize in the early phase. Radioactive minerals such as uranothorite and thorite have been recorded in the investigated granite. They were of syngenetic origin during the emplacement of the alkali feldspar granite. The total oxide content of the analyzed uranothorite and thorite from alkali feldspar granite in the study area is approximately close to ~90 wt% (Tables 2 and 3), which could be related to the hydration processes and/or extensive metamictization [45,46]. Some thorite grains are spread in hematite (Figure 9b,e), which could indicate an extensive alteration. The presence of REEs in uranothorite and thorite could be related to their similarity in ionic radii compared to U and Th. This is responsible for allowing the REEs to occupy the same structural sites in these minerals. The crystallization of monazite may remove thorium from the residual melts, and the substitution in A-site could involve a marked replacement of P and REEs by Si and Th and/or replacement of REEs by Ca and Th as recorded by Williams et al. [47] as follows:

$$P^{5+} + REE^{3+} \leftrightarrow Si^{4+} + Th^{4+}$$

$$2REE^{3+} \leftrightarrow Ca^{2+} + Th^{4+}$$

It is important to know the relationship occurring between the metallogenitic stages of valuable rare metals mineralization. Nb-, Ta-rich minerals are often associated together, whereas Nb^{+5} and Ta^{+5} are very close to each other. Hydrothermal fluids rich in Nb and Ta could be responsible for the formation of columbite. The Ta/(Ta + Nb) ratios of the analyzed columbite spots ranged from 0.01 to 0.03, whereas the Mn/(Mn + Fe) ratios showed a much wider range, varying from 0.02 to 0.28 (Table 4). These ratios indicate a marked enrichment of iron in the analyzed columbite, revealing its typical ferrocolumbite composition according to Černý and Ercit [48].

Zircons are considered valuable carriers of rare metal ores such as Nb, Ta, Y, REEs, Hf, U, and Th. It is well noted that some thorite and xenotime grains occur in the form of fine grains along the rims of zircon crystals (Figures 9c,d and 10c,d). The spatial association of zircon with thorite and xenotime could suggest that they were crystallized from a highly fractionated fluid-rich magma. Additionally, xenotime (Y) is found in association with zircon and can vary in morphology from euhedral, igneous-like grains to irregular secondary grains formed by the dissolution of zircon in the presence of phosphorus-bearing fluids [49]. The average hafnium content in zircon (approximately 3.6; Table 5) is relatively higher than the normal value (approximately 1%). The presence of hafnium-rich zircons in the El Gergab granite is significant because it could be suggested that these rocks have undergone extensive differentiation and fractionation processes that occurred during their formation. This could be related to prolonged magma cooling and crystallization, as well as fractional crystallization [15]. EPMA data for zircons (Table 5) and their pleochroic halos indicate their enrichment in radioactive elements, especially U and Th. However, a detailed study of zircon grains revealed multistage zonation accompanied by changes in morphology. Some interstitial zircons associated with xenotime are evidence of subsequent formation by hydrothermal fluids (Figures 9 and 10). The presence of a combination of zircon, xenotime, and thorite could suggest an intermediate solid solution reaction, whereas the presence of zircon, columbite, and xenotime with anhedral habits could be related to the effect of fluorine on high-field strength elements, which increases their solubility [18].

Monazite and xenotime are the main sources of phosphatic-REE minerals. Igneous and hydrothermal monazite have been distinguished by their ThO_2 contents, vary from 3 to >5 wt% for igneous monazite and <1 wt% for hydrothermal ones [50]. The EPMA data show that the Th content is consistent with its igneous or magmatic origin (Table 6). Xenotime is considered the main source of Y and HREEs, and is distinguished by its Th concentration, which is similar to that of Cínovec granite [51].

7.5. Geodynamic Evolution

The geodynamic evolution of granites remains a subject of debate, and many studies have been conducted to understand the origin of different granitic rock types. Several petrogenetic schemes have been reconstructed to explain the origin of granites, including the following: (1) dehydration melting of tonalite-granodiorte rocks [52]; (2) fractional crystallization of basaltic magma that is derived from the mantle [53]; (3) partial melting of the residual sources after I-type granite extraction [33]; (4) low-pressure melting of the calc–alkaline magmas at the upper crust [54]; and (5) hybridization of the mantle-derived magmas with those of crustal melts [55].

In the ANS region, several models have been proposed in which A-type granites could originate from mantle derived magmas and the partial melting of the lower continental crust. In addition, extreme fractionation of the basaltic source magma has been proposed for A-type granites from El Ineigi [56]. In general, the lithospheric delamination process could have played a markedly useful role in the formation of A-type post-collision granite in response to the upwelling of asthenospheric magmas [20,39]. The heat required for partial melting can arise from various sources, and fluids such as water or carbon dioxide can lower the melting temperature of rocks, which could facilitate the partial melting processes. Once the magma forms, it could undergo further compositional changes during its ascent, contributing to the diversity of granites.

Figure 14 shows a simple petrogenetic model for the Homret El Gergab alkali feldspar A-type granite. This model suggests that the continental lithosphere underwent extension, resulting in upwelling, stretching, and thinning of both the mantle and crust. This model proposes the following: (1) The continental lithosphere underwent a markedly wide extension, which is responsible for the rising, stretching, and thinning of both the mantle and lower crust. This process is accompanied by basaltic magma injection, which is derived from the mantle into the continental lower crust. (2) The transfer of heat and fluids rich in volatiles percolation from basaltic magma into the continental lower crust could have led to a partial melting process of the lower crust materials. (3) The A-type granite could have resulted from the intrusion of the basaltic magma from the mantle into the lower crust [57–59].

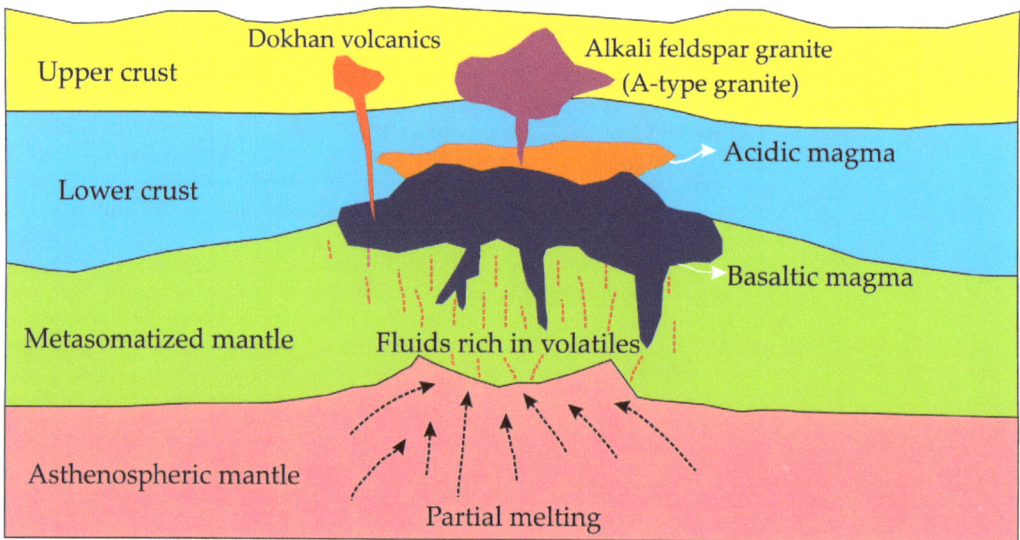

Figure 14. Simplified geodynamic model for the origin of the Homret El Gergab alkali feldspar granite modified after Tavakoli et al. [59].

8. Conclusions

Detailed petrological, geochemical, and mineralogical investigations indicate that the Homret El Gergab alkali feldspar granite is peraluminous, calc–alkaline series high-K, and emplaced within-plate environments. The alkali feldspar rare metal granite in the study area is distinguished by the higher content of alkali feldspar minerals, which form at higher temperatures and are more insoluble than others, indicating a greater degree of differentiation.

Mineralogical studies have identified various minerals, including uranothorite, thorite, columbite, zircon, monazite, and xenotime, as well as other accessory minerals. The close proximity of zircon to thorite and xenotime indicates that they could be formed from fluid-rich magmas and underwent a significant degree of fractionation.

Basaltic magmas play two main roles: (1) they supply the necessary heat to melt the Earth's crust, and (2) they provide volatile substances that seep into the lower continental layers, resulting in the formation of A-type granite by partial melting of the crust.

Author Contributions: Conceptualize, M.M.G., A.S.M. and A.E.A.G.; methodology, M.M.G., M.D., H.A.E.-D. and E.G.P.; software, M.M.G. and A.E.A.G.; validation, M.M.G., A.E.A.G., M.D., H.A.E.-D., E.G.P., A.S.M. and M.A.E.-L.; formal analysis, M.M.G. and E.G.P.; investigation, M.M.G., A.E.A.G., M.D., H.A.E.-D., E.G.P., A.S.M. and M.A.E.-L.; resources, M.M.G., M.D., H.A.E.-D., A.S.M. and M.A.E.-L.; data curation, A.E.A.G., M.M.G. and A.S.M.; Writing—original draft preparation: M.M.G., A.S.M. and A.E.A.G.; writing—review and editing, A.E.A.G., M.M.G. and A.S.M.; visualization, A.E.A.G., M.M.G., E.G.P., M.D., H.A.E.-D. and A.S.M.; supervision, M.M.G., E.G.P., A.E.A.G. and M.D.; funding acquisition, A.E.A.G. and E.G.P. All authors have read and agreed to the published version of the manuscript.

Funding: This research received no external funding.

Data Availability Statement: Data are contained within the article.

Conflicts of Interest: The authors declare no conflicts of interest.

References

1. Stern, R.J.; Hedge, C.E. Geochronologic and isotopic constraints on late Precambrian crustal evolution in the Eastern Desert of Egypt. *Am. J. Sci.* **1985**, *285*, 97–172. [CrossRef]
2. Liégeois, J.P.; Stern, R.J. Sr–Nd isotopes and geochemistry of granite–gneiss complexes from the Meatiq and Hafafit domes, Eastern Desert, Egypt: No evidence for pre-Neoproterozoic crust. *J. Afr. Earth Sci.* **2010**, *57*, 31–40. [CrossRef]
3. Helba, H.; Trumbull, R.B.; Morteani, G.; Khalil, S.O.; Arslan, A. Geochemical and petrographic studies of Ta mineralization in the Nuweibi albite granite complex, Eastern Desert, Egypt. *Miner. Depos.* **1997**, *32*, 164–179. [CrossRef]
4. Abu El-Rus, M.A.; Mohamed, M.A.; Lindh, A. Mueilha rare metals granite, Eastern Desert of Egypt: An example of a magmatic-hydrothermal system in the Arabian- Nubian Shield. *Lithos* **2017**, *294*, 362–382. [CrossRef]
5. Abuamarah, B.A.; Azer, M.K.; Asimow, P.D.; Ghrefat, H.; Mubarak, H.S. Geochemistry and petrogenesis of late Ediacaran rare-metal albite granites of the Arabian-Nubian Shield. *Acta Geol. Sin.* **2021**, *95*, 459–480. [CrossRef]
6. Abdel Gawad, A.E.; Skublov, S.G.; Levashova, E.V.; Ghoneim, M.M. Geochemistry and U–Pb Age dating of zircon as a petrogenetic tool for magmatic and hydrothermal processes in Wadi Ras Abda Syenogranite, Eastern Desert, Egypt. *Arab. J. Sci. Eng.* **2021**, *47*, 7351–7365. [CrossRef]
7. Skublov, S.G.; Abdel Gawad, A.E.; Levashova, E.V.; Ghoneim, M.M. U–Pb geochronology, REE and trace element geochemistry of zircon from El Fereyid monzogranite, south Eastern Desert, Egypt. *J. Miner. Pet. Sci.* **2021**, *116*, 220–233. [CrossRef]
8. Ghoneim, M.M.; Panova, E.G.; Abdel Gawad, A.E.; Yanson, S.Y. Morphological and geochemical features of zircon from intrusive rocks of El Sela area, Eastern Desert, Egypt. *N. Ural. State Min. Univ.* **2020**, *3*, 7–18. [CrossRef]
9. Surour, A.A.; Omar, S.A.M. Historiography and FTIR spectral signatures of beryl crystals from some ancient Roman sites in the Eastern Desert of Egypt. *Environ. Earth Sci.* **2020**, *79*, 520. [CrossRef]
10. Ali, M.A.; Abdel Gawad, A.E.; Ghoneim, M.M. Geology and mineral chemistry of uranium and thorium bearing minerals in rare-metal (NYF) pegmatites of Um Solimate, South Eastern Desert of Egypt. *Acta Geol. Sin.* **2021**, *95*, 1568–1582. [CrossRef]
11. Ghoneim, M.M.; Panova, E.G.; Abdel Gawad, A.E. Natural radioactivity and geochemical aspects of radioactive mineralisation in El Sela, South Eastern Desert, Egypt. *Int. J. Environ. Anal. Chem.* **2023**, *103*, 2338–2350. [CrossRef]
12. Abdel Gawad, A.E.; Ene, A.; Skublov, S.G.; Gavrilchik, A.K.; Ali, M.A.; Ghoneim, M.M.; Nastavkin, A.V. Trace element geochemistry and genesis of beryl from Wadi Nugrus, South Eastern Desert, Egypt. *Minerals* **2022**, *12*, 206. [CrossRef]

13. Sami, M.; Osman, H.; Ahmed, A.F.; Zaky, K.S.; Abart, R.; Sanislav, I.V.; Abdelrahman, K.; Fnais, M.S.; Xiao, W.; Abbas, H. Magmatic Evolution and Rare Metal Mineralization in Mount El-Sibai Peralkaline Granites, Central Eastern Desert, Egypt: Insights from Whole-Rock Geochemistry and Mineral Chemistry Data. *Minerals* **2023**, *13*, 1039. [CrossRef]
14. Alekseev, V.I.; Alekseev, I.V. The Presence of Wodginite in Lithium–Fluorine Granites as an Indicator of Tantalum and Tin Mineralization: A Study of Abu Dabbab and Nuweibi Massifs (Egypt). *Minerals* **2023**, *13*, 1447. [CrossRef]
15. Abdalla, H.M.; Helba, H.; Matsueda, H. Chemistry of zircon in rare metal granitoids and associated rocks, Eastern Desert, Egypt. *Resour. Geol.* **2009**, *59*, 51–68. [CrossRef]
16. Abdallah, S.E.; Azer, M.K.; El Shammari, A.S. The Petrological and geochemical evolution of Ediacaran rare-metal bearing A-type granites from Jabal Aja complex, northern Arabian Shield, Saudi Arabia. *Acta Geol. Sin. Engl. Ed.* **2020**, *94*, 743–762. [CrossRef]
17. Abdelkader, M.A.; Watanabe, Y.; Shebl, A.; El-Dokouny, H.A.; Dawoud, M.; Csámer, Á. Effective delineation of rare metal-bearing granites from remote sensing data using machine learning methods: A case study from the Umm Naggat Area, Central Eastern Desert, Egypt. *Ore Geol. Rev.* **2022**, *1*, 105184. [CrossRef]
18. Abdel-Azeem, M.M. Genesis of the rare metals mineralization in Um Safi acidic volcanics, Central Eastern Desert, Egypt. *Appl. Earth Sci.* **2024**, *30*, 25726838231225051.
19. Basta, E.Z.; Kotb, H.; Awadalla, M.F. Petrochemical and geochemical characteristics of the Dokhan Formation at the type locality, Jabal Dokhan, Eastern Desert, Egypt. *Inst. Appl. Geol. Jeddah Bull.* **1980**, *3*, 121–140.
20. Eliwa, H.A.; El-Bialy, M.Z.; Murata, M. Edicaran postcollisional volcanism in the Arabian-Nubian Shield: The high-K calc-alkaline Dokhan Volcanics of Gabal Samr El-Qaa (592 ± 5 Ma), North Eastern Desert, Egypt. *Precambrian Res.* **2014**, *246*, 180–207. [CrossRef]
21. Khamis, H.A.; El-Bialy, M.; Hamimi, Z.; Afifi, A.; Abdel Wahed, S.A. Petrographic investigation of the Precambrian basement rocks of Esh El Mallaha area, North Eastern Desert, Egypt. *Alfarama J. Basic Appl. Sci.* **2023**, *4*, 340–354.
22. Abd El-Hadi, A.M. Mineralogical and Geochemical Prospection and Radioelements Distribution in Hamrat Al Jirjab Stream Sediments, Esh El Melaha Range, North Eastern Desert, Egypt. Ph.D. Thesis, Faculty of Science, Benha University, Benha, Egypt, 2013; 251p.
23. Barker, F. Trondhjemite: Definition, environment and hypotheses of origin. In trondhjemites, dacites, and related rocks. *Dev. Petrol.* **1979**, *6*, 1–12.
24. Middlemost, E.A.K. Naming materials in the magma/igneous rock system. *Earth Sci. Rev.* **1994**, *37*, 215–224. [CrossRef]
25. Sun, S.S.; McDonough, W.F. Chemical and isotopic systematics of oceanic basalts: Implications for mantle composition and processes. In *Magmatism in Ocean Basins*; Saunders, A.D., Norry, M.J., Eds.; Geological Society London Special Publications: London, UK, 1989; Volume 42, pp. 313–345.
26. Taylor, S.R.; McLennan, S.M. Chemical composition and element distribution in the Earth's crust. *Encycl. Phys. Sci. Technol.* **2001**, *312*, 697–719.
27. Watt, G.R. High-thorium monazite-(Ce) formed during disequilibrium melting of metapelites under granulite-facies conditions. *Min. Mag.* **1995**, *59*, 735–743. [CrossRef]
28. Abd El Ghaffar, N.I. Enrichment of rare earth and radioactive elements concentration in accessory phases from alkaline granite, South Sinai- Egypt. *J. Afr. Earth Sci.* **2018**, *147*, 393–401. [CrossRef]
29. Abdel Gawad, A.E. Mineral chemistry (U, Th, Zr, REE) in accessory minerals from Wadi Rod Elsayalla granitoids, South Eastern Desert, Egypt. *Arab. J. Geosci.* **2021**, *14*, 1996. [CrossRef]
30. Maniar, P.D.; Piccoli, P.M. Tectonic discrimination of granitoids. *Geol. Soc. Am. Bull.* **1989**, *101*, 635–643. [CrossRef]
31. Rickwood, P.C. Boundary line within petrologic diagrams which use oxides of major and minor elements. *Lithos* **1989**, *22*, 247–263. [CrossRef]
32. Pearce, J.A.; Harris, N.B.W.; Tindle, A.G. Trace element discrimination diagrams for the tectonic interpretation of granitic rocks. *J. Petrol.* **1984**, *25*, 956–983. [CrossRef]
33. Whalen, J.B.; Currie, K.L.; Chappell, B.W. A-type granites: Geochemical characteristics, discrimination and petrogenesis. *Contrib. Min. Petrol.* **1987**, *95*, 407–419. [CrossRef]
34. Qiu, J.T.; Qiu, L. Geochronology and magma oxygen fugacity of Ehu S-type granitic pluton in Zhe-Gan-Wan region, SE China. *Geochemistry* **2016**, *76*, 441–448. [CrossRef]
35. Qiu, L.; Yan, D.P.; Zhou, M.F.; Arndt, N.T.; Tang, S.L.; Qi, L. Geochronology and geochemistry of the Late Triassic Longtan pluton in South China: Termination of the crustal melting and Indosinian orogenesis. *Int. J. Earth Sci. (Geol. Rundsch)* **2013**, *103*, 649–666. [CrossRef]
36. Qiu, L.; Li, X.; Li, X.; Yan, D.P.; Ren, M.; Zhang, L.; Cheng, G. Petrogenesis of early cretaceous intermediate to felsic rocks in Shanghai, South China: Magmatic response to Paleo-Pacific plate subduction. *Tectonophysics* **2022**, *838*, 229469. [CrossRef]
37. O' Halloran, D.A. Ras ed Dom migrating complex: A-type granites and syenites from the Bauda Desert, Sudan. *J. Afr. Earth Sci.* **1985**, *3*, 61–75.
38. Abdel Gawad, A.E.; Eliwa, H.; Ali, K.G.; Alsafi, K.; Murata, M.; Salah, M.S.; Hanfi, M.Y. Cancer Risk Assessment and Geochemical Features of Granitoids at Nikeiba, Southeastern Desert, Egypt. *Minerals* **2022**, *12*, 621. [CrossRef]
39. Khedr, M.Z.; Khashaba, S.M.; El-Shibiny, N.H.; Takazawa, E.; Hassan, S.M.; Azer, M.K.; Whattam, S.A.; El-Arafy, R.A.; Ichiyama, Y. Integration of remote sensing and geochemical data to characterize mineralized A-type granites, Egypt: Implications for origin and concentration of rare metals. *Int. J. Earth Sci.* **2023**, *112*, 1717–1745. [CrossRef]

40. Shaw, I.; Bunbury, J.A. Petrological Study of the Emerald Mines in the Egyptian Eastern Desert. In *Lthics at the Millannium*; Moloney, N., Shott, M.J., Eds.; Archaeopress: London, UK, 2003; pp. 203–213.
41. Vidal, P.; Dosso, L.; Bowden, P.; Lameyre, J. Strontium isotope geochemistry in syenite-alkaline granite complexes. In *Origin and Distribution of the Elements (2nd Symposium)*; Ahrens, L.H., Ed.; Pergamon Press: Oxford, UK, 1979; pp. 223–231.
42. Mason, B. *Principles of Geochemistry*, 3rd ed.; John Wiely: New York, NY, USA, 1966; 310p.
43. Watson, E.B.; Harrison, T.M. Zircon saturation revisited: Temperature and composition effects in a variety of crustal magma types. *Earth Planet. Sci. Lett.* **1983**, *64*, 295–304. [CrossRef]
44. Abdel-Karim, A.M. REE-rich accessory minerals in granites from southern Sinai, Egypt: Mineralogy, geochemistry and petrogenetic implications. In Proceedings of the 4th International Conference on Geochemistry, Bab Sharqi, Egypt, 15–16 September 1999; pp. 83–100.
45. Abd El-Naby, H.H. High and low temperature alteration of uranium and thorium minerals, Um Ara granites, south Eastern Desert, Egypt. *Ore Geol. Rev.* **2009**, *35*, 436–446. [CrossRef]
46. Abdel Gawad, A.E.; Ghoneim, M.M.; El-Taher, A.; Ramadan, A.A. Mineral chemistry aspects of U-, Th-, REE-, Cu-bearing minerals at El-Regeita shear zone, South Central Sinai, Egypt. *Arab. J. Geosci.* **2021**, *14*, 1–13. [CrossRef]
47. Williams, M.L.; Jercinovic, M.J.; Hetherington, C.J. Microprobe monazite geochronology: Understanding geologic processes by integrating composition and chronology. *Annu. Rev. Earth Planet Sci.* **2007**, *35*, 137–175. [CrossRef]
48. Černý, P.; Ecrit, T.S. Some recent advances in the mineralogy and geochemistry of Nb and Ta in rare-element granitic pegmatites. *Bull. Min.* **1985**, *108*, 499–532. [CrossRef]
49. Xie, L.W.; Yang, J.H.; Yin, Q.Z.; Yang, Y.H.; Liu, J.B.; Huang, C. High spatial resolution in situ U–Pb dating using laser ablation multiple ion counting inductively coupled plasma mass spectrometry (LA-MIC-ICP-MS). *J. Anal. At. Spectrom.* **2017**, *32*, 975–986. [CrossRef]
50. Schandl, E.S.; Gorton, M.P. A textural and geochemical guide to the identification of hydrothermal monazite: Criteria for selection of samples for dating epigenetic hydrothermal ore deposits. *Econ. Geol.* **2004**, *99*, 1027–1035. [CrossRef]
51. Johan, Z.; Johan, V. Accessory minerals of the Cínovec (Zinnwald) granite cupola, Czech Republic: Indicators of petrogenetic evolution. *Min. Petrol.* **2005**, *83*, 113–150. [CrossRef]
52. King, P.L.; White, A.J.R.; Chappell, B.W.; Allen, C.M. Characterization and origin of aluminous A-type granites from the Lachlan Fold Belt, southeastern Australia. *J. Petrol.* **1997**, *38*, 371–391. [CrossRef]
53. Christiansen, E.H.; Best, M.G.; Radebaugh, J. The origin of magma on planetary bodies. *Planet. Volcanism Across Sol. Syst.* **2022**, *1*, 235–270.
54. Skjerlie, K.P.; Johnston, A.D. Vapor-absent melting at 10 kbar of a biotite-and amphibole-bearing tonalitic gneiss: Implications for the generation of A-type granites. *Geology* **1992**, *20*, 263–266. [CrossRef]
55. Yang, J.H.; Wu, F.Y.; Wilde, S.A.; Xie, L.W.; Yang, Y.H.; Liu, X.M. Tracing magma mixing in granite genesis: In situ U–Pb dating and Hf-isotope analysis of zircons. *Contrib. Min. Petrol.* **2007**, *153*, 177–190. [CrossRef]
56. Mohamed, F.H.; El-Sayed, M.M. Postorogenic and anorogenic A-type fluorite-bearing granitoids, Eastern Desert, Egypt: Petrogenetic and geotectonic implications. *Geochemistry* **2008**, *68*, 431–450. [CrossRef]
57. Annen, C.; Sparks, R.S. Effects of repetitive emplacement of basaltic intrusions on thermal evolution and melt generation in the crust. *Earth Planet. Sci. Lett.* **2002**, *203*, 937–955. [CrossRef]
58. Abd El-Fatah, A.A.; Surour, A.A.; Azer, M.K.; Madani, A.A. Integration of Whole-Rock Geochemistry and Mineral Chemistry Data for the Petrogenesis of A-Type Ring Complex from Gebel El Bakriyah Area, Egypt. *Minerals* **2023**, *13*, 1273. [CrossRef]
59. Tavakoli, N.; Shabanian, N.; Davoudian, A.R.; Azizi, H.; Neubauer, F.; Asahara, Y.; Bernroider, M.; Lee, J.K. A-type granite in the Boein-Miandasht Complex: Evidence for a Late Jurassic extensional regime in the Sanandaj-Sirjan Zone, western Iran. *J. Asian Earth Sci.* **2021**, *213*, 104771. [CrossRef]

Disclaimer/Publisher's Note: The statements, opinions and data contained in all publications are solely those of the individual author(s) and contributor(s) and not of MDPI and/or the editor(s). MDPI and/or the editor(s) disclaim responsibility for any injury to people or property resulting from any ideas, methods, instructions or products referred to in the content.

Article

Petrogenesis of the Newly Discovered Neoproterozoic Adakitic Rock in Bure Area, Western Ethiopia Shield: Implication for the Pan-African Tectonic Evolution

Junsheng Jiang [1,2], Wenshuai Xiang [1,*], Peng Hu [1,*], Yulin Li [3], Fafu Wu [1], Guoping Zeng [1], Xinran Guo [4], Zicheng Zhang [5] and Yang Bai [6]

1. Wuhan Center, China Geological Survey (Central South China Innovation Center for Geosciences), Wuhan 430205, China; jsjiang0818@163.com (J.J.)
2. Research Center for Petrogenesis and Mineralization of Granitoid Rocks, China Geological Survey, Wuhan 430205, China
3. Institute of Geological Survey, China University of Geosciences, Wuhan 430074, China; a942186394@gmail.com
4. The Geological Science Education Center of Guangdong, Guangzhou 510000, China; xinran5566@163.com
5. China National Geological & Mining Corporation, Beijing 100029, China; zhangzicheng@chinagm.com.cn
6. Geophysical Exploration Brigade HuBei Geological Bureau, Wuhan 430056, China
* Correspondence: xiangwenshuai@mail.cgs.gov.cn (W.X.); hpeng@mail.cgs.gov.cn (P.H.)

Abstract: The Neoproterozoic Bure adakitic rock in the western Ethiopia shield is a newly discovered magmatic rock type. However, the physicochemical conditions during its formation, and its source characteristics are still not clear, restricting a full understanding of its petrogenesis and geodynamic evolution. In this study, in order to shed light on the physicochemical conditions during rock formation and provide further constraints on the petrogenesis of the Bure adakitic rock, we conduct electron microprobe analysis on K-feldspar, plagioclase, and biotite. Additionally, we investigate the trace elements and Hf isotopes of zircon, and the Sr-Nd isotopes of the whole rock. The results show that the K-feldspar is orthoclase (Or = 89.08~96.37), the plagioclase is oligoclase (Ab = 74.63~85.99), and the biotite is magnesio-biotite. Based on the biotite analysis results, we calculate that the pressure during rock formation was 1.75~2.81 kbar (average value of 2.09 kbar), representing a depth of approximately 6.39~10.2 km (average value of 7.60 km). The zircon thermometer yields a crystallization temperature of 659~814 °C. Most of the $(Ce/Ce^*)_D$ values in the zircons plotted above the Ni-NiO oxygen buffer pair, and the calculated magmatic oxygen fugacity ($\log fO_2$) values vary from −18.5 to −4.9, revealing a relatively high magma oxygen fugacity. The uniform contents of FeO, MgO, and K_2O in the biotite suggest a crustal magma source for the Bure adakitic rock. The relatively low $(^{87}Sr/^{86}Sr)_i$ values of 0.70088 to 0.70275, positive $\varepsilon_{Nd}(t)$ values of 3.26 to 7.28, together with the positive $\varepsilon_{Hf}(t)$ values of 7.64~12.99, suggest that the magma was sourced from a Neoproterozoic juvenile crust, with no discernable involvement of a pre-Neoproterozoic continental crust, which is coeval with early magmatic stages in the Arabian Nubian Shield elsewhere. Additionally, the mean Nd model ages demonstrate an increasing trend from the northern parts (Egypt, Sudan, Afif terrane of Arabia, and Eritrea and northern Ethiopia; 0.87 Ga) to the central parts (Western Ethiopia shield; 1.03 Ga) and southern parts (Southern Ethiopia Shield, 1.13 Ga; Kenya, 1.2 Ga) of the East African Orogen, which indicate an increasing contribution of pre-Pan-African crust towards the southern part of the East African Orogen. Based on the negative correlation between MgO and Al_2O_3 in the biotite, together with the Lu/Hf-Y and Yb-Y results of the zircon, we infer that the Bure adakitic rock was formed in an arc–arc collision orogenic environment. Combining this inference with the whole rock geochemistry and U-Pb age of the Bure adakitic rock, we further propose that the rock is the product of thickened juvenile crust melting triggered by the Neoproterozoic Pan-African Orogeny.

Keywords: Bure adakitic rock; mineralogical characteristics; Sr-Nd-Hf isotopes; Western Ethiopia Shield

Citation: Jiang, J.; Xiang, W.; Hu, P.; Li, Y.; Wu, F.; Zeng, G.; Guo, X.; Zhang, Z.; Bai, Y. Petrogenesis of the Newly Discovered Neoproterozoic Adakitic Rock in Bure Area, Western Ethiopia Shield: Implication for the Pan-African Tectonic Evolution. *Minerals* **2024**, *14*, 408. https://doi.org/10.3390/min14040408

Academic Editors: Ignez de Pinho Guimarães and Jefferson Valdemiro De Lima

Received: 21 February 2024
Revised: 4 April 2024
Accepted: 9 April 2024
Published: 16 April 2024

Copyright: © 2024 by the authors. Licensee MDPI, Basel, Switzerland. This article is an open access article distributed under the terms and conditions of the Creative Commons Attribution (CC BY) license (https://creativecommons.org/licenses/by/4.0/).

1. Introduction

The East African Orogen (EAO) has recorded a complex history of intra-oceanic and continental margin magmatic and tectono–thermal events from the Neoproterozoic to the Early Cambrian. It mainly consists of the juvenile Arabian Nubian Shield (ANS) and the largely older continental crust of the Mozambique Belt (MB) from north to south [1–3]. The Western Ethiopian Shield (WES) is situated in a key location, relatively close to the transition between the Arabian Nubian Shield and the Mozambique Belt. It is also adjacent to and east of the 'Eastern Saharan Meta-craton' [4]. It is a metamorphic terrane that includes high-grade gneisses and low-grade metavolcanic and metasedimentary rocks with associated intrusions. The granitoid rocks, which have either intruded into greenschist facies volcano–sedimentary sequences or been emplaced at the contact between low- and high-grade terranes, constitute a significant proportion of plutonic rocks in the Precambrian rocks of the WES. Many researchers have focused on the granitoid rocks in the WES, significantly advancing our understanding of regional tectonic evolution [5–11]. However, the magma source of these granitoid rocks in the WES, especially those intruding into the low- and high-grade rock associations within the eastern part of the WES, remains unclear. Also unclear is whether the magma source derived from mixing with pre-Neoproterozoic crustal material or not.

The Bure granite in the WES formed in the Pan-African Orogeny Period (750–650 Ma; [5]). It has an LA-ICP-MS U-Pb age of 773.8 ± 8.1 Ma and is characterized by high Sr (310~401 ppm), Sr/Y (64.9~113.6), and La/Yb (25.7~51.6), and low MgO (0.27~0.41 wt%), Y (2.71~4.78 ppm), and Yb (0.20~0.31 ppm) values [7]. Based on the study by Xu et al. [12], Jiang et al. [7] defined the Bure granite as an adakitic rock. This rock is called the Bure adakitic rock in this study. It is a newly discovered magmatic rock type in this area. However, the magma source and physicochemical conditions of the Bure adakitic rock remain unknown, hindering a comprehensive understanding of its petrogenesis and geodynamic evolution during the Pan-African period. Its mineralogical and isotopic compositions vary significantly depending on the type of precursor rocks and/or igneous processes during the evolution of its parental magma. Thus, knowledge of the mode of origin of these rocks contributes to our understanding of the Neoproterozoic evolutionary history of the WES.

This study investigates the major elements of typical minerals (K-feldspar, plagioclase, and biotite), trace elements and Hf isotopes of zircon, and Sr-Nd isotopes of the whole rock from the Bure adakitic rock in the eastern part of the WES. Combined with local and regional geological, geochemical, and geochronological data, the results shed light on the degree of pre-Neoproterozoic crustal material involvement in the source magmas, and the Neoproterozoic geological evolution of the WES.

2. Geological Setting

The EAO is a Neoproterozoic to early Cambrian mobile belt that reflects the collision between Neoproterozoic India and the African Neoproterozoic continents [1,2,13,14]. Based on its lithological and metamorphic characteristics, the EAO can be broadly subdivided into two terranes, the Arabian Nubian Shield in the north and the Mozambique Belt in the south. The ANS is dominated by low-grade volcano–sedimentary rocks associated with plutons and ophiolitic remnants [4,15–19], and represents the juvenile terrane. However, the MB in the south part of the EAO is a tract of largely older continental crust that was extensively deformed and metamorphosed in the Neoproterozoic/Cambrian ([10,20–22]; Figure 1).

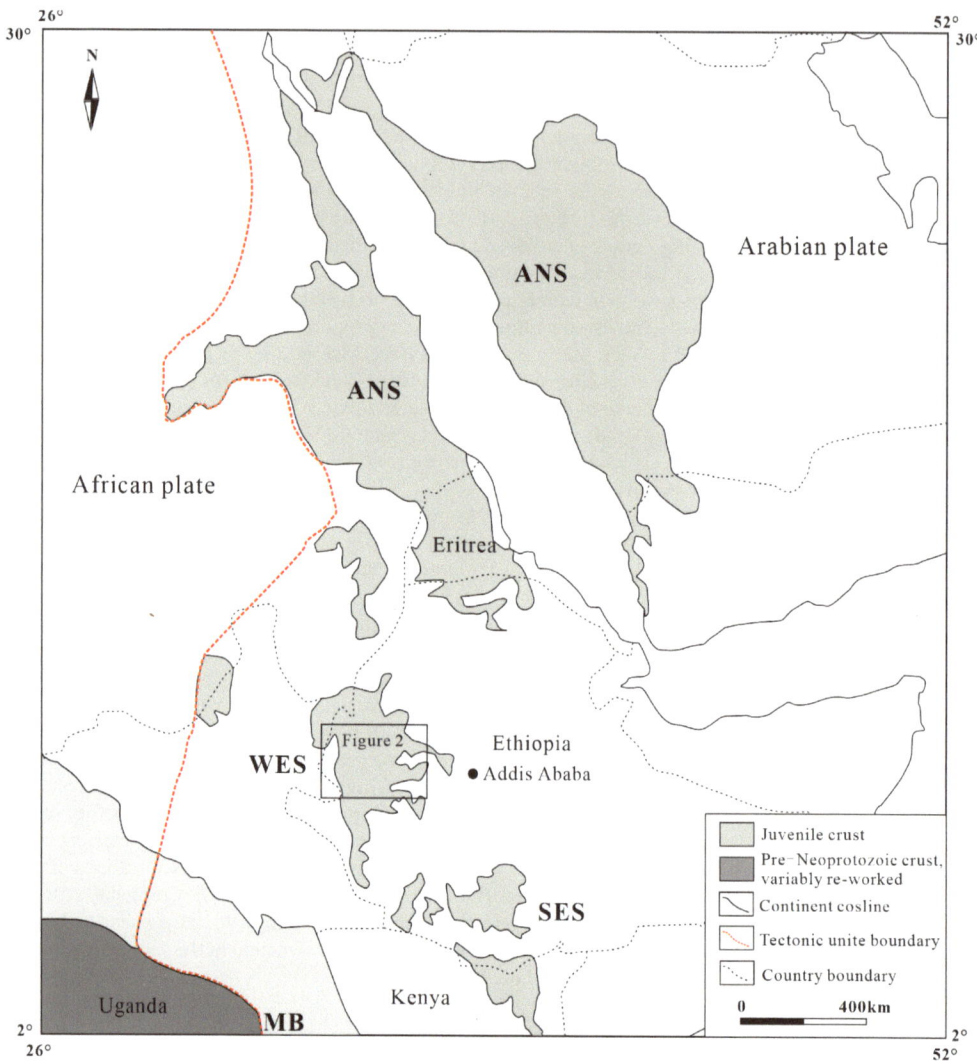

Figure 1. Geological map of the Arabian-Nubian Shield, northeast Africa (after [14]).

The WES is also called the Tuludimtu Orogenic Belt, which is understood to have formed during the amalgamation of western Gondwana before the final closure of the Mozambique Ocean [15]. It can be subdivided into five litho-tectonic domains from west to east—the Daka, Sirkole, Dengi, Kemashi, and Didesa Domains. The Daka Domain lies in the southwest corner of the WES (Figure 2) and consists of pre-Neoproterozoic basement gneisses representing the western basement margin of the Tuludimtu Belt. The Sirkole Domain, composed of gneissic and volcano–sedimentary rocks intruded by granites, is located in the northwestern portion of the WES that extends into Sudan. The Dengi Domain is characterized by a deformed and metamorphosed volcano–sedimentary sequence and the Jamoa-Ganti orthogneiss; there are several intrusive bodies in this domain. It is generally thought to be a volcanic arc sequence related to the closure of the ocean represented by the Tuludimtu Ophiolite to the east. The Kemashi Domain consists of a sequence of metasedimentary rocks and abundant mafic to ultra-mafic volcanic material that has been metamorphosed to upper greenschist/epidote-amphibolite facies. The nature of these ultra-mafic/mafic plutonic rocks within the Kemashi Domain is controversial, with some scholars holding that they represent an ophiolite sequence [4,15,23], named the Tuludimtu Ophiolite. However, others [24–26] hold that these ultra-mafic/mafic plutonic rocks represent Alaskan-type, concentrically zoned intrusions, which were emplaced into an extensional arc or back-arc environment. The Didesa Domain within the eastern boundary of the WES is characterized by amphibolite facies paragneiss and orthogneiss intruded by Neoproterozoic intrusive rocks. It is located in the transition between the Arabian Nubian Shield and the Mozambique Belt.

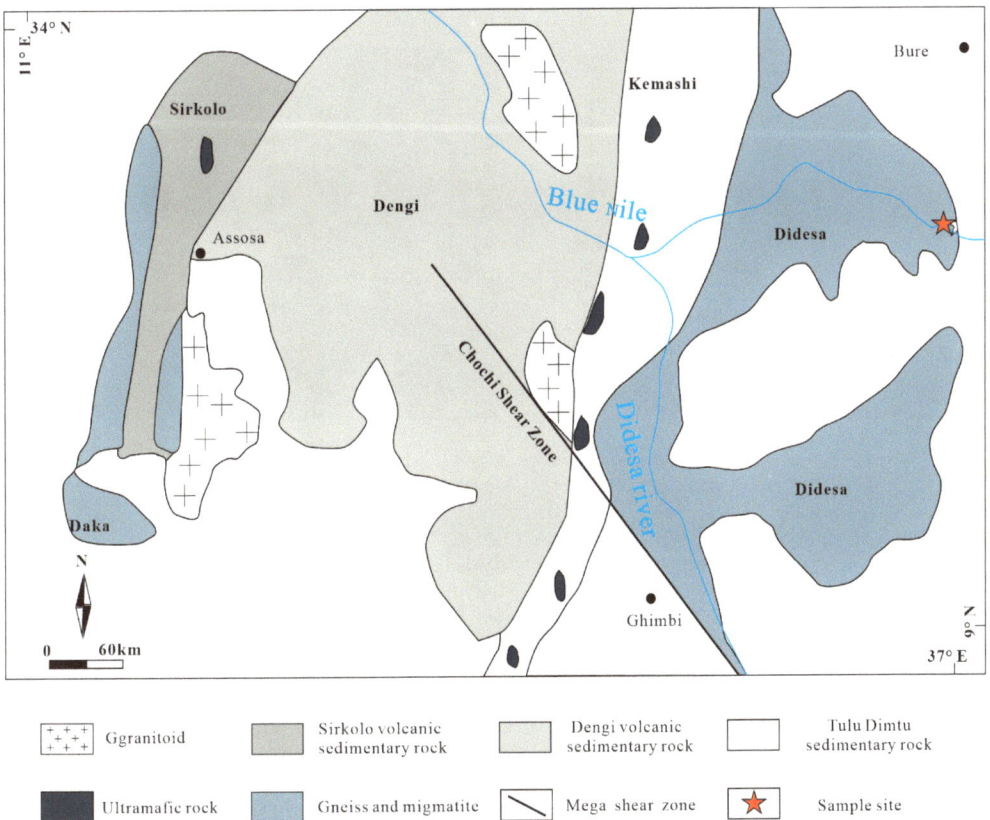

Figure 2. Sketch of the regional geology of the western Ethiopian terrain (after [15]).

Three generations of magmatism at ca. 850–810 Ma, 780–700 Ma, and 650–550 Ma [5,8–10,27,28], which represent pre-, syn-, and post-tectonic environments, respectively, have been recognized by previously limited ages from elsewhere in the WES [10,21]. These intrusions are usually present as strains and dikes and are developed as ductile fault contact or intrusive contact with the surrounding rock. The main types of intrusions are granite, granodiorite, monzogranite, and tonalite. The Bure adakitic rock is located at the eastern boundary of the Didesa Domain, with the surrounding rocks comprising gneisses. This rock assemblage suggests that it not only inherited the unique rock assemblages of the Arabic-Nubian Shield but also developed the typical middle-high grade metamorphic rocks of the Mozambique Belt.

3. Samples and Analytical Methods

3.1. Petrography

The Bure adakitic rock appears light gray in the field, with a fine granitic texture. It is mainly composed of K-feldspar (45–48 wt%), plagioclase (20–23 wt%), quartz (23–25 wt%), biotite (4–5 wt%), and minor amounts of muscovite (1–2 wt%) (Figure 3a). The K-feldspar is heteromorphic granular, with a size of 0.2–1.5 mm, some of which show slight kaolinization on the surface. The plagioclase is granular and 0.1 to 1 mm in size, with characteristics of polysynthetic twins and Carlsbadal bite compound twins. The surface of the plagioclase is usually altered, displaying light sericitization. The quartz is xenomorphic-granular, with a size of 0.05–0.7 mm (Figure 3b).

Figure 3. Hand specimen photograph (**a**) and microphotograph (**b**) for the Bure adakitic rock. Qtz—quartz; Bt—biotite; Kfs—K-feldspar.

3.2. Analytical Methods

Electron microprobe analysis (EMPA) was performed on the K-feldspar, plagioclase, and biotite at the Zhongnan Mineral Resources Supervision and Test Center for Geoanalysis, Wuhan Center, China Geological Survey. During the analysis, a 10-μm spot size was used for the plagioclase and K-feldspar, and a 1-μm spot size was used for the biotite, with an accelerating voltage of 20 kV and a beam current of 20 nA. The integration times for the Ti and Mn peaks were 20 s and that for the remaining elements was 10 s. The SPI and ZBA mineral standards and ZAF calibration were employed for all minerals.

Trace element analyses of zircon were conducted synchronously using LA–ICP-MS at the Wuhan Sample Solution Analytical Technology Co., Ltd. Laser sampling was performed using a GeolasPro laser ablation system consisting of a COMPexPro 102 ArF excimer laser (wavelength of 193 nm and maximum energy of 200 mJ) and a MicroLas optical system. An Agilent 7700e ICP-MS instrument was used to acquire ion-signal intensities. Zircon 91,500 and glass NIST610 were used as external standards for trace element calibration. Helium was applied as a carrier gas. Argon was used as the make-up gas and mixed with the carrier gas via a T-connector before entering the ICP. The spot size and frequency of the laser were set to 32 μm and 10 Hz, respectively. Each analysis incorporated a background acquisition of approximately 20–30 s followed by 50 s of data acquisition from the sample. Excel-based software ICPMSDataCal 11.8 was used to perform quantitative calibration for trace element analysis [29].

About 0.1–0.2 g of whole rock powder of each sample was dissolved in digestion bombs with a mixture of double distilled HNO_3, HF, and $HClO_4$. They were then placed in an electric oven and heated to 190 °C for 48 h. Columns of DoweAG50WX8 and HDEHP resin were used successively for the separation and purification of rare earth elements (REEs) and finally for the separation of Nd and Sm by HCl eluant. The Sr-Nd isotopic measurements were performed using the Triton Ti thermal ionization mass spectrometer (TIMS) at the Laboratory of Isotope Geochemistry, Wuhan Center of China Geological Survey. ^{143}Nd/^{144}Nd and ^{87}Sr/^{86}Sr ratios were normalized to ^{143}Nd/^{144}Nd = 0.7219 and ^{87}Sr/^{86}Sr = 8.375209, respectively. Measurements of the La Jolla and SRM NBS987 standards during this course gave average ^{143}Nd/^{144}Nd and ^{87}Sr/^{86}Sr ratios of 0.511847 ± 3 (2σ, n = 25) and 0.710254 ± 8 (2σ, n = 22), respectively. ^{147}Sm/^{144}Nd and ^{87}Rb/^{86}Sr ratios of the samples were calculated using Sm, Nd, Rb and Sr concentrations as measured by the ICP-MS, and their relative uncertainties are ∼0.3% and ∼1%, respectively, based on USGS standard analyses [30].

In situ Hf isotope ratio analysis was conducted using a Neptune Plus MC-ICP-MS (Thermo Fisher Scientific, Dreieich, Germany) in combination with a Geolas HD excimer ArF laser ablation system (Coherent, Göttingen, Germany) at the Wuhan Sample Solution Analytical Technology Co., Ltd., Hubei, China. A single spot ablation mode at a spot size of 44 μm was used, and the energy density of the laser ablation was ~7.0 J·cm^{-2}. Each measurement consisted of 20 s of acquisition of the background signal followed by 50 s of ablation signal acquisition. The detailed operating conditions of the laser ablation system and the MC-ICP-MS instrument and analytical method are the same as described by [31]. The normalized ^{179}Hf/^{177}Hf = 0.7325 and ^{173}Yb/^{171}Yb = 1.132685 were used to calculate the mass bias of Hf (βHf) and Yb (βYb), respectively [32]. The interference of ^{176}Yb on ^{176}Hf was corrected by measuring the interference-free ^{173}Yb isotope and using ^{176}Yb/^{173}Yb = 0.79639 to calculate ^{176}Yb/^{177}Hf [31]. Similarly, the relatively minor interference of ^{176}Lu on ^{176}Hf was corrected by measuring the intensity of the interference-free ^{175}Lu isotope and using the recommended ^{176}Lu/^{175}Lu = 0.02656 to calculate ^{176}Lu/^{177}Hf. Off-line selection and integration of analyte signals and mass bias calibrations were performed using ICPMSDataCal [33]. In order to ensure the reliability of the analysis data, three international zircon standards of Plešovice, 91,500, and GJ-1 were analyzed simultaneously with the actual samples. Plešovice was used for the external standard calibration to further optimize the analysis and test results. 91,500 and GJ-1 were used as the second standard to monitor the quality of data correction. The external precision (2SD) of Plešovice, 91,500, and GJ-1 was better than 0.000020. The test value is consistent with the recommended value within the error range. At the same time, we used the internationally recognized high Yb/Hf ratio standard sample, Temora 2, to monitor the test data of the high Yb/Hf ratio zircon. The Hf isotopic compositions of Plešovice, 91,500, and GJ-1 have been reported by Zhang et al. [34].

4. Results

4.1. Mineral Compositions

4.1.1. K-Feldspar

The K-feldspar crystals of the Bure adakitic rock show relatively uniform compositional variation in the major elements (Table 1), with 11.23–16.66 wt% of K_2O (average value of 15.74 wt%), 0.37–1.20 wt% of Na_2O (average value of 0.62 wt%), and 10.81–19.36 wt% of Al_2O_3 (average value of 18.51 wt%). The low contents of CaO, MgO, TiO_2, and MnO indicate that there is less isomorphism and the formation temperature of K-feldspar is low [35]. The orthoclase (Or) value is high (89.08–96.37), the albite (Ab) value is low (3.57–10.59), and the anorthite (An) value is almost negligible (0–0.38), suggesting that the K-feldspar in this area is orthoclase (Figure 4a). The Or and Al_2O_3 values in the K-feldspar crystal show the same zigzag variation trend from the core to the edge, but the content of the whole porphyry is relatively stable (Figure 5a,b). This shows that the physical and chemical conditions during the formation of the potassium feldspar did not change much.

Table 1. Electron microprobe composition of K-felspar (wt%) for the Bure adakitic rock.

Ele.	Grt8-3-fs01	Grt8-3-fs02	Grt8-3-fs04	Grt8-5-fs01	Grt8-5-fs02	Grt8-5-fs04	Grt8-5-fs05	Grt8-5-fs06	Grt8-5-fs07	Grt8-5-fs08	Grt8-5-fs09	Grt3-1-fs01	Grt3-1-fs02	Grt3-1-fs03	Grt3-1-fs04	Grt3-1-fs05	Grt3-1-fs06	Grt3-5-fs07	Grt3-5-fs08
CaO		0.042	0.022	0.016	0.041		0.042	0.035	0.012	0.011		0.01	0.02	0.03	0.03		0.03	0.03	0.01
Na₂O	0.648	0.629	0.858	0.555	0.462	0.658	0.586	0.537	0.607	0.387	0.507	0.70	0.74	0.68	0.60	0.79	0.84	0.66	0.75
K₂O	16.085	15.769	15.465	16.152	16.66	16.136	15.93	16.026	14.227	15.874	15.892	16.28	16.07	15.74	16.34	15.91	15.61	16.10	15.02
SrO																			
TFeO																			
MgO																			
MnO																			
SiO₂	64.565	63.241	64.184	65.094	63.424	63.986	63.244	63.48	67.125	63.714	66.579	64.10	64.08	63.72	63.11	64.49	65.81	65.34	64.98
Al₂O₃	19.002	18.788	18.486	19.02	19.106	19.144	18.528	19.086	19.045	19.023	17.944	18.65	17.95	18.16	18.24	18.43	18.65	18.83	18.55
BaO																			
Total	100.30	98.47	99.02	100.84	99.69	99.92	98.33	99.16	101.02	99.01	100.92	99.74	98.87	98.34	98.32	99.61	100.94	100.95	99.30

Number of cation on basis of 8 oxygens

Si	5.950	5.938	5.979	5.963	5.909	5.926	5.952	5.923	6.047	5.942	6.070	5.954	6.000	5.987	5.957	5.983	6.003	5.977	6.008
Al	1.548	1.559	1.522	1.540	1.573	1.567	1.541	1.574	1.517	1.568	1.446	1.531	1.486	1.509	1.522	1.511	1.504	1.523	1.516
Mg	0.000	0.000	0.000	0.000	0.000	0.000	0.000	0.000	0.000	0.000	0.000	0.000	0.000	0.000	0.000	0.000	0.000	0.000	0.000
Fe	0.000	0.000	0.000	0.000	0.000	0.000	0.000	0.000	0.000	0.000	0.000	0.000	0.000	0.000	0.000	0.000	0.000	0.000	0.000
Mn	0.000	0.000	0.000	0.001	0.002	0.002	0.002	0.002	0.001	0.001	0.000	0.000	0.001	0.002	0.001	0.000	0.001	0.001	0.000
Ca	0.000	0.002	0.001	0.025	0.021	0.030	0.027	0.024	0.027	0.017	0.022	0.031	0.034	0.031	0.028	0.035	0.037	0.029	0.034
Na	0.029	0.029	0.039	0.025	0.495	0.477	0.478	0.477	0.409	0.472	0.462	0.482	0.480	0.472	0.492	0.471	0.454	0.470	0.443
K	0.473	0.472	0.459	0.472	0.000	0.000	0.000	0.000	0.000	0.000	0.000	0.000	0.000	0.000	0.000	0.000	0.000	0.000	0.000
Sr	0.000	0.000	0.000	0.000	0.000	0.000	0.000	0.000	0.000	0.000	0.000	0.000	0.000	0.000	0.000	0.000	0.000	0.000	0.000
Ba	0.0	0.2	0.1	0.1	0.2	0.0	0.2	0.2	0.1	0.1	0.0	0.0	0.1	0.2	0.1	0.0	0.1	0.1	0.0
An	5.8	5.7	7.8	5.0	4.0	5.8	5.3	4.8	6.1	3.6	4.6	6.1	6.5	6.2	5.3	7.0	7.5	5.8	7.0
Or	94.2	94.1	92.1	95.0	95.8	94.2	94.5	95.0	93.8	96.4	95.4	93.9	93.4	93.7	94.6	93.0	92.3	94.0	92.9

Ele.	Grt3-4-fs01	Grt3-4-fs02	Grt3-4-fs03	Grt3-4-fs04	Grt3-4-fs05	Grt3-4-fs06	Grt3-4-fs07	Grt3-3-fs05	Grt3-3-fs06	Grt3-3-fs07	Grt3-3-fs08	Grt3-3-fs09	Grt3-5-fs01	Grt3-5-fs02	Grt3-5-fs03	Grt3-5-fs04	Grt3-5-fs05	Grt3-5-fs06	Grt3-5-fs07	Grt3-5-fs08
CaO	0.01	0.03	0.01	0.01	0.02	0.02		0.02	0.02	0.03	0.01	0.03	0.07	0.03	0.01	0.03	0.03	0.07	0.03	0.01
Na₂O	0.55	0.51	0.58	0.57	0.61	0.60	0.55	0.52	0.61	0.68	0.51	0.46	1.20	0.70	0.56	0.62	0.84	0.70	0.66	0.75
K₂O	16.19	15.61	16.46	16.29	16.16	16.55	16.19	15.96	16.44	15.78	16.43	16.35	15.37	15.88	16.02	15.91	15.44	15.75	16.10	15.02
SrO																				
TFeO																				
MgO																				
SiO₂	64.79	62.78	64.62	65.35	66.01	65.06	65.05	65.06	65.78	64.51	64.68	64.55	62.67	64.95	64.78	65.80	63.77	62.82	65.34	64.98
MnO																				
Al₂O₃	18.80	18.78	18.75	18.91	18.53	18.53	19.25	18.81	18.81	18.31	18.43	18.55	18.99	19.05	19.08	19.07	18.33	18.93	18.83	18.55
Total	100.34	97.71	100.40	101.13	101.33	100.76	101.04	100.36	101.67	99.31	100.06	99.93	98.30	100.61	100.44	101.43	98.40	98.27	100.95	99.30

Number of cation on basis of 8 oxygens

Si	5.969	5.935	5.961	5.972	6.011	5.981	5.948	5.980	5.983	5.996	5.985	5.977	5.900	5.958	5.955	5.979	5.980	5.916	5.977	6.008
Al	1.531	1.570	1.529	1.527	1.492	1.506	1.556	1.528	1.513	1.504	1.507	1.518	1.580	1.545	1.550	1.531	1.519	1.575	1.523	1.516
Mg	0.000	0.000	0.000	0.000	0.000	0.000	0.000	0.000	0.000	0.000	0.000	0.000	0.000	0.000	0.000	0.000	0.000	0.000	0.000	0.000
Fe	0.000	0.000	0.000	0.000	0.000	0.000	0.000	0.000	0.000	0.000	0.000	0.000	0.000	0.000	0.000	0.000	0.000	0.000	0.000	0.000
Mn	0.000	0.000	0.000	0.001	0.001	0.001	0.000	0.001	0.001	0.001	0.001	0.000	0.003	0.003	0.000	0.001	0.001	0.004	0.001	0.000
Ca	0.025	0.024	0.026	0.025	0.027	0.027	0.024	0.023	0.027	0.031	0.023	0.020	0.055	0.031	0.025	0.027	0.038	0.032	0.029	0.034
Na	0.476	0.470	0.484	0.475	0.469	0.485	0.472	0.468	0.477	0.468	0.485	0.483	0.461	0.465	0.470	0.461	0.462	0.473	0.470	0.443
K	0.000	0.000	0.000	0.000	0.000	0.000	0.000	0.000	0.000	0.000	0.000	0.000	0.000	0.000	0.000	0.000	0.000	0.000	0.000	0.000
Sr	0.000	0.000	0.000	0.000	0.000	0.000	0.000	0.000	0.000	0.000	0.000	0.000	0.000	0.000	0.000	0.000	0.000	0.000	0.000	0.000
Ba	0.0	0.2	0.1	0.1	0.1	0.1	0.1	0.1	0.1	0.1	0.1	0.1	0.3	0.1	0.0	0.1	0.1	0.3	0.1	0.0
An	4.9	4.8	5.0	5.0	5.4	5.2	4.9	4.7	5.3	6.2	4.5	4.1	10.6	6.3	5.0	5.6	7.6	6.3	5.8	7.0
Or	95.0	95.1	94.9	94.9	94.5	94.7	95.1	95.2	94.6	93.7	95.4	95.8	89.1	93.6	95.0	94.3	92.3	93.3	94.0	92.9

Note: Blank space is below the detection limit.

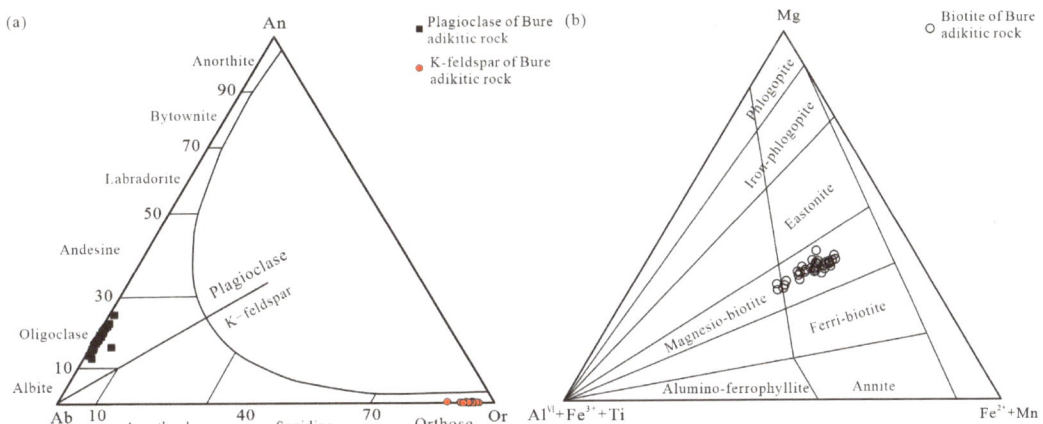

Figure 4. Ternary classification diagram for feldspar (**a**), [36]); Mg–(Al^{VI} + Fe^{3+} + Ti)–(Fe^{2+} + Mn) classification diagram for biotite (**b**), [37]).

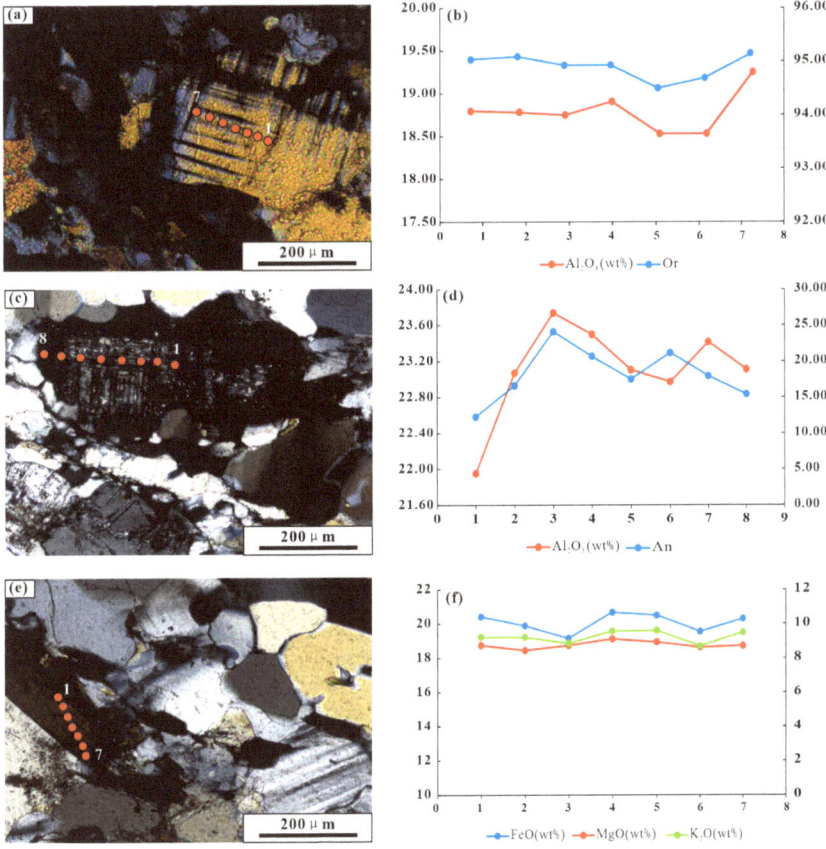

Figure 5. Electron microprobe line profile analysis of K-feldspar (**a**,**b**), plagioclase (**c**,**d**) and biotite (**e**,**f**) for the Bure adakitic rock.

4.1.2. Plagioclase

The major elements in the plagioclase crystals of the Bure adakitic rock show a small range of compositions (Table 2). The SiO_2 content is relatively high, ranging from 62.57 to 67.75 wt% (average value of 65.18 wt%), with small variations of 6.71–10.02 wt% of Na_2O (average value of 8.88 wt%), 0.09–0.74 wt% of K_2O (average value of 0.16 wt%), and 2.52–4.28 wt% of CaO (average value of 3.51 wt%). In addition, the contents of FeO, MnO, and MgO in the plagioclase are below the detection limits. The Ab has high values of 74.63–85.99 (average value = 81.14), while the Or values are almost negligible (0.56–4.87, with an average of 1.00). The An values range from 12.31–24.15, with an average of 17.86. Thus, all the plagioclases of the Bure adakitic rock are macro-feldspar (Figure 4a). In the plagioclase porphyry of the Bure adakitic rock, the content of An and Al_2O_3 has a relatively coupled synchronous change trend (Figure 5c,d). The contents of An are higher in the core and mantle, with an increasing trend from the core to the mantle, and a decreasing trend from the mantle to the edge.

4.1.3. Biotite

The Fe^{2+} and Fe^{3+} in the biotite of the Bure adakitic rock were adjusted using the method proposed by [38], and the number of cations and related parameters of the biotite were calculated using 22 oxygen atoms as the unit. In the major element content of the Bure adakitic rock, there is 35.19–39.67 wt% of SiO_2, with an average value of 37.65 wt%. The biotite has relatively high contents of FeO (16.11–20.99 wt%; average value of 19.12 wt%), Al_2O_3 (13.96–19.13 wt%; average value of 15.77 wt%), and TiO_2 (2.01–3.84 wt%; average value of 2.90 wt%). In comparison, the MgO, K_2O, Na_2O, and CaO contents in biotite are relatively low, with values of 6.54–9.59 wt% of MgO (average value of 8.33 wt%), 7.07–9.85 wt% of K_2O (average value of 8.33 wt%), 0.01–0.16 wt% of Na_2O (average value of 0.08 wt%), and 0.08–0.33 wt% of CaO (average value of 0.16 wt%) (Table 3; Figure 4b).

The low Ca content of the biotite indicates that it was not, or only rarely, affected by chlorite and sericite alteration caused by primary metamorphism after the magmatic stage [39]. In addition, the Ti atomic numbers of the biotite in this study range from 0.13 to 0.22 (mean of 0.17), which is consistent with the fact that the Ti atomic number in the magmatic biotite is less than 0.55. The $Fe^{2+}/(Mg + Fe^{2+})$ ratio in the biotite presents a small variation (0.26–0.52, with an average value of 0.38), also suggesting that the biotite is of magmatic origin. The FeO, MgO, and K_2O contents from the core to the edge of the biotite fluctuate slightly, showing a gentle trend and indicating that there was no mixing of basic magmatic components during crystallization (Figure 5e,f). Generally, the substitution modes of Mg^{2+} and Al^{3+} are crucial in calc-alkaline and peraluminous magmatic systems. The obvious negative correlation of MgO and Al_2O_3 in the biotite implies that the displacement reaction of Mg^{2+} and Al^{3+} may have occurred during the crystallization process of the calc-alkaline and peraluminous magmatic system (Figure 6; [40]).

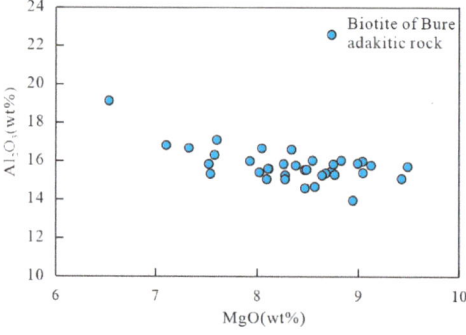

Figure 6. Diagram of the chemical variation of Al_2O_3 vs. MgO in the biotite.

Table 2. Electron microprobe composition of plagioclase (wt%) for the Bure adakitic rock.

	Grt8-1-fs02	Grt8-1-fs03	Grt8-1-fs04	Grt8-1-fs05	Grt8-1-fs06	Grt8-1-fs07	Grt8-1-fs08	Grt8-1-fs09	Grt8-1-fs01	Grt8-1-fs02	Grt8-1-fs03	Grt8-1-fs04	Grt8-1-fs06	Grt8-2-fs01	Grt8-2-fs02	Grt8-2-fs03	Grt8-2-fs04	Grt8-2-fs05	Grt8-2-fs06	Grt8-2-fs07	Grt8-2-fs08
CaO	3.13	3.07	3.28	3.39	3.65	2.68	2.99	2.92	3.65	3.65	3.61	3.26	3.62	2.52	3.54	3.93	4.00	3.75	3.94	3.42	2.78
Na₂O	9.68	9.50	10.02	9.80	7.11	9.69	9.49	9.55	9.52	9.44	9.66	8.40	8.60	9.69	9.72	6.71	8.37	9.63	7.98	8.56	7.96
K₂O	0.12	0.18	0.19	0.10	0.16	0.15	0.17	0.14	0.16	0.13	0.14	0.13	0.09	0.33	0.13	0.17	0.15	0.17	0.18	0.13	0.74
FeO																					
SrO																					
MgO																					
SiO₂	64.99	64.61	65.70	64.23	65.06	65.05	65.12	64.05	64.36	64.72	64.10	65.01	65.56	64.62	65.70	66.41	66.00	65.07	66.07	65.81	64.70
MnO																					
Al₂O₃	22.68	22.89	23.32	23.55	24.11	22.59	23.24	22.46	23.25	23.66	22.67	23.58	24.13	21.95	23.07	23.74	23.50	23.11	22.97	23.41	23.11
BaO																					
Total	100.59	100.25	102.52	101.07	100.10	100.16	101.00	99.13	100.94	101.62	100.17	100.38	102.01	99.10	102.15	100.95	102.01	101.72	101.13	101.33	99.29
								Number of cation on basis of 8 oxygens												
Si	5.686	5.671	5.652	5.607	5.668	5.706	5.668	5.684	5.625	5.615	5.648	5.671	5.638	5.733	5.668	5.725	5.676	5.645	5.720	5.690	5.706
Al	1.754	1.776	1.774	1.817	1.857	1.752	1.788	1.762	1.796	1.815	1.765	1.818	1.834	1.722	1.759	1.809	1.787	1.772	1.758	1.789	1.802
Mg	0.000	0.000	0.000	0.000	0.000	0.000	0.000	0.000	0.000	0.000	0.000	0.000	0.000	0.000	0.000	0.000	0.000	0.000	0.000	0.000	0.000
Fe	0.000	0.000	0.000	0.000	0.000	0.000	0.000	0.000	0.000	0.000	0.000	0.000	0.000	0.000	0.000	0.000	0.000	0.000	0.000	0.000	0.000
Mn	0.147	0.144	0.000	0.158	0.170	0.126	0.139	0.139	0.171	0.170	0.170	0.152	0.167	0.120	0.163	0.181	0.184	0.174	0.182	0.159	0.132
Ca	0.410	0.404	0.418	0.415	0.300	0.412	0.400	0.411	0.403	0.397	0.412	0.355	0.358	0.417	0.406	0.280	0.349	0.405	0.335	0.359	0.340
Na	0.003	0.005	0.005	0.003	0.005	0.004	0.004	0.004	0.004	0.004	0.004	0.004	0.003	0.009	0.003	0.005	0.004	0.005	0.005	0.003	0.021
K	0.000	0.000	0.000	0.000	0.000	0.000	0.000	0.000	0.000	0.000	0.000	0.000	0.000	0.000	0.000	0.000	0.000	0.000	0.000	0.000	0.000
Sr																					
Ba																					
An	15.0	15.0	15.2	15.9	21.9	13.1	14.7	14.3	17.3	17.5	17.0	17.5	18.8	12.3	16.6	24.1	20.7	17.5	21.2	18.0	15.4
Ab	84.2	84.0	83.8	83.5	77.0	86.0	84.3	84.8	81.8	81.8	82.3	81.6	80.6	85.8	82.7	74.6	78.4	81.5	77.7	81.3	79.7
Or	0.7	0.8	1.0	0.6	1.2	0.9	1.0	0.8	0.9	0.8	0.8	0.9	0.6	1.9	0.7	1.2	0.9	0.9	1.1	0.8	4.9

	Grt3-3-fs01	Grt3-3-fs02	Grt3-3-fs03	Grt3-3-fs04	Grt3-6-fs01	Grt3-6-fs02	Grt3-6-fs03	Grt3-6-fs04	Grt3-6-fs06	Grt3-8-fs01	Grt3-8-fs02	Grt3-8-fs03	Grt3-8-fs04	Grt3-8-fs05	Grt3-8-fs06	Grt3-8-fs07
CaO	3.58	3.54	3.33	3.17	3.69	3.52	3.60	3.62	3.59	4.00	3.79	3.88	4.28	3.97	3.87	3.91
Na₂O	8.47	7.73	8.91	9.19	8.83	9.85	7.93	8.26	8.90	8.18	8.89	7.54	9.31	8.90	8.83	9.52
K₂O	0.15	0.12	0.12	0.12	0.18	0.10	0.14	0.16	0.13	0.18	0.16	0.18	0.15	0.12	0.16	0.20
FeO																
SrO																
MgO																
SiO₂	66.55	66.89	67.75	66.29	65.18	64.62	65.97	65.88	66.04	65.32	64.39	65.27	63.68	63.52	63.94	63.39
MnO																
Al₂O₃	23.09	23.42	22.32	22.36	22.92	23.03	23.15	23.14	23.20	23.31	23.22	23.57	23.63	23.21	23.13	23.37
BaO																
Total	101.84	101.70	102.42	101.14	100.79	101.12	100.78	101.07	101.84	100.97	100.46	100.44	101.05	99.72	99.92	100.39
								Number of cation on basis of 8 oxygens								
Si	5.723	5.737	5.789	5.749	5.683	5.639	5.721	5.708	5.692	5.675	5.642	5.682	5.573	5.614	5.634	5.584
Al	1.755	1.776	1.686	1.714	1.766	1.777	1.774	1.772	1.768	1.790	1.798	1.814	1.828	1.814	1.802	1.820
Mg	0.000	0.000	0.000	0.000	0.000	0.000	0.000	0.000	0.000	0.000	0.000	0.000	0.000	0.000	0.000	0.000
Fe	0.000	0.000	0.000	0.000	0.000	0.000	0.000	0.000	0.000	0.000	0.000	0.000	0.000	0.000	0.000	0.000
Mn	0.165	0.162	0.152	0.147	0.172	0.164	0.167	0.168	0.166	0.186	0.178	0.181	0.200	0.188	0.182	0.185
Ca	0.353	0.321	0.369	0.386	0.373	0.417	0.333	0.347	0.372	0.344	0.378	0.318	0.395	0.381	0.377	0.406
Na	0.004	0.003	0.003	0.003	0.005	0.003	0.004	0.005	0.003	0.005	0.004	0.005	0.004	0.003	0.004	0.005
K	0.000	0.000	0.000	0.000	0.000	0.000	0.000	0.000	0.000	0.000	0.000	0.000	0.000	0.000	0.000	0.000
Sr																
Ba																
An	18.8	20.0	17.0	15.9	18.6	16.4	19.9	19.3	18.1	21.1	18.9	21.9	20.1	19.7	19.3	18.3
Ab	80.3	79.2	82.3	83.4	80.4	83.1	79.2	79.7	81.2	77.8	80.2	77.0	79.1	79.7	79.8	80.6
Or	0.9	0.8	0.8	0.7	1.1	0.6	0.9	1.0	0.9	1.1	0.9	1.2	0.8	0.7	0.9	1.1

Note: Blank space is below the detection limit.

Table 3. Electron microprobe composition of biotite (wt%) for the Bure adakitic rock.

	Grt3-1-ms01	Grt3-1-ms02	Grt3-1-ms03	Grt3-1-ms04	Grt3-1-ms05	Grt3-1-ms06	Grt3-1-ms07	Grt3-1-ms08	Grt3-3-ms02	Grt3-3-ms04	Grt3-3-ms06	Grt3-3-ms07	Grt3-3-ms08	Grt3-3-ms09	Grt3-3-ms10	Grt3-3-ms11	Grt3-3-ms12	Grt3-3-ms13	Grt3-6-ms01	Grt3-6-ms02	Grt3-6-ms03	Grt3-6-ms04
SiO_2	37.512	36.417	37.188	36.342	37.424	37.667	36.166	36.344	38.646	37.178	37.884	37.184	38.548	35.189	37.295	36.984	35.891	37.385	38.104	38.663	37.11	37.509
TiO_2	2.547	2.994	2.657	2.731	2.663	2.574	2.866	2.399	2.314	2.645	3.06	3.024	2.596	2.058	2.673	2.73	2.919	2.736	3.586	3.817	3.479	3.536
Al_2O_3	15.299	15.548	15.843	15.789	13.959	15.272	15.474	15.725	15.99	15.1	15.565	15.06	16.669	19.128	16.605	15.403	14.672	15.376	16.037	15.856	15.883	16.038
FeO	20.427	19.903	19.169	20.682	20.507	20.307	20.37	20.457	18.94	18.25	19.449	20.196	18.327	19.128	19.864	20.587	19.835	20.987	19.921	18.693	20.427	19.627
MnO	0.226	0.218	0.214	0.213	0.28	0.222	0.252	0.224	0.233	0.19	0.251	0.267	0.218	0.181	0.27	0.249	0.283	0.262	0.252	0.211	0.198	0.162
MgO	8.765	8.47	8.752	9.126	8.945	8.639	8.734	9.49	9.04	9.43	8.488	8.277	8.045	6.535	8.341	9.045	8.568	8.675	8.546	8.262	8.993	8.829
CaO	0.118	0.18	0.182	0.143	0.139	0.265	0.1	0.193	0.31	0.332	0.129	0.316	0.273	0.18	0.082	0.142	0.29	0.202	0.276	0.211	0.135	0.081
Na_2O	0.097	0.054	0.085	0.096	0.089	0.065	0.089	0.065	0.078	0.089	0.095	0.081	0.108	0.033	0.076	0.11	0.159	0.121	0.104	0.073	0.078	0.078
K_2O	9.25	9.239	8.879	9.583	9.627	8.713	9.516	9.699	9.025	8.992	9.404	9.195	8.297	7.696	9.059	9.231	8.835	9.448	9.141	9.302	9.852	9.425
Total	94.241	93.023	92.969	94.705	93.633	92.977	93.504	94.596	94.576	92.206	94.325	93.6	93.081	87.111	94.265	94.481	91.452	95.192	95.967	94.958	96.155	95.285
										Number of cation on basis of 22 oxygens												
Si	2.833	3.458	2.898	2.853	2.889	2.812	2.927	2.927	2.832	2.815	2.937	2.907	2.909	2.895	2.952	2.851	2.865	2.858	2.865	2.873	2.875	2.927
Al^{IV}	1.167	0.542	1.102	1.147	1.111	1.188	1.073	1.073	1.168	1.185	1.063	1.093	1.091	1.105	1.048	1.149	1.135	1.142	1.135	1.127	1.125	1.073
Al^{VI}	0.028	1.648	0.292	0.288	0.34	0.252	0.214	0.326	0.26	0.251	0.37	0.298	0.318	0.277	0.456	0.677	0.369	0.26	0.245	0.266	0.301	0.341
Ti	0.129	0	0.148	0.176	0.155	0.159	0.157	0.151	0.169	0.14	0.132	0.156	0.177	0.177	0.15	0.125	0.154	0.159	0.175	0.158	0.204	0.217
Fe^{3+}	0.129	0.124	0.28	0.281	0.323	0.211	0.24	0.341	0.233	0.189	0.343	0.303	0.322	0.3	0.44	0.489	0.322	0.254	0.268	0.255	0.344	0.397
Fe^{2+}	0.638	0	1.04	1.023	0.922	1.127	1.101	0.931	1.097	1.136	0.861	0.89	0.927	1.015	0.734	0.603	0.954	1.076	1.056	1.093	0.913	0.787
Mn	0.003	0	0.015	0.014	0.014	0.014	0.019	0.015	0.017	0.015	0.015	0.013	0.016	0.018	0.014	0.012	0.018	0.016	0.019	0.017	0.016	0.014
Mg	1.841	0.211	1.01	0.989	1.014	1.053	1.043	1.001	1.02	1.096	1.024	1.099	0.972	0.961	0.918	0.789	0.955	1.042	1.02	0.994	0.961	0.932
Ca	0.025	0	0.01	0.015	0.015	0.012	0.012	0.022	0.008	0.016	0.025	0.028	0.011	0.026	0.022	0.016	0.007	0.012	0.025	0.017	0.022	0.007
Na	0.032	0.039	0.015	0.008	0.013	0.014	0.013	0.01	0.014	0.016	0.011	0.013	0.014	0.012	0.016	0.005	0.011	0.016	0.025	0.018	0.015	0.011
K	0.913	0.888	0.912	0.923	0.88	0.946	0.961	0.864	0.951	0.959	0.875	0.897	0.921	0.913	0.81	0.795	0.888	0.91	0.9	0.926	0.88	0.898

	Grt8-1-ms01	Grt8-1-ms02	Grt8-1-ms03	Grt8-1-ms04	Grt8-1-ms05	Grt8-1-ms06	Grt8-1-ms07	Grt8-1-ms08	Grt8-3-ms01	Grt8-3-ms02	Grt8-3-ms03	Grt8-3-ms04	Grt8-3-ms05	Grt8-3-ms06	Grt8-5-ms01	Grt8-5-ms02	Grt8-5-ms03	Grt8-5-ms04	Grt8-5-ms05	Grt8-5-ms06	Grt8-5-ms07	Grt8-5-ms08
SiO_2	46.568	47.913	45.736	46.328	45.586	46.436	47.622	47.476	36.007	37.312	38.717	39.669	38.79	39.312	38.683	37.77	37.651	39.147	38.832	38.474	38.011	38.31
TiO_2	1.106	1.021	1.089	0.988	1.444	0.712	0.435	0.41	2.567	2.603	2.652	2.43	2.555	2.708	2.49	3.608	2.887	3.377	3.615	2.806	3.821	3.836
Al_2O_3	29.682	29.113	29.725	29.342	29.037	29.447	30.081	30.942	14.599	15.062	15.274	16.682	15.338	16.827	17.12	15.606	15.604	15.778	15.419	16	16.315	15.844
FeO	4.451	5.048	4.95	5.009	4.875	5.623	5.254	5.515	18.37	18.689	18.977	17.017	17.374	17.353	18.587	19.489	20.461	17.878	18.553	17.535	17.189	18.617
MnO	0.046	0.016	0.072	0.036	0.038	0.048	0.019	0.064	0.064	0.314	0.234	0.228	0.305	0.223	0.301	0.306	0.283	0.266	0.287	0.236	0.2	0.217
MgO	1.586	1.746	1.411	1.418	1.519	1.659	1.582	1.662	8.473	8.096	8.276	7.325	7.537	7.104	7.598	8.116	8.108	8.382	8.021	7.928	7.576	7.522
CaO	0.005	0.008	0.023	0.019				0.005	0.299	0.13	0.081	0.105	0.076	0.133	0.118	0.086	0.168	0.096	0.123	0.102	0.134	0.141
Na_2O	0.189	0.13	0.173	0.171	0.168	0.145	0.083	0.121	0.088	0.082	0.079	0.061	0.054	0.042	0.056	0.04	0.07	0.082	0.072	0.053	0.014	0.014
K_2O	11.513	11.665	11.654	11.521	11.346	11.47	11.491	11.355	8.591	9.037	8.939	8.219	8.931	7.072	8.933	9.389	9.161	9.474	8.885	8.665	8.48	9.057
Total	95.146	96.66	94.833	94.832	94.013	95.54	96.567	97.55	89.122	91.325	93.229	91.736	90.96	90.774	93.886	94.41	94.393	94.48	93.807	91.799	91.74	93.683
										Number of cation on basis of 22 oxygens												
Si	3.183	3.229	3.152	3.187	3.165	3.179	3.209	3.169	2.920	2.950	2.985	3.046	3.039	3.036	2.947	2.899	2.901	2.966	2.967	2.984	2.944	2.938
Al^{IV}	0.817	0.771	0.848	0.813	0.835	0.821	0.791	0.831	1.080	1.050	1.015	0.954	0.961	0.964	1.053	1.101	1.099	1.034	1.033	1.016	1.056	1.062
Al^{VI}	1.574	1.541	1.567	1.567	1.54	1.555	1.597	1.604	0.316	0.354	0.372	0.556	0.456	0.568	0.484	0.31	0.318	0.375	0.356	0.447	0.433	0.370
Ti	0.057	0.052	0.056	0.051	0.075	0.037	0.022	0.021	0.157	0.155	0.154	0.14	0.151	0.157	0.143	0.208	0.167	0.193	0.208	0.164	0.223	0.221
Fe^{3+}	0.254	0.284	0.285	0.288	0.283	0.322	0.296	0.308	0.323	0.345	0.387	0.534	0.448	0.608	0.42	0.35	0.321	0.399	0.431	0.446	0.491	0.422
Fe^{2+}	0	0	0	0	0	0	0	0	0.923	0.891	0.837	0.558	0.691	0.513	0.764	0.901	0.997	0.733	0.755	0.691	0.622	0.772
Mn	0.003	0.001	0.004	0.002	0	0.003	0.001	0.004	0.021	0.021	0.015	0.015	0.02	0.015	0.019	0.02	0.018	0.017	0.019	0.016	0.013	0.014
Mg	0.162	0.175	0.145	0.145	0.157	0.169	0.159	0.165	1.024	0.954	0.951	0.839	0.88	0.818	0.863	0.929	0.931	0.947	0.914	0.917	0.875	0.860
Ca	0	0.001	0.002	0.001	0	0	0	0	0.011	0.011	0.007	0.009	0.006	0.011	0.01	0.007	0.014	0.008	0.01	0.008	0.011	0.012
Na	0.025	0.017	0.023	0.023	0.023	0.019	0.011	0.016	0.014	0.013	0.012	0.009	0.008	0.006	0.008	0.006	0.01	0.012	0.011	0.008	0.002	0.021
K	1.004	1.003	1.025	1.011	1.005	1.002	0.988	0.967	0.889	0.912	0.879	0.805	0.893	0.697	0.868	0.919	0.9	0.916	0.866	0.857	0.838	0.886

Note: Blank space is below the detection limit.

4.2. Trace Element Compositions of Zircon

The zircon trace elements and calculated oxygen fugacity parameters from the Bure adakitic rock are shown in Table 4, respectively. They are depleted in LREEs and enriched in HREEs, with significant positive Ce anomalies and weak negative Eu anomalies in the chondrite-normalized REE patterns (Figure 7), indicating that they are magmatic zircons [41]. The magmatic crystallization temperatures of the Bure adakitic rock calculated based on Ti-in-zircon thermometry [42] vary from 659 to 814 °C (mean of 705 °C). The corresponding logfO_2 values of the zircons from the Bure adakitic rock range from −11.5 to −5.2, with a median of −8.6 [43–45].

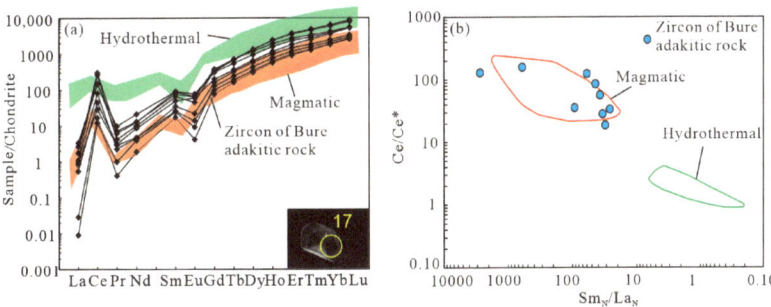

Figure 7. Chondrite-normalized REE patterns (**a**) and Ce/Ce* vs. Sm_N/La_N (**b**); [41]). The Chondrite data for the normalization and plotting are from [46].

Table 4. Trace element compositions of zircon (ppm) for the Bure adakitic rock.

No.	La	Ce	Pr	Nd	Sm	Eu	Gd	Tb	Dy	Ho	Er	Tm	Yb	Lu	Y	Ti	t/°C [42]	(Ce/Ce*)$_D$ [43]	logfO_2 [45]	ΔFMQ [45]
BR0101 Grt1-05	0.01	10.39	0.04	0.87	2.66	0.24	15.51	6.08	77.30	32.03	150.74	32.78	307.81	63.94	957.16	4.49	675.1	383.1	−9.6	7.9
BR0101 Grt1-06	0.22	49.94	0.21	2.13	5.36	1.27	32.86	11.94	146.11	55.28	240.64	46.74	413.50	84.28	1513.41	3.63	659.0	560.7	−9.1	8.9
BR0101 Grt1-07	0.13	27.75	0.29	3.99	7.62	3.49	40.70	13.41	163.41	64.89	299.34	64.48	615.10	132.45	1890.88	21.77	814.2	190.1	−5.2	9.0
BR0101 Grt1-15	0.63	88.37	0.94	10.03	13.87	4.47	62.97	20.56	239.48	89.35	382.39	74.31	658.28	135.13	2380.80	8.13	723.0	177.5	−9.8	6.4
BR0101 Grt1-17	0.26	161.18	0.39	4.73	9.97	2.61	71.27	26.37	336.42	132.64	588.23	115.67	1005.19	199.90	3637.32	3.06	646.3	971.1	−7.8	10.5
BR0101 Grt1-26	0.41	184.47	0.69	6.19	11.69	3.74	66.89	25.61	325.23	134.75	609.07	121.47	1079.09	217.81	3838.12	5.38	689.1	874.7	−5.7	11.5
BR0101 Grt1-32	0.20	17.83	0.27	2.20	3.99	0.80	20.21	7.06	86.07	34.79	161.39	33.70	315.24	68.03	1017.32	8.03	721.8	224.5	−9.0	7.3
BR0101 Grt1-35	0.00	7.49	0.09	1.79	3.76	0.53	25.91	8.91	110.30	42.79	192.76	39.88	359.63	74.73	1224.50	7.90	720.5	116.6	−11.5	4.8
BR0101 Grt1-39	0.80	93.02	0.58	6.05	13.51	3.77	77.29	26.39	298.01	108.37	448.54	84.63	703.25	137.54	2797.50	5.60	692.3	284.1	−9.7	7.3

4.3. Zircon Lu-Hf Isotopes and Whole-Rock Sr-Nd Isotopes

Ten Lu-Hf isotopic analyses were conducted on the zircons of the Bure adakitic rock sample, yielding $^{176}Hf/^{177}Hf$ ratios of 0.282572~0.282734, and $\varepsilon_{Hf}(t)$ values from 7.64 to 12.99 (average value = 11; Table 5). On the Age-$\varepsilon_{Hf}(t)$ diagram, the corresponding two-stage Hf model ages vary from 802–1161 Ma (Figure 8a). The Sr-Nd isotopic results of the Bure adakitic rock are shown in Table 6. The $^{87}Sr/^{86}Sr$ ratios ranging from 0.707381 to 0.70745 (average value = 0.70741) are higher than that of the current original mantle value ($^{87}Sr/^{86}Sr$ = 0.7045; Table 6). Correspondingly, the calculated ($^{87}Sr/^{86}Sr$)i ratios vary from 0.70088 to 0.70275 (average value = 0.70184), and the $\varepsilon_{Nd}(t)$ values have a relatively large variation of 3.26 to 7.28 (average value = 4.72; Figure 8b). Their two-stage Nd model ages range from 820 to 1210 Ma.

Table 5. Zircon Hf isotopic data for the Bure adakitic rock.

No.	^{176}Hf/^{177}Hf	1σ	^{176}Lu/^{177}Hf	1σ	^{176}Yb/^{177}Hf	1σ	Age (Ma) [7]	ε$_{Hf}$(t) [47]	TDM$_2$ (Ma)
BR0101Grt1-02	0.282734	0.000020	0.001766	0.000051	0.064510	0.001820	660	12.44	803
BR0101Grt1-05	0.282623	0.000013	0.000892	0.000018	0.029611	0.000544	753	10.91	974
BR0101Grt1-07	0.282680	0.000021	0.001743	0.000019	0.055813	0.000457	743	12.29	877
BR0101Grt1-12	0.282702	0.000017	0.002429	0.000022	0.070191	0.000389	743	12.73	849
BR0101Grt1-17	0.282599	0.000020	0.003626	0.000051	0.129672	0.002161	744	8.50	1122
BR0101Grt1-21	0.282689	0.000022	0.002049	0.000063	0.067411	0.002036	696	11.48	893
BR0101Grt1-26	0.282718	0.000021	0.003151	0.000071	0.109087	0.002564	746	12.99	834
BR0101GRT1-10	0.282631	0.000019	0.002470	0.000074	0.056882	0.001625	715	9.63	1027
BR0101GRT1-15	0.282645	0.000020	0.002279	0.000038	0.060088	0.001038	728	10.46	983
BR0101GRT1-22	0.282667	0.000017	0.000706	0.000009	0.017882	0.000247	729	12.02	883
BR0101GRT1-28	0.282657	0.000022	0.001282	0.000020	0.033778	0.000571	735	11.54	919
BR0101GRT1-31	0.282644	0.000021	0.001278	0.000002	0.034658	0.000082	713	10.61	962
BR0101GRT1-32	0.282620	0.000023	0.000766	0.000016	0.020989	0.000557	731	10.38	991
BR0101GRT1-35	0.282644	0.000020	0.001313	0.000004	0.033838	0.000049	753	11.44	940
BR0101GRT1-39	0.282572	0.000029	0.002650	0.000025	0.072997	0.000613	724	7.64	1161

Table 6. Sr–Nd isotopic data for the Bure adakitic rock.

Sample No.	^{87}Rb/^{86}Sr	^{87}Sr/^{86}Sr	±2σ	(^{87}Sr/^{86}Sr)$_i$	^{147}Sm/^{144}Nd	^{143}Nd/^{144}Nd	±2σ	ε$_{Nd}$(t)	TDM$_2$ (Ma)
BR0101Grt1	0.523	0.707381	0.000006	0.701898	0.122	0.512463	0.000009	3.62	1140
BR0101Grt2	0.623	0.707409	0.000006	0.700880	0.115	0.512618	0.000007	7.28	820
BR0101Grt3	0.449	0.707450	0.000010	0.702746	0.130	0.512485	0.000007	3.26	1210

Note: ε$_{Nd}$(t) = [(^{143}Nd/^{144}Nd)$_{sample(t)}$/(^{143}Nd/^{144}Nd)$_{CHUR(t)}$ − 1] × 10$^{−4}$; T$_{DM2}$ = 1/λ × {1 + [(^{143}Nd/^{144}Nd)$_{sample}$ − ((^{147}Sm/^{144}Nd)$_{sample}$ − (^{147}Sm/^{144}Nd)$_{crust}$) × (e^{λt}−1) − (^{143}Nd/^{144}Nd)$_{DM}$]/((^{147}Sm/^{144}Nd)$_{crust}$ − (^{147}Sm/^{144}Nd)$_{DM}$)}. (^{147}Sm/^{144}Nd)$_{CHUR}$ = 0.1967, and (^{143}Nd/^{144}Nd)$_{CHUR}$ = 0.512638 [48]; (^{147}Sm/^{144}Nd)$_{DM}$ = 0.2136, and (^{143}Nd/^{144}Nd)$_{DM}$ = 0.51315 [49]; (^{147}Sm/^{144}Nd)$_{crust}$ = 0.118 [50].

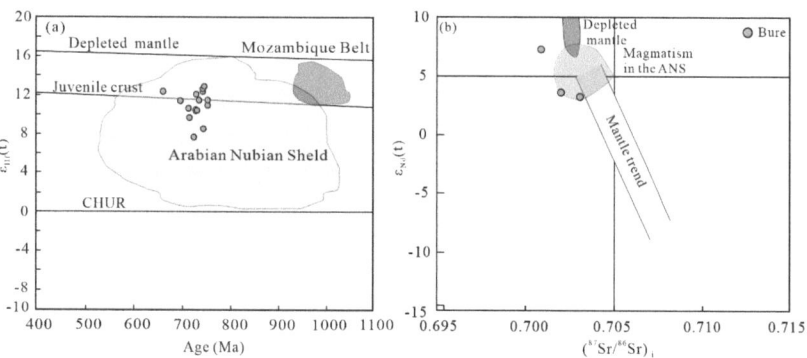

Figure 8. Diagrams of Hf (**a**) and Sr-Nd isotopes (**b**) for the Bure adakitic rock. Zircon Hf isotope-age data obtained from the Arabian Nubian Shield [51]; Mozambique Belt [52]; ranges for depleted mantle (DM), chondritic uniform reservoir (CHUR), and juvenile crust from Griffin et al. [53]. Sr-Nd isotopic data of the Depleted Mantle [54] and the Arabian Nubian Shield [10,28,55].

5. Discussion

5.1. Physicochemical Condition of Magma Crystallization

Zircon, a mineral that typically crystallizes early in acidic magma, usually at temperatures close to the magma formation temperature, serves as an indicator of initial crystallization in granitoids. Thus, the magmatic crystallization temperature of the Bure adakitic rock calculated based on Ti-in-zircon thermometry varies from 659 to 814 °C, with a mean of 705 °C. In conclusion, we propose that the crystallization temperature of the Bure adakitic rock was concentrated between 659 to 814 °C.

Emplacement pressure can be estimated from biotite compositions using the empirical formula of the biotite all-aluminum manometer in granitoids based on the hornblende manometer: $p \times 100 = 3.03 \times T^{Al} - 6.53 (\pm 0.33)$ [56]. The estimated pressures show a range from 1.75×10^5 to 2.81×10^5 Pa (mean 2.09×10^5 Pa) for the Bure adakitic rock. The calculated emplacement depth of the Bure adakitic rock is 6.39~10.2 km (mean 7.60 km) according to the empirical formula $p = \rho g h$ (ρ =2800 kg/m^3; g = 9.8 m/s^2), which indicates that the magmatic emplacement depth was relatively deep.

Generally, the Fe^{3+}, Fe^{2+}, and Mg^{2+} values in biotite can be used to estimate the oxygen fugacity during crystallization. The electron probe data of the biotite in the Bure adakitic rock projected into the correlation diagram of biotite composition and oxygen buffer pairs show that all the data fall between the Ni-NiO and Fe_2O_3-Fe_3O_4 buffer lines and all are close to the Ni-NiO buffer lines, implying that the biotite in the Bure adakitic rock crystallized in a high oxygen fugacity environment (Figure 9). The presence of the variable valence elements of Ce and Eu in zircon makes it an ideal candidate for calculating the oxygen fugacity in coexisting magmas [42]. Unlike most rare earth elements, which exist in the +3 valence, the Ce element can exist in the form of Ce^{4+} in magmas. The similar radius of Ce^{4+} and Zr^{4+} leads to Ce^{4+} being more likely than Ce^{3+} to enter the zircon lattice due to isomorphism. Thus, Ballard et al. [43] proposed that the positive Ce anomaly of zircon can reflect the oxidation state in magma. Most of the points of the Bure adakitic rock are in the FMQ–HM range, and nearly half of the calculated zircon points reach the magmatic oxygen fugacity level of MH, suggesting a high oxygen fugacity of the magma (Figure 10a,b).

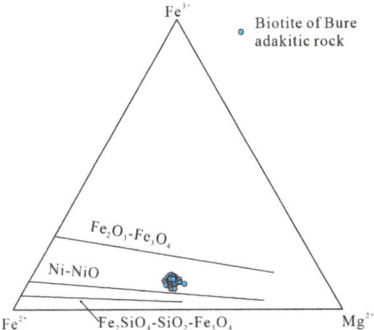

Figure 9. Correlative diagram between biotite composition and oxygen buffer-reagents [57].

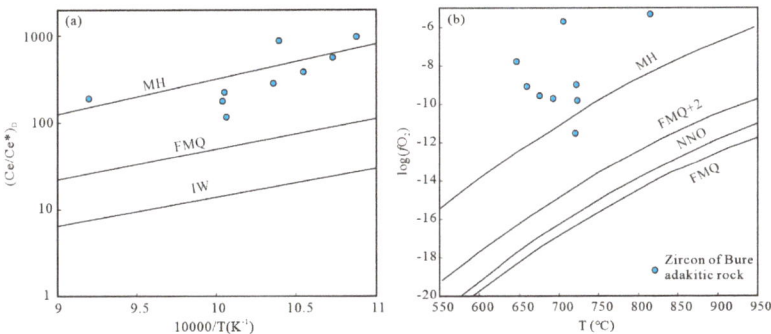

Figure 10. (Ce/Ce*)$_D$ of the zircons vs. 10,000/T (**a**); [58]) and logfO_2 vs. T (**b**); [59]) diagrams for the Bure adakitic rock.

5.2. Magma Source and Genesis

The relationship between the MB and ANS, collectively referred to as the EAO by Stern [4], is not well understood. The inherited zircons of Mesoproterozoic age reported from the different granitic populations in the contrasting low- and high-grade terranes by Kebede et al. [8,9] indicate a contribution of pre-Neoproterozoic crustal material to the source magmas of these rocks. In eastern Ethiopia, Teklay et al. [60] suggested pre-Neoproterozoic crustal reworking based on Paleoproterozoic zircon inheritance and Mesoproterozoic to Archean crust residence ages for the granitoids. Kröner and Sassi [61] also reported a Mesoproterozoic to Paleoproterozoic crystalline basement intruded by Neoproterozoic granitoids in northern Somalia. Farther north in the ANS, studies [62–64] rule out the involvement of pre-Neoproterozoic crust. These studies seem to indicate the increasing importance of pre-Neoproterozoic crust southwards in the EAO, but detailed and systematic investigations are necessary to fully understand the issue.

As mentioned above, the biotite in the Bure adakitic rock enriched in iron and aluminum [7], together with the major elements plotted onto the MgO–FeO–Al$_2$O$_3$ and TFeO/(TFeO + MgO)–MgO diagrams, suggest that the rock is a calc-alkaline orogenic granite (Figure 11a), with a crustal magmatic source affinity (Figure 11b). The positive $\varepsilon_{Hf}(t)$ values > 7 (ranging from 7.64 to 12.99) of the Bure adakitic rock fall above the Hf isotope evolution line of the chondrites, and completely fall into the ANS area [51], implying generation from a juvenile source. The Sr-Nd isotope results show that the Bure adakitic rock has low (^{87}Sr/^{86}Sr)i values of 0.70088–0.70275 and positive $\varepsilon_{Nd}(t)$ values of 3.26 to 7.28, suggesting that the rock was sourced from a juvenile crust rather than lithospheric mantle material [54]. The (^{87}Sr/^{86}Sr)$_i$-$\varepsilon_{Nd}(t)$ map shows that the Bure adakitic rock is consistent with the magmatic rocks in the ANS [10,28,55], which further indicates that the magma was derived from a juvenile crust. Although the Nd isotope depleted mantle model age of 820 Ma to 1210 Ma (average age = 1060 Ma) of the Bure adakitic rock is older than that of the crystallization age of 733.8 Ma [7], it is obviously younger than the Mesoproterozoic and Archaean ancient crust. This result further demonstrates that the Arab-Nubian Shield in the Neoproterozoic was characterized by a juvenile crust. The mean Nd model age for the WES is 1.03 Ga, which is between those calculated by Stern [22] based on existing Nd isotopic data from northern Ethiopia and Eritrea (mean value of 0.87 Ga; [22,55,65]) and the Southern Ethiopia Shield (1.13 Ga), respectively. This indicates that the transition between northern and southern Ethiopia lies in the Western Ethiopia Shield, reflecting a gradual transition between the northern ANS and the southern MB of the EAO. Additionally, the mean Nd model ages from the northern parts (Egypt, Sudan, Arabia Shield, and Eritrea and northern Ethiopia) to the central parts (western Ethiopia shield) and southern parts (southern Ethiopia shield, Kenya) of the EAO show an increasing trend, which indicates an increasing contribution of pre-Pan-African crust towards the southern part of the EAO (Figure 12).

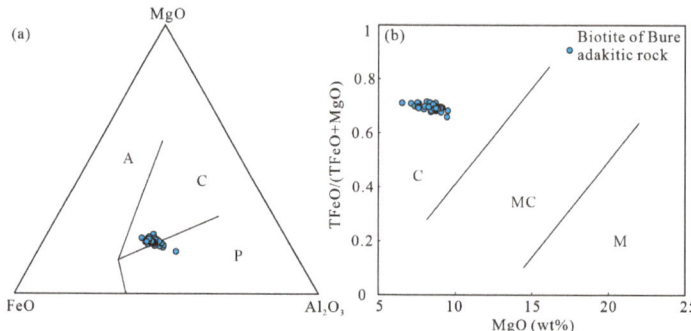

Figure 11. MgO–FeO–Al$_2$O$_3$ discrimination diagram of the tectonic setting (**a**); [66]) and TFeO/(TFeO + MgO) vs. MgO diagram (**b**); [67]) of biotite. A: anorogenic alkaline suites; C: calc-alkaline orogenic suites; P: peraluminous suites; C: crustal source; M: mixing source between crust and mantle; M: mantle source.

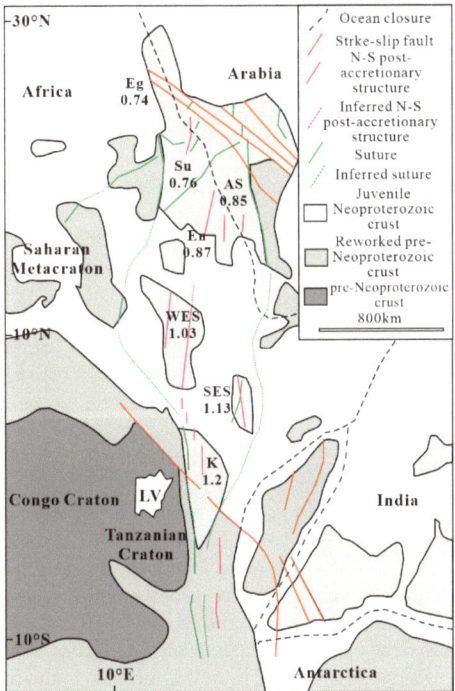

Figure 12. The mean Nd-model ages of the EAO in Africa [22]. Eg—Egypt; Su—Sudan; As—Arabian Shield; En—Eritrea and northern Ethiopia; SES—Southern Ethiopia Shield; K—Kenya.

5.3. Tectonic Environment

The plagioclase in the Bure adakitic rock shows no distinct zonal structure, indicating that the magma chamber was almost undisturbed, and the original molten slurry was in a balanced crystalline environment. In general, the crystallized minerals from the molten slurry easily reacted with the melt to form a uniform composition of minerals, leading to no zonal characteristics in the crystallized minerals. In the Lu/Hf-Y and Yb-Y diagrams of zircon, the trace elements of zircon from the Bure adakitic rock fall into the volcanic arc environment (VAB) and the area towards the within plate environment (WPB; Figure 13a,b). As mentioned above, the zircon U-Pb age of 750~710 Ma from the Bure adakitic rock [7] corresponds to the tectono–thermal event of approximately 780–700 Ma measured in previous studies of other locations in the ANS. This suggests a syn-tectonic environment [5,8–10,22]. In addition, the high SiO_2 (72.26–72.78 wt%), Al_2O_3 (14.91–15.82 wt%), Sr (310–401 ppm), Sr/Y (64.9–113.6), and La/Yb (25.7–51.6), low MgO (0.27–0.41 wt%), Y (2.71–4.78 ppm), and Yb (0.20–0.31 ppm), and Na_2O/K_2O values of 1.13–1.38 [7] of the Bure adakitic rock suggest that it was mainly formed by the partial melting of a thickened juvenile lower crust. Consequently, we propose that the Bure adakitic rock is the product of thickened juvenile crust melting triggered by the Pan-African Orogeny during the Neoproterozoic [68].

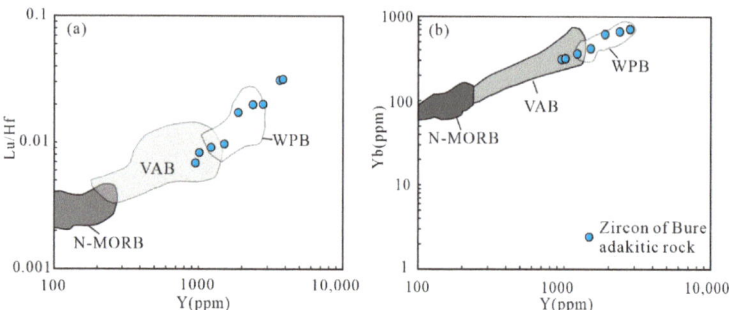

Figure 13. Lu/Hf vs. Y (**a**) and Yb vs. Y (**b**) diagrams of zircons [69] for the Bure adakitic rock. N-MORB: normal mid-ocean ridge basalt; VAB: volcanic arc basalt; WPB: within-plate basalt.

6. Conclusions

The petrological, mineralogical, and geochemical features of the Bure adakitic rock lead to the following conclusions:

(1) The crystallization temperature of the Bure adakitic rock ranges from 659 to 814 °C, and its calculated emplacement depth was 6.39~10.2 km (average of 7.60 km). The Fe^{3+}, Fe^{2+}, and Mg^{2+} values of biotite, and the positive Ce anomaly and calculated magmatic oxygen fugacity values of zircon reveal a high oxygen fugacity of the magma.

(2) The major elements of biotite and the Sr-Nd-Hf isotopes indicate that the Bure adakitic rock was derived from juvenile crustal materials. Additionally, the mean Nd model ages progressively increase from the northern to the central and southern parts of the EAO, which indicates an increasing contribution of the pre-Pan-African crust towards the southern part of the EAO.

(3) The Bure adakitic rock is the product of thickened juvenile crust melting triggered by the Pan-African Orogeny during the Neoproterozoic.

Author Contributions: J.J., W.X. and P.H. conceived this contribution and conducted all field and analytical work, assisted by Y.L., F.W., G.Z., X.G., Z.Z. and Y.B. The manuscript was written by J.J., W.X. and P.H., with contribution from Y.L., F.W., G.Z., X.G., Z.Z. and Y.B. All authors have read and agreed to the published version of the manuscript.

Funding: J.S.J. acknowledges support from the Natural Science Foundation of Hubei Province of China (2022CFB850), National Natural Science Foundation of China (42202092), Open Fund of the Research Center for Petrogenesis and Mineralization of Granitoid Rocks, China Geological Survey (No. PMGR202018), and the China Geological Survey (DD20221802, DD20230575).

Data Availability Statement: Data are contained within the article.

Acknowledgments: Staff at Wuhan Sample Solution Analytical Technology Co., Ltd., Zhongnan Mineral Resources Supervision and Test Center for Geoanalysis, Wuhan Center, China Geological Survey are gratefully acknowledged for assistance with instrument operation. We thank the reviewers for the journal Minerals.

Conflicts of Interest: Zicheng Zhang is employee of China National Geological & Mining Corporation. The paper reflects the views of the scientists and not the company. The remaining authors declare that the research was conducted in the absence of any commercial or financial relationships that could be construed as a potential conflict of interest.

References

1. Collins, A.S.; Pisarevsky, S.A. Amalgamating eastern Gondwana: The evolution of the Circum-Indian Orogens. *Earth-Sci. Rev.* **2005**, *71*, 229–270. [CrossRef]
2. Fritz, H.; Abdelsalam, M.; Ali, K.A.; Bingen, B.; Collins, A.S.; Fowler, A.R.; Ghebreab, W.; Hauzenberger, C.A.; Johnson, P.R.; Kusky, T.M.; et al. Orogen styles in the East African Orogen: A review of the Neoproterozoic to Cambrian tectonic evolution. *J. Afr. Earth Sci.* **2013**, *86*, 65–106. [CrossRef] [PubMed]

3. Stern, R.J. Arc-assembly and continental collision in the Neoproterozoic African Orogen: Implications for the consolidation of Gondwanaland. *Annu. Rev. Earth Planet. Sci.* **1994**, *22*, 319–351. [CrossRef]
4. Abdelsalam, M.; Stern, R. Sutures and shear zones in the Arabian-Nubian Shield. *J. Afr. Earth Sci.* **1996**, *23*, 289–310. [CrossRef]
5. Ayalew, T.; Peccerillo, A. Petrology and geochemistry of the Gore-Gambella plutonic rocks: Implications for magma genesis and the tectonic setting of the Pan-African Orogenic Belt of western Ethiopia. *J. Afr. Earth Sci.* **1998**, *27*, 397–416. [CrossRef]
6. Bowden, S.; Gani, N.D.; Alemu, T.; O'Sullivan, P.; Abebe, B.; Tadesse, K. Evolution of the Western Ethiopian Shield revealed through U-Pb geochronology, petrogenesis, and geochemistry of syn- and post-tectonic intrusive rocks. *Precambrian Res.* **2020**, *338*, 105588. [CrossRef]
7. Jiang, J.S.; Hu, P.; Xiang, W.S.; Wang, J.X.; Lei, Y.J.; Zhao, K.; Zeng, G.P.; Wu, F.F.; Xiang, P. Geochronology, geochemistry and its implication for regional tectonic evolution of adakite-like rock in the Bure area, western Ethiopia. *Acta Geol. Sin.-Engl.* **2021**, *95*, 1260–1272. (In Chinese with English Abstract)
8. Kebede, T.; Koeberl, C.; Koller, F. Geology, geochemistry and petrogenesis of intrusive rocks of the Wallagga area, western Ethiopia. *J. Afr. Earth Sci.* **1999**, *29*, 715–734. [CrossRef]
9. Kebede, T.; Koeberl, C.; Koller, F. Magmatic evolution of the suqii-wagga garnet-bearing two-mica granite, wallagga area, western Ethiopia. *J. Afr. Earth Sci.* **2001**, *32*, 193–221. [CrossRef]
10. Woldemichael, B.W.; Kimura, J.; Dunkley, D.J.; Tani, K.; Ohira, H. SHRIMP U–Pb zircon geochronology and Sr-Nd isotopic systematic of the Neoproterozoic Ghimbi-Nedjo mafic to intermediate intrusions of Western Ethiopia: A record of passive margin magmatism at 855 Ma? *Int. J. Earth Sci.* **2010**, *99*, 1773–1790. [CrossRef]
11. Xiang, W.S.; Jiang, J.S.; Lei, Y.J.; Zhao, K. Petrogenesis of A-type granite and geological significance of Bure area, western Ethiopia. *Earth Sci.* **2021**, *46*, 2299–2310. (In Chinese with English Abstract).
12. Xu, J.F.; Shinji, R.; Defant, M.J.; Wang, Q.; Rapp, R.P. Origin of Mesozoic adakitic intrusive rocks in the Ningzhen area of east China: Partial melting of delaminated lower continental crust? *Geology* **2002**, *30*, 1111–1114. [CrossRef]
13. Jacobs, J.; Thomas, R.J. Himalayan-type indenter-escape tectonics model for the southern part of the late Neoproterozoic-early Palaeozoic East African Antarctic orogen. *Geology* **2004**, *32*, 721–724. [CrossRef]
14. Johnson, P.; Andresen, A.; Collins, A.S.; Fowler, A.; Fritz, H.; Ghebreab, W.; Kusky, T.; Stern, R. Late Cryogenian–Ediacaran history of the Arabian-Nubian Shield: A review of depositional, plutonic, structural, and tectonic events in the closing stages of the northern East African Orogen. *J. Afr. Earth Sci.* **2011**, *61*, 167–232. [CrossRef]
15. Allen, A.; Tadesse, G. Geological setting and tectonic subdivision of the Neoproterozoic orogenic belt of Tuludimtu, western Ethiopia. *J. Afr. Earth Sci.* **2003**, *36*, 329–343. [CrossRef]
16. Cox, G.M.; Lewis, C.J.; Collins, A.S.; Halverson, G.P.; Jourdan, F.; Foden, J.; Nettle, D.; Kattan, F. Ediacaran terrane accretion within the Arabian–Nubian Shield. *Gondwana Res.* **2012**, *21*, 341–352. [CrossRef]
17. Kröner, A.; Linnebacher, P.; Stern, R.; Reischmann, T.; Manton, W.; Hussein, I. Evolution of Pan-African island arc assemblages in the southern Red Sea Hills, Sudan, and in southwestern Arabia as exemplified by geochemistry and geochronology. *Precambrian Res.* **1991**, *53*, 99–118. [CrossRef]
18. Robinson, F.; Foden, J.; Collins, A.; Payne, J. Arabian Shield magmatic cycles and their relationship with Gondwana assembly: Insights from zircon U–Pb and Hf isotopes. *Earth Planet. Sci. Lett.* **2014**, *408*, 207–225. [CrossRef]
19. Shackleton, R. The final collision zone between East and West Gondwana: Where is it? *J. Afr. Earth Sci.* **1996**, *23*, 271–287. [CrossRef]
20. Meert, J.G. A synopsis of events related to the assembly of eastern Gondwana. *Tectonophysics* **2003**, *362*, 1–40. [CrossRef]
21. Woldemichael, B.W.; Kimura, J.I. Petrogenesis of the Neoproterozoic Bikilal Ghimbi gabbro, western Ethiopia. *J. Mineral. Petrol. Sci.* **2008**, *103*, 23–46. [CrossRef]
22. Stern, R.J. Crustal evolution in the East African Orogen: A neodymium isotopic perspective. *J. Afr. Earth Sci.* **2002**, *34*, 109–117. [CrossRef]
23. Tadesse, G.; Allen, A. Geology and geochemistry of the Neoproterozoic Tuludimtu Ophiolite suite, western Ethiopia. *J. Afr. Earth Sci.* **2005**, *41*, 192–211. [CrossRef]
24. Braathen, A.; Grenne, T.; Selassie, M.; Worku, T. Juxtaposition of Neoproterozoic units along the Baruda–Tulu Dimtu shear-belt in the East African Orogen of western Ethiopia. *Precambrian Res.* **2001**, *107*, 215–234. [CrossRef]
25. Grenne, T.; Pedersen, R.B.; Bjerkgård, T.; Braathen, A.; Selassie, M.G.; Worku, T. Neoproterozoic evolution of Western Ethiopia: Igneous geochemistry, isotope systematics and U–Pb ages. *Geol. Mag.* **2003**, *140*, 373–395. [CrossRef]
26. Mogessie, A.; Belete, K.; Hoinkes, G. Yubdo-Tulu Dimtu mafic-ultramafic belt, Alaskan-type intrusions in western Ethiopia: Its implication to the ArabianNubian Shield and tectonics of the Mozambique Belt. *J. Afr. Earth Sci.* **2000**, *30*, 62.
27. Blades, M.L.; Collins, A.S.; Foden, J.; Payne, J.L.; Xu, X.; Alemu, T.; Woldetinsae, G.; Clark, C.; Taylor, R.J.M. Age and hafnium isotopic evolution of the Didesa and Kemashi Domains, western Ethiopia. *Precambrian Res.* **2015**, *270*, 267–284. [CrossRef]
28. Kebede, T.; Koeberl, C. Petrogenesis of A-type granitoids from the Wallagga area, western Ethiopia: Constraints from mineralogy, bulk-rock chemistry, Nd and Sr isotopic compositions. *Precambrian Res.* **2003**, *121*, 1–24. [CrossRef]
29. Liu, Y.S.; Hu, Z.C.; Gao, S.; Günther, D.; Xu, J.; Gao, C.G.; Chen, H.H. In situ analysis of major and trace elements of anhydrous minerals by LA-ICP-MS without applying an internal standard. *Chem. Geol.* **2008**, *257*, 34–43. [CrossRef]

30. Qiu, X.-F.; Ling, W.-L.; Liu, X.-M.; Kusky, T.; Berkana, W.; Zhang, Y.-H.; Gao, Y.-J.; Lu, S.-S.; Kuang, H.; Liu, C.-X. Recognition of Grenvillian volcanic suite in the Shennongjia region and its tectonic significance for the South China Craton. *Precambrian Res.* **2011**, *191*, 101–119. [CrossRef]
31. Hu, Z.C.; Liu, Y.S.; Gao, S.; Liu, W.; Yang, L.; Zhang, W.; Tong, X.; Lin, L.; Zong, K.Q.; Li, M.; et al. Improved in situ Hf isotope ratio analysis of zircon using newly designed X skimmer cone and jet sample cone in combination with the addition of nitrogen by laser ablation multiple collector ICP-MS. *J. Anal. At. Spectrom.* **2012**, *27*, 1391–1399. [CrossRef]
32. Fisher, C.M.; Vervoort, J.D.; Hanchar, J.M. Guidelines for reporting zircon Hf isotopic data by LA-MC-ICPMS and potential pitfalls in the interpretation of these data. *Chem. Geol.* **2014**, *363*, 125–133. [CrossRef]
33. Liu, Y.S.; Gao, S.; Hu, Z.C.; Gao, C.G.; Zong, K.Q.; Wang, D.B. Continental and oceanic crust recycling-induced melt-peridotite interactions in the Trans-North China Orogen: U–Pb dating, Hf isotopes and trace elements in zircons of mantle xenoliths. *J. Petrol.* **2010**, *51*, 537–571. [CrossRef]
34. Zhang, W.; Hu, Z. Estimation of Isotopic Reference Values for Pure Materials and Geological Reference Materials. *At. Spectrosc.* **2020**, *41*, 93–102. [CrossRef]
35. Chen, G.Y.; Sun, D.S.; Zhou, X.R.; Shao, W.; Gong, R.T.; Shao, Y. *Mineralogy of Guojialing Granodiorite and Its Relationship to Gold Mineralization in the Jiaodong Peninsula*; Chinese University of Geosciences: Beijing, China, 1993; pp. 1–230.
36. Deer, W.A.; Howie, R.A.; Zussman, J. *An Introduction to the Rock Forming Minerals*, 2nd ed.; Longman Group: Harlow, UK, 1992; pp. 1–232.
37. Foster, M.D. *Interpretation of the Composition of Trioctahedral Mica*; U.S. Geological Survey Professional Paper 354-B; U.S. Government Printing Office: Washington, DC, USA, 1960; pp. 11–48.
38. Lin, W.W.; Peng, L.J. Estimation of Fe3+ and Fe2+ in hornblende and biotite by electron probe analysis data. *J. Chang. Coll. Geol.* **1994**, *24*, 155–162.
39. Kumar, S.; Pathak, M. Mineralogy and geochemistry of biotites from Proterozoic granitoids of western Arunachal Himalaya: Evidence of bimodal granitogeny and tectonic affinity. *J. Geol. Soc. India* **2010**, *75*, 715–730. [CrossRef]
40. Guo, Y.Y.; He, W.Y.; Li, Z.C.; Ji, X.Z.; Han, Y.; Fang, W.K.; Yin, C. Petrogenesis of Ge'erkuohe porphyry granitoid, western Qinling: Constraints from mineral chemical characteristics of biotites. *Acta Petrol. Sin.* **2015**, *31*, 3380–3390. (In Chinese with English Abstract)
41. Hoskin, P.W. Trace-element composition of hydrothermal zircon and the alteration of Hadean zircon from the Jack Hills, Australia. *Geochim. Cosmochim. Acta* **2005**, *69*, 637–648. [CrossRef]
42. Watson, E.B.; Harrison, T.M. Zircon Thermometer Reveals Minimum Melting Conditions on Earliest Earth. *Science* **2005**, *308*, 841–844. [CrossRef]
43. Ballard, J.R.; Palin, M.J.; Campbell, I.H. Relative oxidation states of magmas inferred from Ce (IV)/Ce (III) in zircon: Application to porphyry copper deposits of northern Chile. *Contrib. Mineral. Petrol.* **2002**, *144*, 347–364. [CrossRef]
44. Trail, D.; Watson, E.B.; Tailby, N.D. Ce and Eu anomalies in zircon as proxies for the oxidation state of magmas. *Geochim. Cosmochim. Acta* **2012**, *97*, 70–87. [CrossRef]
45. Li, W.K.; Cheng, Y.Q.; Yang, Z.M. Geo-fO2: Integrated software for analysis of magmatic oxygen fugacity. *Geochem. Geophy. Geosy.* **2019**, *20*, 2542–2555. [CrossRef]
46. Sun, S.S.; McDonough, W.F. Chemical and isotopic systematics of oceanic basalts: Implications for mantle composition and processes. *Geol. Soc. Lond. Spec. Publ.* **1989**, *42*, 313–345. [CrossRef]
47. Bouvier, A.; Blichert-Toft, J.; Vervoort, J.D.; Gillet, P.; Albarède, F. The case for old basaltic shergottites. *Earth Planet. Sci. Lett.* **2008**, *266*, 105–124. [CrossRef]
48. Wasserburg, G.J.; Jacobsen, S.B.; DePaolo, D.J.; McCulloch, M.T.; Wen, T. Precise determination of SmNd ratios, Sm and Nd isotopic abundances in standard solutions. *Geochim. Cosmochim. Acta* **1981**, *45*, 2311–2323. [CrossRef]
49. Liew, T.C.; Hofmann, A.W. Precambrian crustal components, plutonic associations, plate environment of the Hercynian Fold Belt of central Europe: Indications from a Nd and Sr isotopic study. *Contrib. Mineral. Petrol.* **1988**, *98*, 129–138. [CrossRef]
50. Jahn, B.-M.; Condie, K.C. Evolution of the Kaapvaal Craton as viewed from geochemical and Sm-Nd isotopic analyses of intracratonic pelites. *Geochim. Cosmochim. Acta* **1995**, *59*, 2239–2258. [CrossRef]
51. Khan, J.; Yao, H.-Z.; Zhao, J.-H.; Tahir, A.; Chen, K.-X.; Wang, J.-X.; Song, F.; Xu, J.-Y.; Shah, I. Geochronology, geochemistry, and tectonic setting of the Neoproterozoic magmatic rocks in Pan-African basement, West Ethiopia. *Ore Geol. Rev.* **2024**, *164*, 105858. [CrossRef]
52. Manda, B.W.; Cawood, P.A.; Spencer, C.J.; Prave, T.; Robinson, R.; Roberts, N.M.W. Evolution of the Mozambique Belt in Malawi constrained by granitoid U-Pb, Sm-Nd and Lu-Hf isotopic data. *Gondwana Res.* **2018**, *68*, 93–107. [CrossRef]
53. Griffin, W.; Graham, S.; O'Reilly, S.Y.; Pearson, N. Lithosphere evolution beneath the Kaapvaal Craton: Re–Os systematics of sulfides in mantle-derived peridotites. *Chem. Geol.* **2004**, *208*, 89–118. [CrossRef]
54. Pearce, J.A.; Harris, N.B.W.; Tindle, A.G. Trace Element Discrimination Diagrams for the Tectonic Interpretation of Granitic Rocks. *J. Petrol.* **1984**, *25*, 956–983. [CrossRef]
55. Zeng, G.P.; Wang, J.X.; Xiang, W.S.; Zhang, Z.C.; Jiang, J.S.; Xiang, P. The Augaro Arc-type Granite in the Nubia Shield, Western Eritrea: Petrogenesis and Implications for Neoproterozoic Geodynamic Evolution of the East African Orogen. *Northwestern Geol.* **2024**, *57*, 159–173. (In Chinese with English abstract)

56. Uchida, E.; Endo, S.; Makino, M. Relationship Between Solidification Depth of Granitic Rocks and Formation of Hydrothermal Ore Deposits. *Resour. Geol.* **2007**, *57*, 47–56. [CrossRef]
57. Wones, D.R.; Eugster, H.P. Stability of biotite: Experiment, theory, and application. *Am. Mineral.* **1965**, *50*, 1228–1272.
58. Jiang, J.-S.; Zheng, Y.-Y.; Gao, S.-B.; Zhang, Y.-C.; Huang, J.; Liu, J.; Wu, S.; Xu, J.; Huang, L.-L. The newly-discovered Late Cretaceous igneous rocks in the Nuocang district: Products of ancient crust melting trigged by Neo–Tethyan slab rollback in the western Gangdese. *Lithos* **2018**, *308–309*, 294–315. [CrossRef]
59. Loader, M.A.; Nathwani, C.L.; Wilkinson, J.J.; Armstrong, R.N. Controls on the magnitude of Ce anomalies in zircon. *Geochim. Cosmochim. Acta* **2022**, *328*, 242–257. [CrossRef]
60. Teklay, M.; Kröner, A.; Mezger, K.; Oberhänsli, R. Geochemistry, Pb Pb single zircon ages and Nd Sr isotope composition of Precambrian rocks from southern and eastern Ethiopia: Implications for crustal evolution in East Africa. *J. Afr. Earth Sci.* **1998**, *26*, 207–227. [CrossRef]
61. Kröner, A.; Sassi, F.P. Evolution of the northern Somali basement: New constraints from zircon ages. *J. Afr. Earth Sci.* **1996**, *22*, 1–15. [CrossRef]
62. Harris, N.B.W.; Marzouki, F.M.H.; Ali, S. The Jabel Sayid complex, Arabian Shield: Geochemical constraints on the origin of peralkaline and related granites. *J. Geol. Soc.* **1986**, *143*, 287–295. [CrossRef]
63. Stern, R.J.; Kröner, A. Late Precambrian Crustal Evolution in NE Sudan: Isotopic and Geochronologic Constraints. *J. Geol.* **1993**, *101*, 555–574. [CrossRef]
64. Stern, R.J.; Abdelsalam, M.G. Formation of juvenile continental crust in the Arabian-Nubian Shield, evidence from granitic rocks of the Nakasib suture, NE Sudan. *Geol. Rundsch.* **1998**, *87*, 150–160. [CrossRef]
65. Zeng, G.P.; Wang, J.X.; Xiang, W.S.; Tong, X.R.; Shao, X.; Hu, P.; Wu, F.F.; Jiang, J.S.; Xiang, P. Petrogenesis and Geological Significance of the Adi Keyh A-type Rhyolite in Central Eritrea. *South China Geol.* **2022**, *38*, 157–173. (In Chinese with English)
66. Abdel-Rahman, A.F.M. Nature of biotites from alkaline, calcalkaline, and peraluminous magmas. *J. Petrol.* **1994**, *35*, 525–541. [CrossRef]
67. Zhou, Z.X. The origin of intrusive mass in Fengshandong, Hubei Province. *Acta Petrol. Sin.* **1986**, *2*, 59–70. (In Chinese with English)
68. Defant, M.J.; Drummond, M.S. Derivation of some modern arc magmas by melting of young subducted lithosphere. *Nature* **1990**, *347*, 662–665. [CrossRef]
69. Schulz, B.; Klemd, R.; Brätz, H. Host rock compositional controls on zircon trace element signatures in metabasites from the Au-stroalpine basement. *Geochim. Cosmochim. Acta* **2006**, *70*, 697–710. [CrossRef]

Disclaimer/Publisher's Note: The statements, opinions and data contained in all publications are solely those of the individual author(s) and contributor(s) and not of MDPI and/or the editor(s). MDPI and/or the editor(s) disclaim responsibility for any injury to people or property resulting from any ideas, methods, instructions or products referred to in the content.

Article

Lithium-, Phosphorus-, and Fluorine-Rich Intrusions and the Phosphate Sequence at Segura (Portugal): A Comparison with Other Hyper-Differentiated Magmas

Michel Cathelineau [1,*], Marie-Christine Boiron [1], Andreï Lecomte [1], Ivo Martins [2], Ícaro Dias da Silva [2] and Antonio Mateus [2]

[1] Université de Lorraine, CNRS, GeoRessources, 54000 Nancy, France; marie-christine.boiron@univ-lorraine.fr (M.-C.B.); andrei.lecomte@univ-lorraine.fr (A.L.)
[2] Faculdade de Ciências, Instituto Dom Luiz, Universidade de Lisboa, Campo Grande, 1749-016 Lisboa, Portugal; ivojmartins@gmail.com (I.M.); ifsilva@fc.ul.pt (Í.D.d.S.); amateus@ciencias.ulisboa.pt (A.M.)
* Correspondence: michel.cathelineau@univ-lorraine.fr

Citation: Cathelineau, M.; Boiron, M.-C.; Lecomte, A.; Martins, I.; da Silva, Í.D.; Mateus, A. Lithium-, Phosphorus-, and Fluorine-Rich Intrusions and the Phosphate Sequence at Segura (Portugal): A Comparison with Other Hyper-Differentiated Magmas. *Minerals* **2024**, *14*, 287. https://doi.org/10.3390/min14030287

Academic Editors: Ignez de Pinho Guimarães and Jefferson Valdemiro De Lima

Received: 25 January 2024
Revised: 19 February 2024
Accepted: 4 March 2024
Published: 8 March 2024

Copyright: © 2024 by the authors. Licensee MDPI, Basel, Switzerland. This article is an open access article distributed under the terms and conditions of the Creative Commons Attribution (CC BY) license (https://creativecommons.org/licenses/by/4.0/).

Abstract: Near the Segura pluton, hyper-differentiated magmas enriched in F, P, and Li migrated through shallowly dipping fractures, which were sub-perpendicular to the schistosity of the host Neoproterozoic to Lower Cambrian metasedimentary series, to form two swarms of low-plunging aplite–pegmatite dykes. The high enrichment factors for the fluxing elements (F, P, and Li) compared with peraluminous granites are of the order of 1.5 to 5 and are a consequence of the extraction of low-viscosity magma from the crystallising melt. With magmatic differentiation, increased P and Li activity yielded the crystallisation of the primary amblygonite–montebrasite series and Fe-Mn phosphates. The high activity of sodium during the formation of the albite–topaz assemblage in pegmatites led to the replacement of the primary phosphates by lacroixite. The influx of external, post-magmatic, and Ca-Sr-rich hydrothermal fluids replaced the initial Li-Na phosphates with phosphates of the goyazite–crandallite series and was followed by apatite formation. Dyke emplacement in metasediments took place nearby the main injection site of the muscovite granite, which plausibly occurred during a late major compression event.

Keywords: peraluminous aplite and pegmatite; phosphorus; lithium; fluorine; eosphorite; montebrasite; Raman spectroscopy

1. Introduction

Lithium- and phosphorus-rich pegmatites are often evidence of the transition between the magmatic and hydrothermal stages in volatile-rich magmatic systems (F, B) active in orogenic belts. Their relationships with a nearby batholith are highly debated, with two main genetic processes being frequently opposed as follows: (i) the fractional crystallisation of a parent S-type granitic magma with a progressive enrichment in incompatible elements during differentiation [1–3] or (ii) anatectic-type melting [4]. Such spatial but not necessarily genetic associations are also found in the Segura region in the southern part of the Central Iberian Zone (CIZ) in Portugal, which is close to the Portugal–Spain border. This swarm of aplites and pegmatites was described by [5] in a pioneering study of the main features of these intrusive bodies. Detailed maps of the Segura region enabled a thorough sampling of the granite facies and nearby aplite/pegmatites. They motivated a more detailed study of the mineral succession from the magmatic to hydrothermal stage. There, the pegmatite and aplite dykes are hosted by the surrounding metapelitic schist series but are close to the westernmost edge of the Cabeza de Araya batholith [6–9]. The perphosphorus feature of the dyke magmas is particularly significant, indicating the role of P, F, and Li as fluxing agents. The abundance of these elements results in a specific mineralogy, especially that of phosphates, which are then particularly sensitive to the evolution of fluids during magma cooling and to the late introduction of external chemical components by hydrothermal fluids.

In the present work, we describe the wide variety of Li, Na, Fe-Mn, and Ca-Sr phosphates found in the aplite dykes and associated pegmatites of the Segura region, with the following objectives: (1) to study the petrographic, mineralogical, textural, and chemical characteristics of the phosphates and establish a paragenetic sequence of the phosphate phases in relation to the magmatic–hydrothermal evolution; (2) to define the geochemical characteristics of the magmas in terms of P, Li, and F evolution; (3) to discuss the origin and behaviour of phosphorus in these magmatic–hydrothermal stages and to understand why several distinct phosphates crystallised in addition to apatite in all the facies at the magmatic stage; and (4) to compare the characteristics of the Segura granites with other similar intrusive suites. The mineralogical and geochemical characteristics of these aplites and pegmatites display similarities to other peraluminous and leucocratic intrusive bodies in the region, such as the Argemela intrusion which is located 30 km to the east [10], as well as with the reference intrusions which are described in detail, such as Beauvoir in the French Massif Central [11] and Tres Arroyos in Spain [3,12,13]. Therefore, a comparison was mainly focused on the range of phosphorus content and its potential role with fluorine on melt mobility.

2. Local Geology

The Segura region is in the southern domain of the CIZ in the easterly part of the Góis–Panasqueira–Argemela–Segura strip, which covers an area rich in granitic intrusions and hydrothermal manifestations associated with different Sn-W-Nb-Ta-Li mineralisation types (Figure 1a). Many studies have reported evidence of Variscan polyphase deformation, metamorphism, and multi-stage magmatism throughout the CIZ [14–17]). However, the bulk of the magmatic intrusions were formed between 320 and 305 Ma, which was roughly concurrent with the Late Variscan deformation and retrograde metamorphism (D3-M3) following the main compressional Late Devonian event and Barrovian metamorphism (D1-M1) as well as the Mississippian regional extensional tectonic event synchronous with Buchan-type metamorphism (D2-M2), which are particularly well described in the northern part of the CIZ. The Late Variscan pluton intrusion was contemporaneous with the D3-M3 upright folding and steep axial plane foliation, thereby defining the present-day regional NW-SE Variscan trend marked by the Ordovician–Silurian synclines and by the conjugated WNW-ESE and ENE-WSW late-D3 regional transcurrent shear zone activity and related structures [15–19].

The CIZ is characterised by the high volume of peraluminous granitic rocks [20] that are syn-to-post kinematic [21–23], and it includes strongly peraluminous, Ca-poor, and variably P-enriched two-mica leucogranites, which are interpreted as being derived from the melting of a metasedimentary source along with several series of metaluminous to weakly peraluminous granites [23,24]. In the Góis–Panasqueira–Argemela–Segura strip, many granite intrusions are peraluminous, like those of Segura, Penamacor, Panasqueira, Orca, and Castelo Branco (Figure 1a). Most of these granites are dated 299–312 Ma with two stages of Li pegmatites at 310 ± 5 and 301 ± 3 Ma ([21]; synthesis in [25]). Some are backed by older intrusions, recording an important magmatic event at the Cambrian-Early Ordovician transition, such as the Fundão, Oledo-Idanha-a-Nova, and Zebreira granites and granodiorites [26,27].

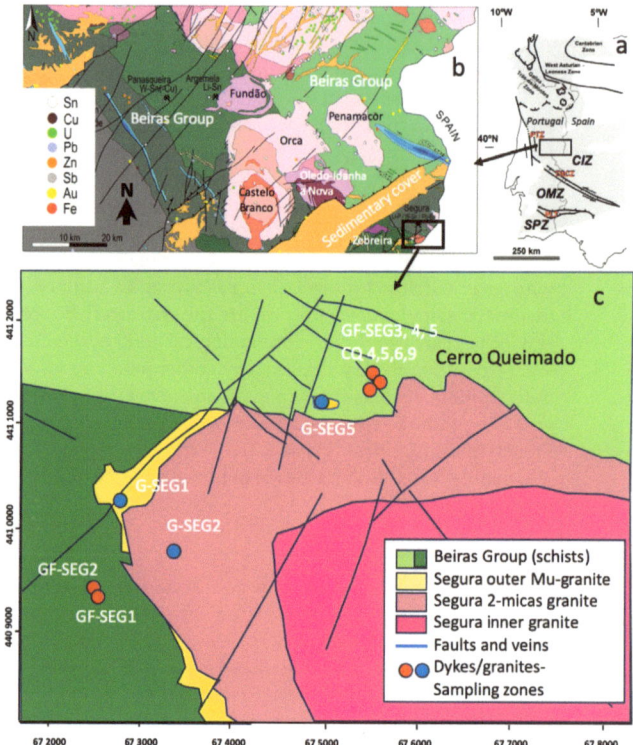

Figure 1. (**a**) Map of main geological units in Portugal and Spain; CIZ—Central Iberian Zone; OMZ—Ossa–Morena Zone; SPZ—South Portuguese Zone; PLT—Pulo do Lobo Terrane (contiguous to the SW Iberian suture); PTZ—Porto–Tomar Shear Zone; TBCSZ—Tomar–Badajoz–Córdoba Shear Zone. (**b**) Simplified map of the Góis–Panasqueira–Argemela–Segura Sn-W strip, which is south of the Central Iberian Zone (CIZ, Portugal) according to the 1:500.000 geological map and the SIORMINP mineral occurrence map of the Portuguese Geological Survey (LNEG) [28]. Different shades of green represent the Beiras Group formations, and the blues the Ordovician–Silurian units defining the upright D3 synclines in this sector of the CIZ NE-SW faults represent late-D3 shear zones that were systematically reactivated in the Alpine orogenic cycle, thus controlling the shapes and disposition of the Cenozoic basins (in orange). The remaining colours represent Cadomian (fuchsia), Ordovician (cherry and dark pink), and Late Variscan (320–300 Ma) granitoids (dark and light orange and light pink). Coloured small circles indicate mineralisation occurrences as in the official SIORMINP catalogue (LNEG). (**c**) Geological setting of the studied Segura area in the northwestern border of the Segura–Cabeza de Araya Batholith and an indication of the sample location for dykes. The geological map was adapted from the geological maps of Portugal at 1:500,000 [29]; the coordinate system is WGS 84 29T, UTM (zone 29).

The Segura granites form the westernmost tip of the Cabeza de Araya batholith, near the border between Portugal and Spain (Figure 1). The Cabeza de Araya batholith extends more than 60 km on the Spanish side [7–9], with three sub-facies noted as A, B, and C, thus forming a zoned structure [30]. In the Segura region (Portugal), a zoned edge with two outer facies, in particular a two-mica facies and a thin border of the so-called outer muscovite-rich granite [5] (Figure 2a–c). The pluton has induced an aureole comprising the contact metamorphism of up to one kilometre. The muscovite-rich and two-mica granite facies exposed in the Segura region have been dated by TIMS U-Pb on zircon [5] at 312.9 Ma and 311 ± 0.5 Ma, respectively.

Figure 2. Outcrops of the aplite–pegmatite dykes and granites in the Segura region. (**a**) Inner facies of the Cabeza de Araya batholith, with large euhedral cordierite crystals (noted Crd) and predominant biotite; (**b**) inner rim of the Cabeza de Araya batholith, showing a two-mica coarse-grained facies with local tourmaline and showing large retrogressed cordierite crystals; (**c**) medium-to-coarse-grained muscovite leucogranite with tourmaline from the outer rim of the Cabeza de Ayara pluton in Segura region, intruding spotted schists of the Beiras Group (BG); (**d**) low-dipping dyke hosted in BG metasediments in the Cerro Queimado area (see Figure 1b); (**e**) sub-horizontal dyke intrusive in BG metasediments in the Cerro Queimado area; (**f**) aplite dyke from the southern area to the southwest of Segura; (**g**) pegmatite exposure at the level of the river (southwest of the Segura batholith); (**h**,**i**) two other views of the dykes intruding on the metasediments, thereby showing geometric arrays due to the forced intrusion on the schists, which open in several directions and fall as enclaves within the magma.

Considering the hydroxyl apatite crystal's chemistry, as well as the decrease in $^{87}Sr/^{86}Sr$ and the ^{18}O shift to higher values from the two-mica granite to the lepidolite-type aplite–pegmatite dykes, [5] ones concludes that (i) the aplite–pegmatite dykes originate from muscovite granite magma from the fractional crystallisation of quartz, plagioclase, K-feldspar and that (ii) aplites cannot derive from the main two-mica granite and correspond to distinct pulses of magma from the partial melting of heterogeneous metapelitic rocks.

On the Portuguese side, dykes of fine-grained granites (aplites), which are locally associated with pegmatites, intrude on the schist–metagraywacke complex of the Neoproterozoic to Lower Cambrian ages, thereby forming the Beiras Group (BG [31]) (Figure 2d–i). There, the greenschist facies metasedimentary rocks are occasionally interbedded with the metaconglomerate layers. Dykes are located in two distinct areas, both on the northern and southwestern flanks of the westernmost end of the Segura pluton. These bodies, trending NE-SW, show low dipping angles (<45°) and are sub-perpendicular to the schists, which are characterised by a vertical or steeply dipping foliation (Figure 2d–f). They are generally around a metre or are exceptionally several metres thick and vary in length from a several decameters to a few hundred metres. Smaller dykes are also observed (Figure 2g–h).

3. Materials and Methods

The methodology used is original and based on mapping phosphorus minerals at the thin section scale by micro-X-ray fluorescence (micro-XRF), followed by imaging and the semi-quantitative investigation of the phosphates by a Scanning Electron Microscope (SEM) equipped with an Energy Dispersive Spectrometer (EDS). Then, a (i) SEM-WDS (Wavelength Dispersive Spectrometer) is used when the complexity of the textures hampers the use of standard electron microprobe analyses (EMPAs) and (ii) an electron microprobe in all other cases. Micro-XRF mapping, SEM, and EPMA measurements were performed at the SCMEM analytical platform of the GeoRessources laboratory, Nancy, France.

This methodology was applied to eighteen thin sections mapped in transmitted light with a Keyence VHX macroscope before micro-XRF mapping on the selected samples to establish the paragenetic sequence of the hydrothermal mineral assemblages. Micro-XRF maps were made choosing the most representative assemblages to render SEM and electron microprobe investigations easier. Micro-XRF mapping was carried out using a M4 Tornado instrument (Bruker Nano Analytics, Berlin, Germany) equipped with an Rh X-ray tube (Be side window) and polycapillary optics, thus giving an X-ray beam with a 25–30 µm diameter on the sample. The X-ray tube was operated at 50 kV and 200 µA. Two 30 mm^2 Xflash® SDD detectors (Bruker Nano Analytics, Berlin, Germany) measured X-rays with an energy resolution of <135 eV at 250,000 cps. All analyses were carried out in a 2 kPa vacuum. The main elements such as Ca, Mg, Mn, Fe, P, Al, K, Na, and Si were mapped, and the composite chemical images were generated. Eleven samples were mapped to depict the mineral assemblages in the veins and their chemical zoning (phosphates).

SEM investigations were performed on eight samples using a JEOL JSM-7600F Schottky-FEG-SEM (JEOL Ltd, Tokyo, Japan) equipped with an Oxford Instruments 20 mm^2 SDD-type EDS spectrometer (Oxford Instruments plc, Abingdon, UK). All elements (including trace elements) have standard deviations (2σ) of less than 10%.

Quantitative analysis of the major elements was performed with a CAMECA SX100 EPMA (CAMECA SAS, Gennevilliers, France) equipped with five WDS spectrometers using an accelerating voltage of 15 kV, a probe current of 12 nA (an accelerating voltage of 25 kV and a probe current of 150 nA for trace elements), and a beam diameter of 1 µm. The peak and background counting times were 10 and 5 s, respectively. The following crystals were used: TAP (F, Na, Mg, Al, Si, and Rb), LPET (K, Cl, and Nb), LiF (Mn and Fe), and PET (Cs). The standards were natural minerals and synthetic oxides as follows: topaz (FKα), albite (Na, SiKα), olivine (MgKα), andradite (CaKα), Al_2O_3 (AlKα), orthoclase (KKα), $MnTiO_3$ (Mn, TiKα), Fe_2O_3 (FeKα), $RbTiPO_5$ (RbLα), $SrSO_4$ (SrLα), and barite (BaLα).

The phosphate crystals were examined with an HR Horiba Jobin Yvon Raman system (Jobin Yvon, Longjumeau, France) (GeoRessources laboratory, Nancy, France) using a 514.5 nm Ar$^+$ laser emission line at a resolution of 2 cm^{-1} in the 100–4000 cm^{-1} range. Repeated acquisition using the highest magnification was accumulated to improve the signal-to-noise ratio. Raman spectra were generally obtained after five acquisitions of 10 to 20 s each, both in the low-frequency (200–1200 cm^{-1}) and high-frequency ranges (3550–3800 cm^{-1}).

Whole-rock major and trace element analyses were performed on field samples at the «Service d'Analyse des Roches et des Minéraux (SARM)», Centre de Recherches Pétrographiques et Géochimiques (CRPG), Nancy, France. Major elements were analysed by inductively coupled plasma optical emission spectrometry on a Thermo-Fischer ICap 6500 instrument (Thermo-Fischer Scientific, Waltham, MA, USA). Trace elements, including rare earth elements (REEs), were determined by inductively coupled plasma mass spectrometry (Thermo-Elemental X7) (Thermo-Fischer Scientific, Waltham, USA). Detailed analytical procedures are given in [32]. For three samples, data were obtained by an XRF analytical tool at Lisboa University, and the loss on ignition (L.O.I.) was not determined but estimated by difference to 100%.

4. Results

4.1. Field Observations

Two main aplite–pegmatite swarms were distinguished, which was mainly based on field relationships, the mineralogical and geochemical data detailed below, and structural and textural characteristics.

In the northern border of the Segura pluton, there is the so-called "Cerro Queimado" mining area where banded aplite–pegmatite dykes are intrusive and have filled sub-horizontally opened fractures in the metasedimentary sequence. They are sub-perpendicular to the foliation, striking N140° and dipping between 20 and 45° to the south (Figures 1b and 2a,b). Other low-dipping (<30° to the south–southwest), N135°–150° aplite–pegmatite dykes are found in the southern border of the most occidental part of the Segura pluton (Figure 1b). They occur as multi-decimetric-thick to metre-thick intrusions, which are well exposed in the river valley (Figure 2c–f).

In both pegmatite and aplite bodies, albite, K-feldspars, and quartz are the main constituents, with both minerals forming bands parallel to the boundary with the host schist (Figure 3a). Pegmatite develops as infillings of the opened fractures within the aplite dykes, usually displaying growth textures that are perpendicular to the aplite boundary (Figure 3b). A series of phosphates, like amblygonite and other (Fe-Mn, Ca-Sr)-P phases, Nb-Ta oxides, and topaz are abundant. Both aplites and pegmatites are leucocratic magmatic facies with low amounts of black constituents, except Nb-Ta oxides and local tourmaline.

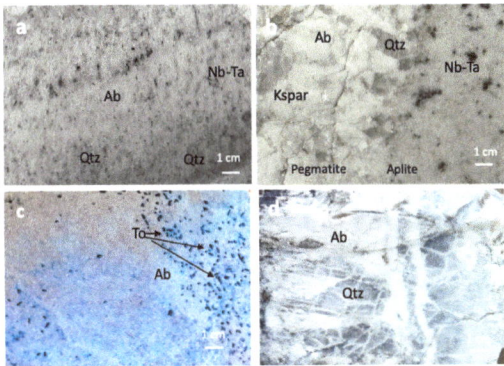

Figure 3. Textures of the Segura dykes. (**a**,**b**) Contact between the aplite and pegmatite in the samples from the northern zone of the Segura batholith (Cerro Queimado). Facies are leucocratic with black dots corresponding to the Nb-Ta oxides (Nb-Ta). The development of pegmatite occurs through the growth of albite (Ab) and quartz (Qtz) as subparallel crystals and minor K-feldspar (Kspar). (**c**) Macroscopic feature of the tourmaline (To) bearing aplite from the southern zone of the Segura dyke swarm. (**d**) Pegmatite with comb quartz and albite development.

Thus, tourmaline is only found in places in the northern swarm and is not abundant. In contrast, tourmaline-rich aplite dykes occur in the southern swarm (Figure 3c).

Some of the dykes are affected by later magmatic–hydrothermal stages, thereby resulting either in the abundant formation of quartz (Figure 3d) or lepidolite that is associated with cassiterite. The latter assemblage is not the subject of the present work.

4.2. Dyke Mineralogy

Detailed composite chemical maps of the mineral assemblages were obtained using micro-XRF and are presented in Figure 4. In Figure 4a,b, the typical texture of the pegmatite developed onto aplite is characterised by the crystallisation of euhedral quartz prisms and tabular albite perpendicular to the aplite boundary. Li-Na phosphates (amblygonite series) form euhedral crystals subparallel to the quartz prisms and tabular albite or are crystallised onto the euhedral K-feldspars and albite. They are also included as small patches within quartz. The mineral assemblage generally ends by filling the remaining space with K-feldspar (Figure 4a) or, most frequently, quartz (Figure 4b–d).

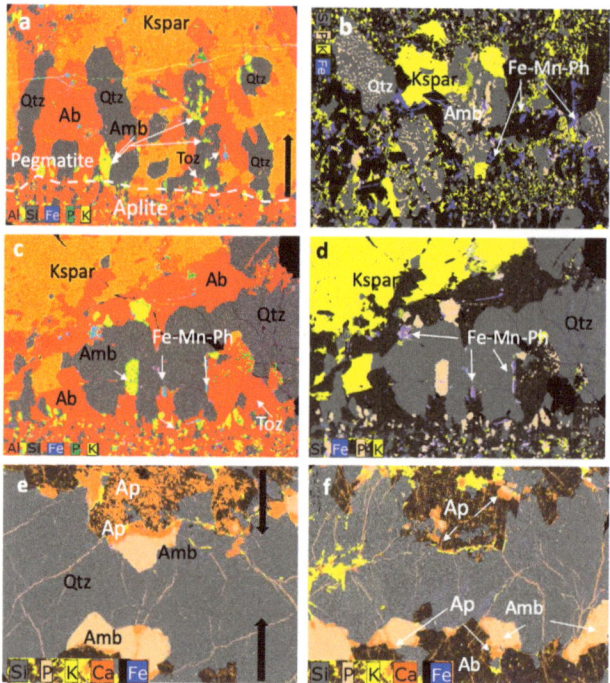

Figure 4. Composite micro-XRF chemical maps of aplite and pegmatite dykes. (**a**) Boundary between aplite and pegmatite showing the growth of albite (in red), quartz (grey), and amblygonite (light yellow) perpendicularly to the boundary, as well as K-feldspar (in orange: Al, K); (**b**) pegmatite texture with phosphates included in quartz; (**c**) amblygonite (in green-yellow), albite, K-feldspar, and quartz; (**d**) same chemical map with the Fe phosphate in violet-blue; (**e**) amblygonite on both sides of the cavities filled with quartz (black arrow indicates how the cavity has been filled) and apatite (in orange) replaces the K-feldspars; (**f**) amblygonite as euhedral crystals formed on albite and cemented by quartz and, in orange, apatite filling microfractures. Toz: topaz; Amb: amblygonite; Fe-Mn Ph: eosphorite–childrenite Fe-Mn phosphate series; Kspar: K-feldspar; Ab: albite; Ap: apatite; Qtz: quartz.

The Fe-Mn phosphates are tiny euhedral crystals found within or on albite, which are frequently at the boundary with the Li-Na-phosphates and seem synchronous with albite. Albite corrodes the quartz that is deposited as the euhedral crystal palisade onto the aplite and is accompanied by topaz. Albite also corrodes the K-feldspars. Synchronous to the albite stage, Na-rich phosphate replaces the Li phosphate from the amblygonite series. Apatite later invades the early texture, thereby forming euhedral needles and microdomains within the albite and earlier phosphate, which frequently occurs along the former grain boundaries (microdomains presented in orange in Figure 4e,f). Identifying the few grains of magmatic apatite is difficult because most of the feldspars are rich in small apatite inclusions or needles linked to hydrothermal alteration.

The Na-Li phosphates show complex textures and relationships. Crystals of the amblygonite series are the first phosphate to crystallise as large and sometimes centimetric crystals (sample GF-SEG3). They are frequently euhedral, and sometimes centimetric prisms are oriented perpendicular to the aplite layers and grow towards the centre of the pegmatite veins (microdomain in the yellow-green colour in Figure 5a–c). They are associated with topaz, which appears in red in the composite micro-XRF maps from Figure 5c,d. These large crystals are partially replaced by the Na-Al phosphates (lacroixite), which form replacement patches at the microscale. Microdomains are a few dozen micrometres long within the initial mass of the Li-phosphate (Figure 5e,f). This pervasive replacement thus yields a mixture of the two phases (montebrasite, lacroixite) at the microscale (Figure 5f).

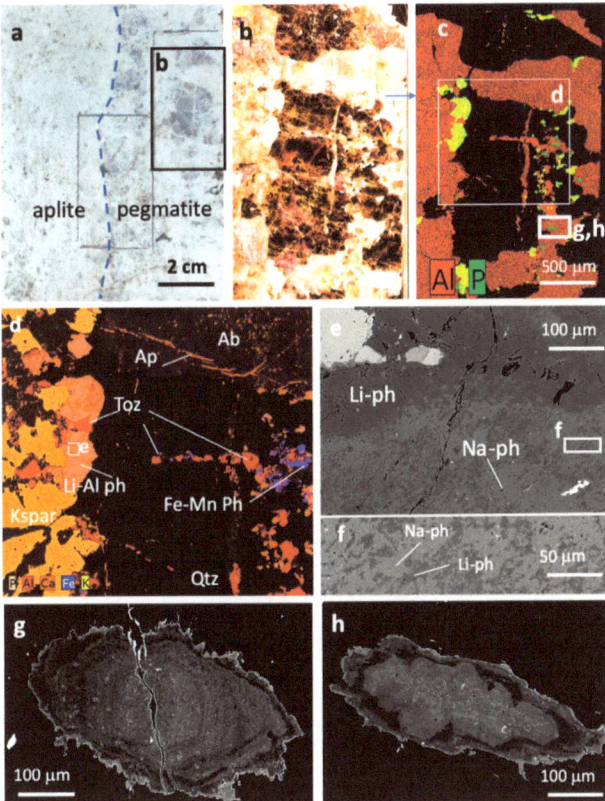

Figure 5. (**a**) Boundary between aplite and pegmatite (macroscopic photograph), with an indication of the zone mapped in figures (**b**,**c**); (**b**) transmitted light microphotograph of the thick section; (**c**) chemical micro-XRF map showing amblygonite (yellow) on both sides of the cavity filled with quartz and the Fe-Mn phosphate (in green); (**d**) magnified detail of zone (**c**) showing the distribution of the amblygonite formed onto K-feldspars and topaz as inclusions in quartz (in red); (**e**) microtextures of Li-phosphates (Li-ph) (amblygonite and mixed lacroixite/amblygonite); (**f**) magnification of the amblygonite/lacroixite at the micron scale; (**g**,**h**) crystals of Fe-Mn phosphates (BSE SEM image),

showing the euhedral growth bands. Toz: topaz; Li-ph: Li phosphates; Amb: amblygonite; Na-ph: Na-phosphate (lacroixite-dominated); Fe-Mn-Ph: eosphorite–childrenite Fe-Mn phosphate series; Kspar: K-feldspar; Ab: albite; Ap: apatite; Qtz: quartz.

In addition to the Li (Al, Na) phosphates, Fe-Mn (Al) phosphates crystallised, apparently relatively early as isolated grains (Figure 5c, inset noted g, h and enlargements in Figure 5g,h). Fe-Mn phosphates show alternated Fe-rich and Mn-rich growth zones with euhedral development, indicating their early formation in the melt before being included in albite crystals. They are coated by a late overgrowth enriched in iron, which finally develops in micro-fissures nearby where insufficient growth space is available, probably during a relatively late stage (Figure 5h). The external rim corresponds to a late overgrowth, which fills minor fractures of the surrounding albite around the initial crystal and is particularly enriched in iron. Raman spectra have shown that the phosphate belongs to the eosphorite–childrenite solid solution (see below Raman and crystal-chemistry data).

Crandallite forms at the grain boundary or, at the expense of earlier phosphates, replaces them, particularly Li-Na-phosphates, as shown in Figure 6a. Figure 6b,c illustrate the microscopic aspect of the amblygonite crystal, showing an intense alteration at the boundary with albite. Microdomains are small, ranging from a few microns to tens of micrometres, and they are very complex at a small scale, as shown by the enlargement from Figure 6d. Along the microfissures affecting the amblygonite (Figure 6d), goyazite–crandallite develops as euhedral crystals with a very large composition between domains characterized by variation in the mean atomic number "Z" (from dark grey to light grey), corresponding to the Ca-Sr exchange (Figure 6e).

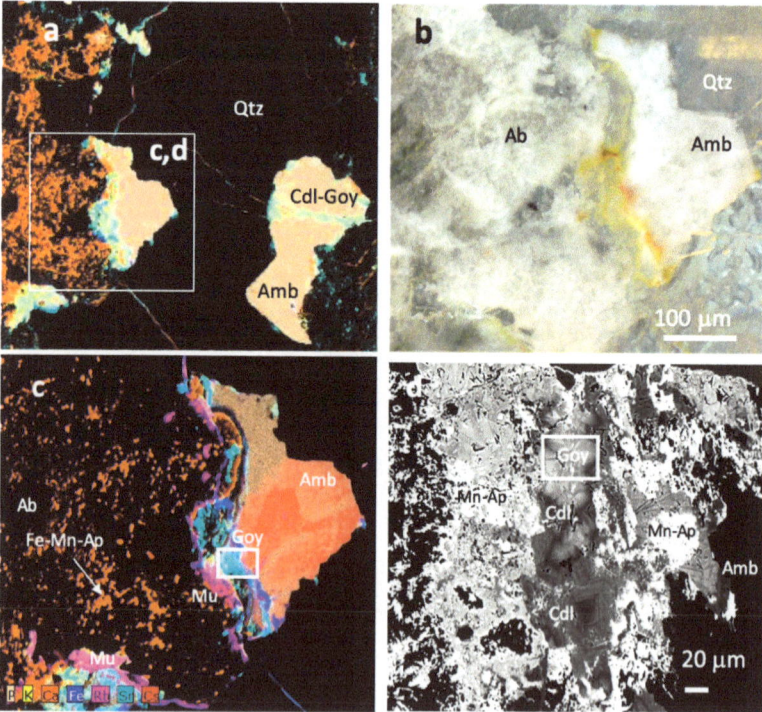

Figure 6. *Cont.*

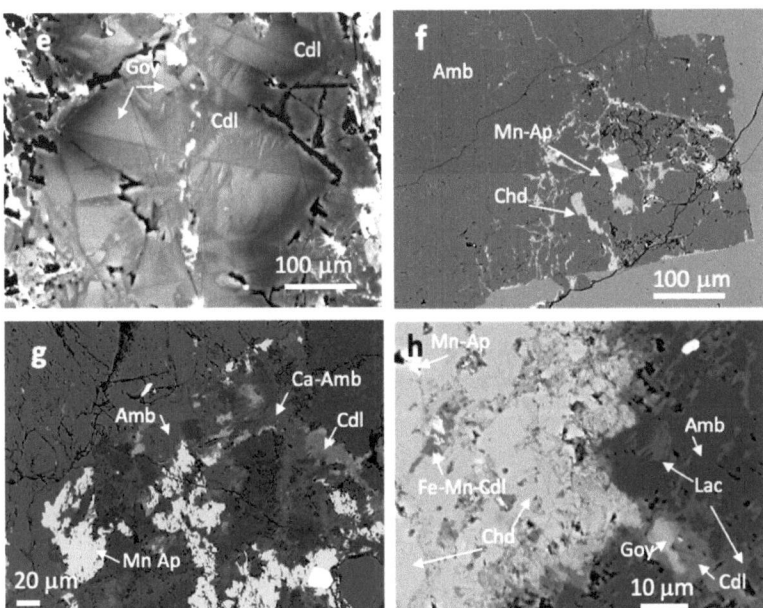

Figure 6. (**a**) Amblygonite crystals developed on the edge of a cavity filled with quartz; (**b**) crandallite and Fe-Mn-rich apatite formed in microfractures; (**c**) detail of the amblygonite crystal of figure (**a**); (**d**) micro-XRF composite chemical map showing the replacement of amblygonite by complex assemblages of phosphates at the boundary between the crystal and feldspars; (**e**) detail of the complex zone of replacement formed by crandallite and goyazite and later Mn-rich apatite; (**f**) detail of the growth bands of the goyazite crystals with the earliest bands of crandallite; (**g**) complex association of phosphates replacing amblygonite. Amblygonite may contain calcium with no clear evidence of the addition of crandallite patches, and apatite is enriched in Fe and Mn; (**h**) association of the four types of phosphates with mutual replacements, with the same abbreviations as those in Figure 4; Goy: goyazite; Cdl: crandallite; Chd: chidrenite; Lac: lacroixite; Mn-Ap: Mn-rich apatite; Qtz: quartz; and Mu: muscovite.

Apatite grains grow as patches in feldspars, especially albite (Figure 6c). They also locally replace all the earlier phosphate types and fill pores and microfractures, thereby developing needles, plumose crystals, or microdomains as grain aggregates (Figure 6f,g). In the tourmaline-rich aplites of the southern swarm, apatite forms poikilitic aggregates around quartz, mica, and tourmaline. The apatite composition is heterogeneous, incorporating significant amounts of Al, Fe, Mn, and traces of Sr. When these element concentrations increase, the fluorine concentration decreases, suggesting an increase in OH at the expense of fluorine when shifting from the ideal F apatite stoichiometry.

4.2.1. Raman Data on Phosphates

All phosphates were analysed by Raman spectroscopy. The three main solid solutions (eosphorite–childrenite, amblygonite-lacroixite, and crandallite–goyazite) are confirmed by comparison with the reference spectra from the international database RRUFF [33]. The most representative spectra are presented in Figure 7. For the eosphorite-chidrenite series ((Fe, Mn) $AlPO_4(OH)_2(H_2O)$), Raman spectroscopy enabled the observation of bands at 969 cm^{-1} and 1018 cm^{-1}, which were assigned to the phosphate bonds. For the other peaks, the assigning proposed by [34] is as follows following: Raman bands at 562 cm^{-1}, 595 cm^{-1}, and 608 cm^{-1} are assigned to the $\nu 4$ bending modes of the PO_4, HPO_4, and H_2PO_4 units; Raman bands at 405, 427, and 466 cm^{-1} are attributed to the $\nu 2$ modes of these units. Hydroxyl and water stretching bands are also observed at 3200–3500 cm^{-1}. Differences between the peaks of eosphorite and childrenite are a function of the abundance of manganese and mostly concern the peak at 253, 557, 618, and 1053 cm^{-1} noted in italics in Figure 7a.

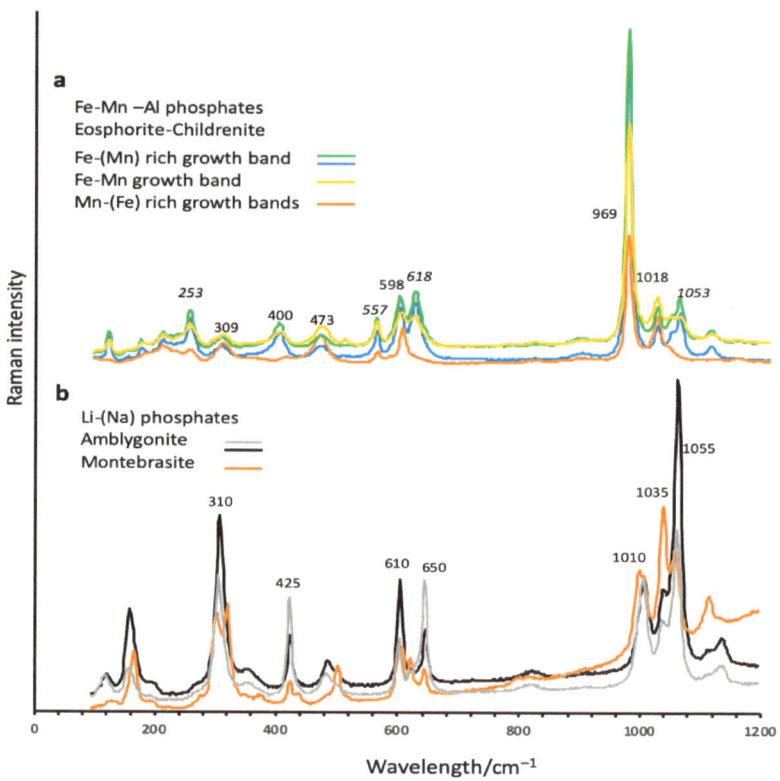

Figure 7. Representative Raman spectra of two phosphate series. (**a**) The eosphorite–childrenite series and (**b**) the Li phosphate series represented by amblygonite and intermediate phases with a spectrum closer to montebrasite.

The Raman spectra for the amblygonite–montebrasite series are presented in Figure 7b. Most spectra display intermediate features between amblygonite and montebrasite, which are typical of the natural series of such minerals. The main changes as a function of the fluorine content concern a few peaks [35], particularly from 599 to 604 cm^{-1} and between 1056 and 1066 cm^{-1}. The spectra present a series of peaks between 1050 and 1056 cm^{-1}, which are closer to the montebrasite following [35]. Significant differences occur between 3379 and 3348 cm^{-1} (not shown in Figure 7).

4.2.2. Crystal Chemistry of Phosphates

The crystal chemistry of the analysed phosphates is shown in Figure 8. The different series are the following: eosphorite–childrenite, the montebrasite–amblygonite–lacroixite series, crandallite–goyazite, and apatite.

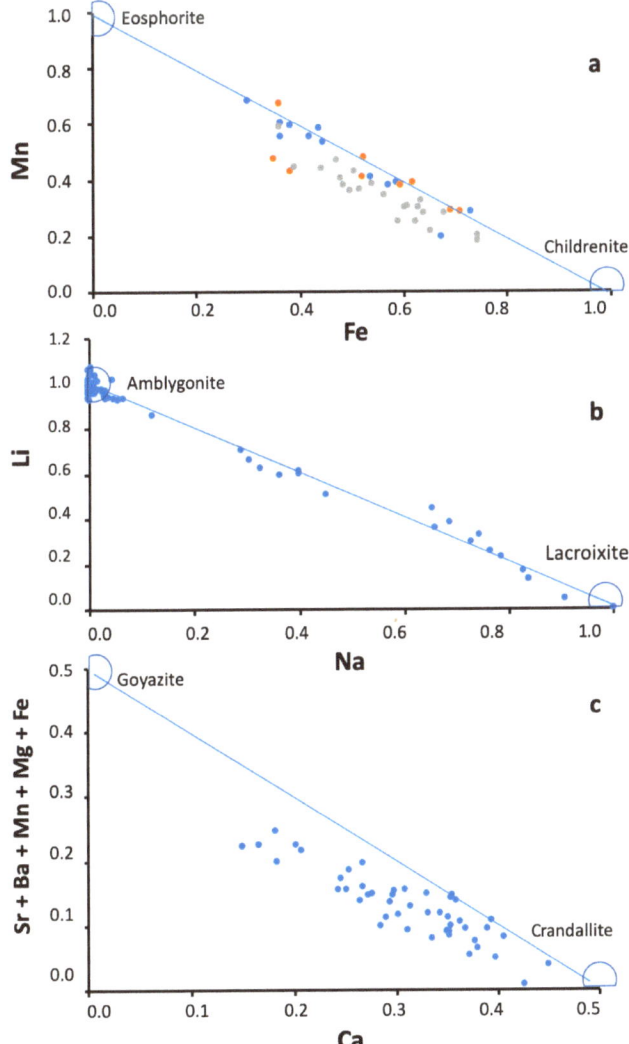

Figure 8. (**a**) Fe and Mn (apfu) in Fe-Mn phosphates (base P = 1) for three distinct crystals represented by three distinct colors. The figure shows that the amplitude of the Fe-Mn exchange covers the same range in each crystal. (**b**) Na versus Li (apfu) diagram for the Na-Li (lacroixite) phosphates (montebrasite–amblygonite) series. The intermediate composition corresponds to the partial replacement of the Na-Li-end-members by the Na-end-member. (**c**) Ca versus the Sr (apfu) plot for the crandallite–goyazite series.

Eosphorite–childrenite displays a significant Fe-Mn substitution (Table 1), which is found in all the crystals, as illustrated by Figure 8a, for three distinct crystals. In the external rim, more than 22% FeO is reached for 4 to 5.3% MnO. They contain low amounts of magnesium (MgO ranges from 0.1 to 1.3%). Still, the main zoning corresponds to a significant Fe and Mn concentration fluctuation with a Fe/Fe + Mn ratio varying from 0.3 to 0.7.

Table 1. Representative analyses and corresponding structural formulae of the Fe-Mn-rich phosphate series (eosphorite–childrenite) from the Segura dykes.

Weight %	Phosphates from the Eosphorite-Childrenite Series							
MgO	0.97	0.18	0.29	n.d.	0.08	0.16	0.18	0.21
Al_2O_3	22.68	22.89	22.99	21.98	22.16	22.18	21.96	22.47
P_2O_5	31.23	30.89	30.27	29.54	30.3	29.81	30.4	30.36
CaO	0.27	0.15	0.26	n.d.	0.02	0.14	0.06	n.d.
FeO	9.51	11.91	13.37	15.64	18.01	18.65	21.90	22.49
MnO	21.35	18.28	17.72	14.24	11.86	11.65	8.72	8.7
total	87.07	85.63	86.39	83.14	84.44	84.67	85.66	86.74
a.p.f.u.	Structural formulae							
Mg	0.05	0.01	0.02	0.00	0.00	0.01	0.01	0.01
Al	1.01	1.03	1.06	1.04	1.02	1.04	1.01	1.03
P	1	1	1	1	1	1	1	1
Ca	0.01	0.01	0.01	0	0	0.01	0	0
Fe	0.30	0.38	0.44	0.52	0.59	0.62	0.71	0.73
Mn	0.68	0.59	0.59	0.48	0.39	0.39	0.29	0.29
Fe + Mn	0.98	0.97	1.01	1.03	0.98	1.01	1.00	1.02
Fe/(Fe + Mn)	0.31	0.39	0.43	0.52	0.60	0.61	0.71	0.72

n.d.: not determined.

The montebrasite–amblygonite series represents the predominant Li-Al phosphate end-members (Figure 8b, Table 2). They are characterised by variable (OH, F) contents and a systematic replacement by lacroixite (Na-Al end-member). Lacroixite is probably physically mixed with the former montebrasite–amblygonite as compositions do not enrich the Na-end-member. The data points with a ratio of Na/Na+ Li that is higher than 0.5 are scarce (Table 2). Mixed analyses cannot be precluded as the lacroixite microdomains are relatively small (Figure 8b).

Table 2. Representative analyses and corresponding structural formulae of the Li-Na-rich phosphate series (amblygonite-lacroixite) from the Segura dykes.

Weight %	Lacroixite			Amblygonite	
Na_2O	14.37	15.63	15.92	0.03	0.04
MgO	0.00	0.00	0.00	0.00	0.04
Al_2O_3	31.62	31.31	32.40	35.33	34.32
P_2O_5	45.13	43.24	43.54	47.55	48.04
CaO	0.00	0.00	0.00	0.06	0.09
FeO	0.00	0.00	0.00	0.08	0.06
MnO	0.00	0.00	0.00	0.04	0.00
F	12.50	12.54	11.74	6.07	6.32
H_2O *	0.00	0.00	0.56	3.78	3.00
Li_2O *	2.57	1.57	1.49	9.99	10.09
Total	105.57	103.97	105.65	102.94	102.01
O=F	5.26	5.28	4.94	2.56	2.66
Total *	100.31	98.69	100.71	100.38	99.35

Table 2. Cont.

a.p.f.u.		Structural formulae			
Na	0.73	0.83	0.84	0.00	0.00
Li	0.27	0.17	0.16	1.00	1.00
Mg	0.00	0.00	0.00	0.00	0.00
Al	0.98	1.01	1.04	1.03	0.99
P	1.00	1.00	1.00	1.00	1.00
Ca	0.00	0.00	0.00	0.00	0.00
Fe	0.00	0.00	0.00	0.00	0.00
Mn	0.00	0.00	0.00	0.00	0.00
OH	0.00	0.00	0.10	0.63	0.49
F	1.03	1.08	1.01	0.48	0.49

*: asterisk corresponds to calculated values taking into account charge balance.

Crandallite–goyazite (Ca-Sr end-member) show a high variation in the Ca or Sr contents and additional cations such as Ba (Figure 8c, Table 3). The divalent cation site is partly vacant, as shown in Figure 8c. The solid solution is mainly in the crandallite field as the ratio of Ca/Ca + Sr is mostly above 0.5. F-apatite, and (F-OH) Al-Fe-Mn-rich apatite end-members are also found with stoichiometry close to apatite but with variable Mn contents.

Table 3. Representative analyses and corresponding structural formulae of the Ca-Sr-rich phosphate series (goyazite–crandallite) from the Segura dykes. *: asterisk corresponds to calculated values taking into account charge balance.

Weight %	Phosphates from the Goyazite-Crandallite Series								
Na_2O	0.00	0.00	0.24	0.00	0.15	0.12	0.21	0.00	0.00
MgO	0.35	0.00	0.26	0.00	0.17	0.17	0.18	0.00	0.00
Al_2O_3	33.99	33.34	34.51	33.40	34.15	33.40	35.50	35.15	35.98
P_2O_5	32.43	31.82	32.12	31.42	31.60	30.61	32.76	34.22	33.26
CaO	4.26	4.62	6.44	7.49	8.83	9.51	9.40	10.75	11.18
FeO	0.00	0.00	0.14	0.00	0.04	0.00	0.37	0.00	0.00
MnO	0.00	0.00	0.00	0.00	0.15	0.01	0.00	0.00	0.00
SrO	14.58	14.08	10.60	8.17	7.69	6.01	5.51	3.61	0.59
BaO	0.00	0.00	0.00	0.00	0.40	0.00	0.05	0.00	0.00
F	3.90	3.35	2.68	4.11	2.22	2.10	1.19	1.97	3.49
O=F	1.64	1.41	1.12	1.72	0.93	0.88	0.50	0.83	1.46
Total	85.61	83.86	87.10	80.48	86.13	83.32	85.29	83.73	81.01
H_2O *	12.75	14.73	14.57	17.80	16.29	19.29	15.57	15.44	17.53
Total *	87.25	85.27	85.43	82.20	83.71	80.71	84.43	84.56	82.47
a.p.f.u.	Structural formulae								
Na	0.00	0.00	0.03	0.00	0.02	0.02	0.03	0.00	0.00
Mg	0.04	0.00	0.03	0.00	0.02	0.02	0.02	0.00	0.00
Al	2.92	2.92	2.99	2.96	3.01	3.04	3.02	2.86	3.01
P	2.00	2.00	2.00	2.00	2.00	2.00	2.00	2.00	2.00

Table 3. Cont.

a.p.f.u.		Structural formulae								
Ca	0.33	0.37	0.51	0.60	0.71	0.79	0.73	0.80	0.85
Fe	0.00	0.00	0.00	0.00	0.00	0.00	0.00	0.00	0.00
Mn	0.00	0.00	0.00	0.00	0.01	0.00	0.00	0.00	0.00
Sr	0.41	0.40	0.31	0.23	0.22	0.18	0.16	0.10	0.02
Ba	0.00	0.00	0.00	0.00	0.01	0.00	0.00	0.00	0.00
R^{2+}	0.78	0.76	0.84	0.84	0.97	0.98	0.91	0.89	0.87
Ca/(Ca + Sr)	0.89	0.96	1.25	1.44	1.52	1.63	1.64	1.78	1.96
F	0.00	0.79	0.62	0.98	0.53	0.51	0.27	0.43	0.78
OH	4.32	3.50	4.07	3.58	4.47	4.59	4.63	3.94	3.99
F + OH	4.32	4.28	4.69	4.56	4.99	5.10	4.90	4.37	4.77

4.3. Geochemistry of Aplite and Pegmatite Dykes from Segura

All the dykes correspond to fractionated aplite–pegmatites enriched in P, Li, F, and to a lesser extent B (for one type of tourmaline-rich aplite). These rocks represent highly differentiated magmas, similar to the LCT facies under the rare element Li subclass of the complex type of classification that is used in [36]. New chemical analyses of granites, aplites, and pegmatites from Segura were obtained on a series of collected samples, as indicated in Figure 1 and reported in Table 4. They were compared with data from previous works (four mean analyses only available from [5]) on the same region and with similar intrusive granite bodies (Argemela: [10], Panasqueira: [37]), Tres Arroyos [3], and Beauvoir (B1 facies, [11]).

Table 4. Representative analyses of pegmatites and aplites from the Segura dyke pairs of aplites and pegmatites noted, with CQ-4, -5, -6, and -9 being adjacent samples and each pair being from the same dyke. bld: below the detection limit. In the three samples, loss on ignition (L.O.I.) was not determined but estimated by difference (*).

	Pegmatite									Aplite			
wt.%	GF SEG1	GF SEG4	Peg CQ-5	Peg CQ-6	Peg CQ-9	Peg CQ-4	GF SEG-3	GF SEG-2	GF SEG-5	Apl CQ-4	Apl CQ-5	Apl CQ-6	Apl CQ-9
SiO_2	76.78	69.19	64.04	71.54	70.57	69.29	70.19	70.99	71.74	69.83	69.53	71.19	69.69
TiO_2	0.01	bdl	bdl	bdl	bdl	bdl	0.005	0.02	0.005	bdl	bdl	bdl	bdl
Al_2O_3	13.43	14.77	18.71	15.12	15.36	16.41	16.85	15.33	16.63	16.47	17.84	16.42	16.12
Fe_2O_3	0.19	0.19	0.413	0.159	0.069	0.073	0.76	1.13	0.76	0.171	0.389	0.268	0.167
MnO	0.006	0.024	0.095	0.021	0.000	0.000	0.078	0.048	0.052	0.020	0.089	0.023	0.017
MgO	bdl	bdl	bdl	bdl	bdl	bdl	0.020	bdl	0.010	bdl	bdl	bdl	bdl
CaO	0.3	0.22	0.31	0.74	0.53	0.53	0.18	0.51	0.22	0.70	0.22	0.53	0.85
Na_2O	6.49	6.06	6.39	5.40	4.96	5.48	6.45	5.03	5.48	6.43	6.45	6.42	6.15
K_2O	1.12	3.85	1.22	2.68	3.72	4.31	2.66	3.5	2.54	2.08	1.00	1.77	2.05
P_2O_5	0.46	1.7	4.76	2.49	3.72	2.63	2.43	1.00	1.94	2.29	2.05	2.05	2.38
L.O.I.	1.21 *	4.00 *	2.11	1.32	1.28	1.19	1.11	n.d.	1.2	1.51	1.61	1.6	1.75
Total	98.79	96.00	98.05	99.47	100.21	99.92	101.43	97.56	100.98	99.50	99.18	100.26	99.17
Li -ppm	n.d.	1160	3278	704	1718	818	n.d.	1000	n.d.	157	1543	304	141
F -ppm	1123	2900	19,100	3300	3900	2800	2805	2100	1806	3100	18200	3400	3200

The SiO_2 contents displayed by the Segura facies range from 72.2 to 74.0 wt% in the granites and from 73.3 to 74.8 wt% SiO_2 in the aplite–pegmatite bodies, being comparable to the values recorded for the Panasqueira granites (72.5–75 wt% SiO_2) but slightly more siliceous than the Argemela granites (66–72 wt% SiO_2). The Segura granites display intermediate features between the rather albitic facies from Argemela and the most evolved, albite-rich Panasqueira granites. The K_2O to Na_2O ratios range between 0.7 and 1.2 and are thus lower than the Panasqueira granites (0.5 and 3.1) and higher than the Argemela granites (0.2 to 0.7). The enrichment in Na$_2$O depicts the differentiation and the more albitic feature of the evolved magmas (Figure 9).

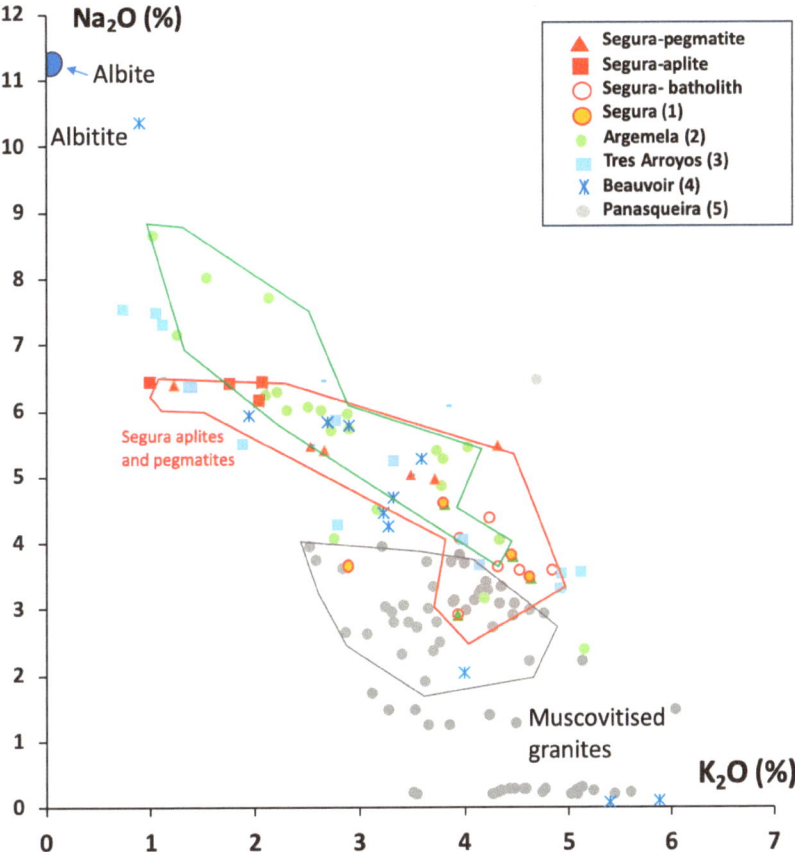

Figure 9. Na$_2$O versus K$_2$O diagram for the Segura aplites, pegmatites, and granites. Data are compared with those from the literature as follows: 1: Segura [5], 2: Argemela [10], 3: Tres Arroyos [3], 4: Beauvoir [11], and 5: Panasqueira [37]. Enveloppe colors correspond to data point colors.

They are similar to the Beauvoir B1 and Tres Arroyos facies and are largely superimposed on most Argemela facies. All these granites are significantly enriched in sodium compared with the Panasqueira granite.

Mafic components are low to very low, with the FeO + MgO + TiO$_2$ sum being comprised between 0.14 and 1.40, being much lower than in the Panasqueira granites (1.71 and 3.12 wt%) and either similar or higher than in the Argemela ones (0.03 to 0.19). All the rocks are magnesium-poor, with the M index (M = 100 Mg/Mg + $\sum$Fe) being between 8 and 45, similarly to the Panasqueira (15 to 31) and Argemela granites (6 to 46). The magmatic trends are more evident to decipher at Segura and Argemela in the geochemical diagram Al – (K + Na + 2Ca) versus the Fe + Mg + Ti diagram from [38], where the parameter A increases as B decreases. All the rocks are distinctly peraluminous (Figure 10).

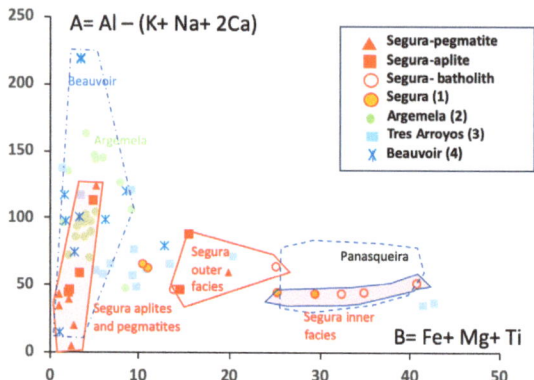

Figure 10. Diagram A (=Al − (K+ Na + Ca)) versus B (=Fe+ Mg+ Mn) from [38] applied to the Segura facies and with a comparison with the reference peraluminous differentiated granites. 1: Segura [5], 2: Argemela [10], 3: Tres Arroyos [3], and 4: Beauvoir [11]. For Panasqueira, only the envelope is shown.

At Panasqueira, the peraluminous parameter Al − (K + Na + 2Ca) is, however, increased by the superimposed alteration effects, which is evident for the most altered facies but also perceptible for many of the less modified granite facies [37]. The Segura aplite and pegmatite dykes are significantly enriched in phosphorous, with the P_2O_5 contents in the 0.36 to 2.23 wt% range being higher than the Panasqueira granites (0.23 to 0.74 wt%). Compared with the Panasqueira granites, which are dominated by a strong correlation between the phosphorous and calcium content meaning that the two elements are both entirely contained in apatite (Figure 11), the Segura aplite–pegmatite veins are characterised by an excess in phosphorus, which is not linked to apatite.

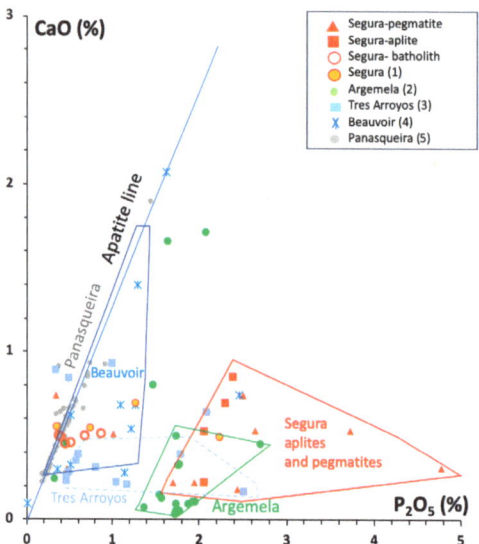

Figure 11. CaO versus P_2O_5 diagram applied to the same series of granites as in Figure 9. It shows the increased phosphorus content of the Segura aplites and pegmatites and the shift from the apatite line, which is high in the Segura and Argemela magmas compared with the other P-rich magmas: 1: Segura [5], 2: Argemela [10], 3: Tres Arroyos [3], 4: Beauvoir [11], and 5: Panasqueira [37].

Thus, data points are located to the right of the apatite line, thereby indicating that for a given calcium content, the phosphorus is connected to another phosphate (Li, Na, or another element such as Sr, for instance, being the dominant phosphate) other than apatite, as confirmed by the petrography and crystal chemistry analyses presented above. The Segura aplites and pegmatites are even more enriched in P_2O_5 than the Argemela granites. The F, Li, and P contents are roughly correlated positively.

5. Discussion
5.1. Comparison with Other Similar Geochemical Suites

Leucocratic peraluminous magmas are formed either by the fractional crystallisation of a parent S-type granitic magma with progressive enrichment in incompatible elements during differentiation [1,3,39] or by anatectic-type melting, which is independent of the formation of the peraluminous granitic plutons that are found in the environment close to pegmatites [4,40]. Low proportions of partial protolith melting could produce low volumes of magmas enriched in incompatible elements, with a mineralogical composition close to the ultimate products of differentiation. This duality between the presence of pegmatites that are often later than the plutons and derived from distinct melting processes at different structural levels in the lithosphere has been evoked in several examples as follows: the Limousin pegmatites (French Massif Central [41]) and the Fregeneda-Almendra pegmatite field in Portugal [42]. The same situation occurs at Segura. The Authors in [5] have already ruled out the hypothesis of a genetic link between the main Cabeza de Araya batholith in Segura and the spatially nearby dykes. What is particularly interesting here is the extreme enrichment in volatiles (F, Li, and P), which may have also influenced the melting temperature of the magma at the migmatisation front.

The abnormal contents in P, Al, and F are much above those expected from peraluminous granite differentiation [12,43–45]. Figure 12 shows that the Segura aplites and pegmatites have around the same F content as the Argemela and Tres Arroyos granites (2500–4000 ppm) but higher P_2O_5 values between 2 and 3% (Figure 12). The reasons for these high enrichments can differ [46].

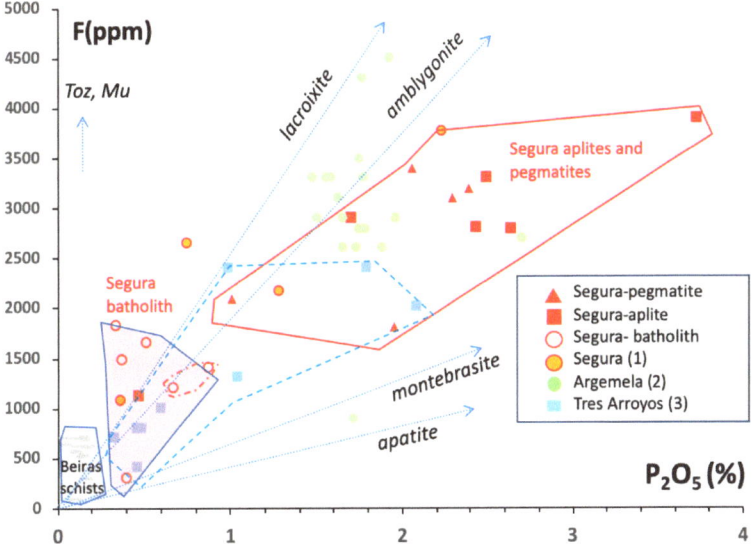

Figure 12. F versus P_2O_5 diagram showing the increased phosphorus content of the Segura aplites and pegmatites (red symbols) as a function of fluorine content. Reference lines in the direction of the phosphate end-members are reported. 1: Segura: [5], 2: Argemela [10], and 3: Tres Arroyos [3].

The studied aplites and pegmatites are particularly enriched in P, F, Li (locally B), Al, and Na and present a positive correlation between all these elements and the aluminium saturation index of ASI = [Al/(Ca − 1.67P + Na + K)] as defined by [47,48] (Figure 13). These enrichments can occur at several stages of the magma genesis, migration, and evolution as follows: from the melting zone, during the upward journey and magmatic differentiation, later when a supercritical fluid is extracted from the cooling magma, and, finally, in the fluid phase acting throughout the magmatic–hydrothermal transition.

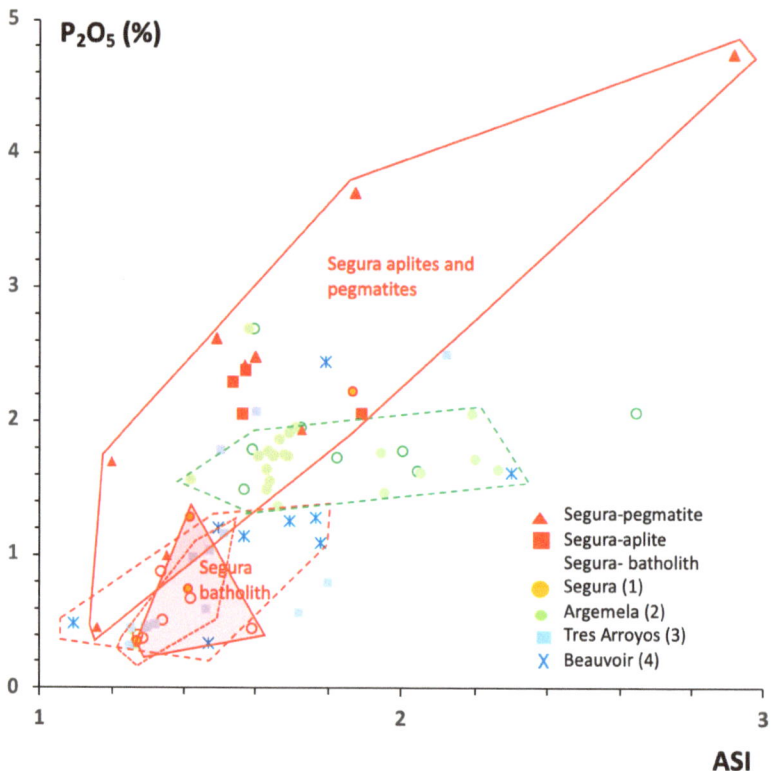

Figure 13. P_2O_5 versus the aluminium saturation index (ASI = [Al/(Ca − 1.67P + Na + K)]) diagram. 1: Segura [5], 2: Argemela [10], 3: Tres Arroyos [3], and 4: Beauvoir [11].

5.2. Causes of Enrichments in Phosphorus, Lithium, and Fluorine

It is difficult to prove that the schists of the Beiras Group, which are the host rocks of the aplite and pegmatite sills, are also the initial sources of the magma. However, the metasedimentary sequence is thick (up to 11 km; [49]) and voluminous, so the melting of metasediments that are compositionally similar to those exposed could be envisaged at a lower level. The P_2O_5, F, and Li content ranges of the Beiras schists are 2000–3000 ppm, 500–700 ppm, and 45–80 ppm, respectively. The F and P_2O_5 concentration ranges in the schists are around two to three times lower than the main batholith granites and five times lower than the outer facies (muscovite granite).

During metapelite anatexis, the phosphorus content of the initial magma is mainly controlled by the apatite content of the source, the degree of melting, and the solubility of the apatite in the melt. The low initial stock of REE and Y explains the absence of monazite or xenotime.

Thomas et al. [50] suggest extreme phosphorus enrichment may be due to metastable sub-liquidus immiscibility. The spontaneous separation into two undercooled melt phases could be linked to thermal fluctuations. Thomas et al. [50] point out that the driving force behind phase separation is a decrease in the free energy of the system.

The very high fluorine content of late-phase liquids could be interpreted as the result of extreme fractionation of granitic magma. The initial content of the parent granitic magma could well be no more than a conventional value of around 0.5–1% F. For phosphorus, Thomas et al. [50] also consider the effect of boiling to be related to decompression after fracturing under the impact of tectonic stress on the formation of phosphorus-enriched globules, which might produce a highly phosphate-rich magma.

5.3. Migration and Extraction of Phosphorus-Rich Magma

The enrichment factor for P_2O_5 is around 2.5 to 3 between the outer muscovite granite and the aplite–pegmatite dykes from Segura and 1.5 to 3 for fluorine concerning the very same group of rocks (Figure 12). The very high levels of F, P, H_2O Li_2O, Rb_2O, Cs_2O, and B in the residual magma liquids related to these rocks may have resulted in a considerable decrease in viscosity, with values of the order of 10 Pa [51,52] and lower being likely. Thanks to this very low viscosity, intergranular magmas can move through granitic matrices according to the model developed by [53]. They can then extract themselves from the magmas and migrate through structural discontinuities (such as shear or fault zones), particularly in the schistose host rock. Such a geometry is typically observed at Segura, i.e., a series of low-dipping dykes filling dilating spaces in the foliation plane or inserting themselves into fractures sub-perpendicular to the foliation, which constitute the most common plane of weakness in the shales.

5.4. The Behaviour of Li, F, and P during the Crystallization of Perphosphorus Magmas

Some phosphorus may be incorporated in crystallising feldspars when calcium is low in the melt. As phosphorus and aluminium form a coupled substitution (the so-called berlinite substitution, $P^{5+} + Al^{3+} = 2Si^{4+}$), the incorporation of phosphorus in feldspar is favoured in peraluminous magmas. Then, considering the strong affinity between P and Li, the phosphorus concentrated in the residual melt is partially removed first by the amblygonite–montebrasite and the eosphorite–childrenite phosphate series formation, which appears first. Amblygonite and eosphorite develop early during the crystallisation of aplites and pegmatites but represent the final stage of the magmatic evolution that began lower down during the crystallisation of the muscovite-rich granite or another non-exposed late intrusive granite. It is difficult to ascribe the Fe-Mn compositional changes to a particular stage of the magma evolution except for its very late crystallisation and a rather low fO_2, allowing for the incorporation of the relatively low amount of iron available in the melt as Fe^{2+} in the eosphorite. At that stage, the crystallisation of the amblygonite–montebrasite series regulates or decreases the fluorine content in the melt [54] and probably the saturation with respect to topaz. During the final stage of the hydrothermal evolution of pegmatites, phosphates from the goyazite–crandallite series form along grain boundaries at the expense of the Li-Na phosphates, after which numerous secondary apatite crystals form and constitute the final expression of the P-bearing phases. When alkali feldspars crystallise out of a perphosphorus and peraluminous magmatic system, they can become the main reservoirs of phosphorus. In the case of Segura, the late apatite grains likely crystallised from this stock, which means that the P concentrations of the feldspars can no longer be used as indicators of the initial P richness of the magma. The overall paragenetic sequence is proposed in Figure 14. According to [55,56], phosphorous can be released secondarily by the Al-Si ordering of alkali feldspars, and fluorine is still high, probably in the remaining fluids. During the magmatic–hydrothermal transition, the crystallisation of these late phosphates needs external hydrothermal inputs (Ca-Sr (Ba), then Ca).

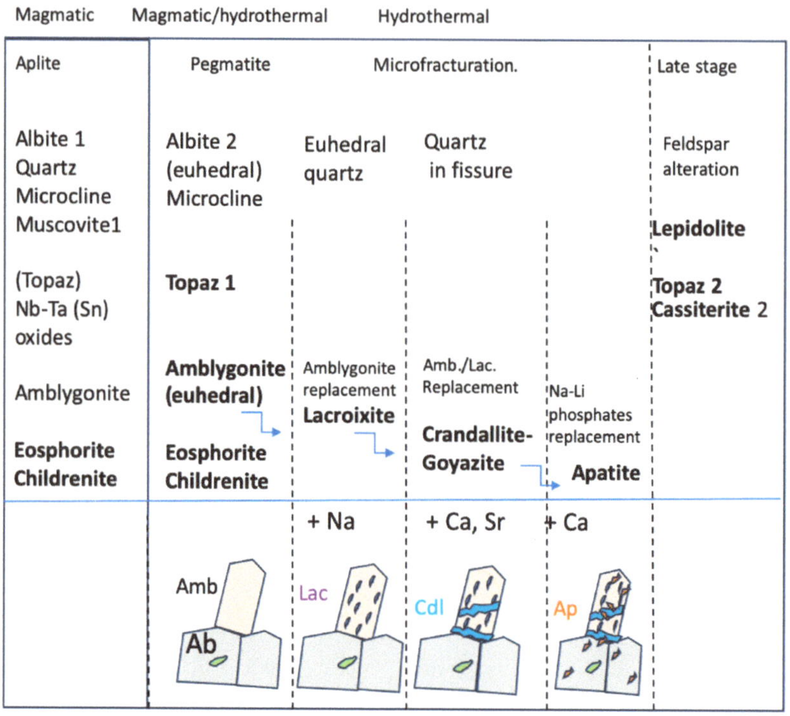

Figure 14. Schematic representation of the mineralogical evolution of the Segura aplites and pegmatites with the time from the magmatic to hydrothermal stages, especially for the phosphates. Amb: amblygonite, Lac: lacroixite, Cdl: crandallite, and Ap: Apatite.

5.5. Conceptual Model of the Melt Injection in the Beiras Schists

A model is proposed for the formation of the aplite–pegmatite dykes at Segura in Figure 15. These dykes cannot be related geochemically to the main inner Cabeza de Araya cordierite monzogranite but only to the external facies, in particular, the muscovite-rich facies, which constitutes a relatively small mass of magmas injected at the boundary between the batholith and the metasedimentary country rocks. The upwelling geometry of the magma that formed the muscovite-rich granite is likely to be independent of that of the main batholith, the centre of which is located more than 40 km to the southeast. As shown by the arrows in Figure 15, the migration pathways are challenging to assess. A fourth magma injection at the boundary with the Cabeza de Araya batholith or a distinct pathway for the muscovite granite could have occurred. The magmas that evolved in the late stages of the Late Variscan evolution appear to have risen through zones of crustal weakness in many places, which could be the case for the Cabeza de Araya batholith [6,30]. In the schematic cross-section of the upper lithosphere thought to represent the situation during the 310–300 Ma period, it is proposed that of a series of intrusive facies ascend from a migmatitic zone at depth. Such a geometry was already suggested by [57] for the G2-G5 including the rare metal granite intrusions at Panasqueira, and the Argemela region (cross-section in [10]) where processes occur under a NE-SW shortening. To explain the formation of dykes in the metasediments near this injection locus, it is once again necessary to refer to the opening conditions of the foliation planes or fractures perpendicular to this plane. The local stress tensor at that time stimulates these openings.

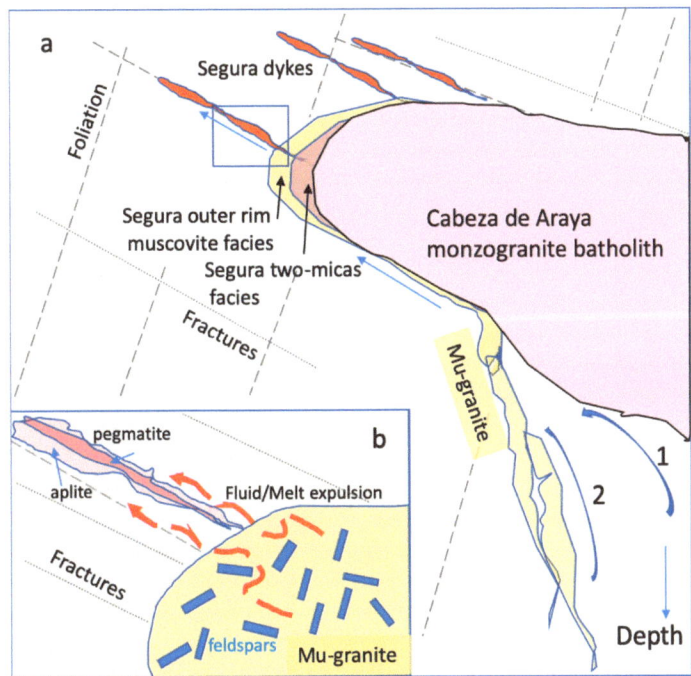

Figure 15. (**a**) Conceptual model of the intrusive dykes (vertical cross-section) in the westernmost sector of the Cabeza de Araya batholith, the intrusion of the muscovite-rich outer facies and dykes in the metasediments. The magma migration pathways are as follows: arrow 1 corresponds to an additional layer of the Cabeza de Araya batholith like the other facies that compose it, with a feeder zone whose centre is located far to the southeast, and arrow 2 to an autonomous magma intrusion that is independent of the initial pathways. (**b**) Inset in figure (**a**), with the extraction of a low-viscosity magma migrating through open tension gashes and fractures in the metasediments.

The formation of magmas hyper-enriched in F, Li, and P was facilitated by the already high concentrations of the muscovite granite facies. Additional processes are required as outlined above, notably filter-pressing mechanisms expulsing these magmas with high water, F, Li, and P contents, and very low viscosity, with these facilitating their extraction from the crystallizing melt when subjected to lateral pressure. Once they have removed themselves from the initial partly crystallised melt, they may migrate into the zones of weakness open in the metasedimentary series. The crystallisation of the aplites and then the pegmatites take place in a sub-closed system. Still, the rise in temperature at the same time triggers convective processes which favour the arrival of fluids in equilibrium with the metamorphic series and induce the elements necessary for the metasomatism of the initial assemblages and, in particular, the crystallisation of the crandallite–goyazite series and the late hydrous apatites. The percolation of late fluids with an essentially metamorphic signature, which is mainly shown by the abundance of methane-rich fluid inclusions, has been identified [58,59]. Unfortunately, this fluid event provoked quartz recrystallisation and has erased all traces of previous magmatic fluids.

6. Conclusions

The aplites and pegmatites of Segura were formed from magma and fluids particularly enriched in phosphorus, fluorine, water, and lithium. Compared with other highly evolved granites, such as those of Argemela, Tres Arroyos, and Beauvoir, the concentrations of P, F, and Li in the aplites and pegmatites are higher than those in the other granites mentioned. Their aluminium saturation index (ASI) is also higher.

The enrichment factors for these elements compared with peraluminous granites or monzogranites are 1.5 to 5. These enrichments were made possible by favourable processes such as extracting the low-viscosity magma from the crystallizing melt.

The formation of the dykes in the metasediments took place close to the main injection site of the muscovite-rich granite, which is off-centre from that of the Cabeza de Araya monzogranite. The opening of the foliation planes or fractures perpendicular to this plane required sub-horizontal stresses during a major compression event linked to the opening of the main muscovite granite feeder drain from the deep zones where the partial melting of the metasediments occurs.

The paragenetic sequences were demonstrated using micro-X-ray fluorescence chemical imaging and show the phosphates crystallise to be in the following order: Fe-Mn phosphate > amblygonite (montebrasite) > lacroixite > crandallite–goyazite > apatite-Mn. The alteration of the initial lithium-bearing phosphate assemblage is consistent with developments already described elsewhere, notably that of the crystallisation of lacroixite at the expense of amblygonite, the formation of the crandallite–goyazite series at the expense of amblygonite, and hydrothermal apatite at the expense of albite and the earlier phosphate phases.

The contribution of calcium and other divalent cations indicates the late involvement of hydrothermal fluids external to the magmatic system.

Author Contributions: M.C., sampling, data acquisition and interpretation, conceptualization, writing (original draft preparation, review and editing), and funding acquisition; M.-C.B., sampling, data acquisition and interpretation, writing (review and editing), and funding acquisition; A.L., data acquisition and writing (review and editing); I.M., fieldwork, geological mapping, sampling, and writing (review and editing); Í.D.d.S., fieldwork, geological mapping, sampling, and writing (review and editing); A.M., fieldwork, geological mapping, sampling, conceptualization, data interpretation, writing (review and editing), and funding acquisition. All authors have read and agreed to the published version of the manuscript.

Funding: This research was funded by ERAMIN2 through the project MOSTMEG (n°: ERA-MIN/0002/2019 (for more details: http://doi.org/10.54499/ERA-MIN/0002/2019; https://mostmeg.rd.ciencias.ulisboa.pt/). This work contributes to a better knowledge of granite-related ore systems, namely those enriched in lithium. It benefits the research carried out within the framework of the programmes of Labex Ressources21 under the reference ANR-10-LABX-21-RESSOURCES21 supported by the Agence Nationale de la Recherche through the national programme "Investissements d'avenir", which contributes to the co-funding of the ERAMIN project. Additional support came from the Portuguese Fundação para a Ciência e a Tecnologia (F.C.T.) I.P./MCTES through the PIDDAC national funds (n°s: UIDB/50019/2020, LA/P/0068/2020; more details at https://doi.org/10.54499/UIDB/50019/2020; https://doi.org/10.54499/LA/P/0068/2020) and PD/BD/142783/2018).

Data Availability Statement: The data presented in this study are available on request from the corresponding author due to the future reseach data based.

Acknowledgments: O. Rouer and M.C. Caumon (GeoRessources) are thanked for their help when acquiring the EMPA and Raman data. Four anonymous reviewers are warmly acknowledged for their suggestions.

Conflicts of Interest: The authors declare no conflicts of interest.

References

1. Breiter, K.; Durisova, J.; Korbelova, Z.; Lima, A.; Vasinova Galiova, M.; Hlozkova, M.; Dosbaba, M. Rock textures and mineral zoning—A clue to understanding rare-metal granite evolution: Argemela stock, Central-Eastern Portugal. *Lithos* **2022**, *410–411*, 106562. [CrossRef]
2. Garate-Olave, I.; Müller, A.; Roda-Robles, E.; Gil-Crespo, P.P.; Pesquera, A. Extreme fractionation in a granite–pegmatite system documented by quartz chemistry: The case study of Tres Arroyos (Central Iberian Zone, Spain). *Lithos* **2017**, *286*, 162–164. [CrossRef]

3. Garate-Olave, I.; Roda-Robles, E.; Gil-Crespo, P.P.; Pesquera, A.; Errandonea-Martin, J. The Tres Arroyos granitic aplite-pegmatite field (Central Iberian Zone, Spain): Petrogenetic constraints from the evolution of Nb-Ta-Sn oxides, whole-rock geochemistry and U-Pb geochronology. *Minerals* **2020**, *10*, 1008. [CrossRef]
4. Webber, K.L.; Simmons, W.B.; Falster, A.U.; Hanson, S.L. Anatectic pegmatites of the Oxford County pegmatite field, Maine, USA. *Can. Mineral.* **2019**, *57*, 811–815. [CrossRef]
5. Antunes, I.M.; Neiva, A.M.; Farinha Ramos, J.M.; Silva, P.B.; Silva, M.M.; Corfu, F. Petrogenetic links between lepidolite-subtype aplite-pegmatite, aplite veins and associated granites at Segura (central Portugal). *Chem. Erde Geochem.* **2013**, *73*, 323–341. [CrossRef]
6. Castro, A. Structural pattern and ascent model in the Central Extremadura batholith, Hercynian belt, Spain. *J. Struct. Geol.* **1986**, *8*, 633–645. [CrossRef]
7. Corretgé, L.G.; Suarez, O. A Garnet-Cordierite Granite Porphyry Containing Rapakivi Feldspars in the Cabeza de Araya Batholith (Extremadura, Spanish Hercynian Belt). *Mineral. Petrol.* **1994**, *50*, 97–111. [CrossRef]
8. Fernandéz, C.; Castro, A. Pluton accommodation at high strain rates in the upper continental crust. The example of the Central Extremadura batholith, Spain. *J. Struct. Geol.* **1999**, *21*, 1143–1149. [CrossRef]
9. García-Moreno, O.; Corretgé, L.G.; Holtz, F.; García-Aria, M.; Rodríguez, C. Phase relations in the Cabeza de Araya cordierite monzogranite, Iberian Massif: Implications for the formation of cordierite in a crystal mush. *Geol. Acta* **2017**, *15*, 337–359.
10. Michaud, J.A.S.; Gumiaux, C.; Pichavant, M.; Gloaguen, E.; Marcoux, E. From magmatic to hydrothermal Sn-Li-(Nb-Ta-W) mineralization: The Argemela area (central Portugal). *Ore Geol. Rev.* **2020**, *116*, 103215. [CrossRef]
11. Cuney, M.; Marignac, C.; Weisbrod, A. The Beauvoir topaz-lepidolite albite granite (Massif Central, France)—The disseminated magmatic Sn-Li-Ta-Nb-Be mineralization. *Econ. Geol.* **1992**, *87*, 1766–1794. [CrossRef]
12. London, D.; Wolf, M.B.; Morgan, G.B.; Gallego-Garrido, M. Experimental Silicate–Phosphate Equilibria in Peraluminous Granitic Magmas, with a Case Study of the Alburquerque Batholith at Tres Arroyos, Badajoz, Spain. *J. Petrol.* **1999**, *40*, 215–240. [CrossRef]
13. Garate-Olave, I.; Roda-Robles, E.; Gil-Crespo, P.P.; Pesquera, A. The phosphate mineral associations from the Tres Arroyos aplite-pegmatites (Badajoz, Spain): Petrography, mineral chemistry and petrogenetic implications. *Can. Mineral.* **2020**, *58*, 747–765. [CrossRef]
14. Ribeiro, A.; Munhá, J.; Dias, R.; Mateus, A.; Pereira, E.; Ribeiro, L.; Fonseca, P.; Araújo, A.; Oliveira, T.; Romão, J.; et al. Geodynamic evolution of the SW Europe Variscides. *Tectonics* **2007**, *26*, TC6009. [CrossRef]
15. Díez Fernández, R.; Pereira, M.F. Extensional orogenic collapse captured by strike-slip tectonics: Constraints from structural geology and U-Pb geochronology of the Pinhel shear zone (Variscan orogen, Iberian Massif). *Tectonophysics* **2016**, *691*, 290–310. [CrossRef]
16. Azor, A.; Dias da Silva, Í.; Gómez Barreiro, J.; González-Clavijo, E.; Martínez Catalán, J.R.; Simancas, J.F.; Martínez Poyatos, D.; Pérez-Cáceres, I.; González Lodeiro, F.; Expósito, I.; et al. Deformation and Structure. In *The Geology of Iberia: A Geodynamic Approach: Volume 2: The Variscan Cycle*; Quesada, C., Oliveira, J.T., Eds.; Springer International Publishing: Cham, Switzerland, 2019; pp. 307–348.
17. Dias da Silva, Í.; González Clavijo, E.; Díez-Montes, A. The collapse of the Variscan belt: A Variscan lateral extrusion thin-skinned structure in NW Iberia. *Int. Geol. Rev.* **2021**, *63*, 659–695. [CrossRef]
18. Dias da Silva, I.; Gomez-Barreiro, J.; Martínez Catalan, J.R.; Ayarza, P.; Pohl, J.; Martinez, E. Structural and microstructural analysis of the Retortillo Syncline (Variscan belt, Central Iberia). Implications for the Central Iberian Orocline. *Tectonophysics* **2017**, *717*, 99–115. [CrossRef]
19. Ribeiro, M.L.; Castro, A.; Almeida, A.; González Menéndez, L.; Jesus, A.; Lains, J.A.; Lopes, J.C.; Martins, H.C.B.; Mata, J.; Mateus, A.; et al. Variscan Magmatism. In *The Geology of Iberia: A Geodynamic Approach, Regional Geology Reviews*; Quesada, C., Oliveira, J.T., Eds.; Springer: Cham, Switzerland, 2019.
20. Villaseca, C.; Barbero, L.; Herreros, V. A re-examination of the typology of peraluminous granite types in intracontinental orogenic belts. *Trans. R. Soc. Edinb. Earth Sci.* **1998**, *89*, 113–119. [CrossRef]
21. Dias, G.; Leterrier, J.; Mendes, A.; Simoes, P.P.; Bertrand, J.M. U–Pb zircon and monazite geochronology of post-collisional Hercynian granitoids from the Central Iberian Zone (Northern Portugal). *Lithos* **1998**, *45*, 349–369. [CrossRef]
22. Lopez Plaza, M.Y.; Martinez Catalan, J.R. Síntesis estructural de los granitoides hercínicos del Macizo Hespérico. In *Geología de los Granitoides y Rocas Asociadas del Macizo Hespérico*; Bea, F., Carnicero, A., Gonzalo, J.C., Lopez Plaza, M., Rodriguez Alonso, M.D., Eds.; Libro homenaje a L.C. García de Figuerola: Editorial Rueda, Madrid, 1987; pp. 195–210.
23. Roda-Robles, E.; Villaseca, C.; Pesquera, A.; Gil-Crespo, P.P.; Vieira, R.; Lima, A.; Garate-Olave, I. Petrogenetic relationships between Variscan granitoids and Li-(F-P)-rich aplite-pegmatites in the Central Iberian Zone: Geological and geochemical constraints and implications for other regions from the European Variscides. *Ore Geol. Rev.* **2018**, *95*, 408–430. [CrossRef]
24. Martins, I.; Mateus, A.; Cathelineau, M.; Boiron, M.C.; Ribeiro da Costa, I.; Dias da Silva, Í.; Gaspar, M. The Lanthanide "Tetrad Effect" as an Exploration Tool for Granite-Related Rare Metal Ore Systems: Examples from the Iberian Variscan Belt. *Minerals* **2022**, *12*, 1067. [CrossRef]
25. Melleton, J.; Gloaguen, E.; Frei, D.; Lima, A.; Vieira, R.; Martins, T. Polyphased rare-element magmatism during late orogenic evolution: Geochronological constraints from NW Variscan Iberia. *Bull. Soc. Geol. Fr.* **2022**, *193*, 7. [CrossRef]
26. Pereira, A.; Pereira, L.; Macedo, C. Os plutonitos da Zebreira (Castelo Branco): Idade e enquadramento estructural. *Mem. Not. Publ. Mus. Lab. Mineral. Geol.* **1986**, *101*, 21–31.
27. Antunes, I.M.; Neiva, A.M.; Silva, M.M.; Corfu, F. The genesis of I- and S-type granitoid rocks of the Early Ordovician Oledo pluton, Central Iberian Zone (central Portugal). *Lithos* **2009**, *111*, 168–185. [CrossRef]

28. Inverno, C.; Carvalho, D.d.; Parra, A.; Reynaud, R.; Filipe, A.; Martins, L. *Carta de Depósitos Minerais de Portugal (Folha 3), à Escala 1:200,000*; LNEG: Lisboa, Portugal, 2020.
29. Instituto Geológico e Mineiro. *Carta Geológica de Portugal, escala 1:500,000*; Serviços Geológicos de Portugal, Instituto Geológico e Mineiro: Lisboa, Portugal, 1992.
30. Vigneresse, J.L.; Bouchez, J.L. Successive granitic magma batches during pluton emplacement: The case of Cabeza de Araya (Spain). *J. Petrol.* **1997**, *38*, 1767–1776. [CrossRef]
31. Sousa, M.B. Considerações sobre a estratigrafia do Complexo Xisto-Grauváquico (CXG) e sua relação com o Paleozóico Inferior. *Cuad. Geol. Ibérica* **1984**, *9*, 9–36.
32. Carignan, J.; Hild, P.; Mevelle, G.; Morel, J.; Yeghicheyan, D. Routine analyses of trace elements in geological samples using flow injection and low pressure on line liquid chromatography coupled to ICP-MS: A study of geochemical reference material B.R., DR-N, EB-N, AN-G and G.H. *Geostand. Newsl.* **2001**, *25*, 187–198. [CrossRef]
33. Lafuente, B.; Downs, R.T.; Yang, H.; Stone, N. The power of databases: The RRUFF project. In *Highlights in Mineralogical Crystallography*; Armbruster, T., Danisi, R.M., Eds.; W. De Gruyter: Berlin, Germany, 2015; pp. 1–30.
34. Frost, R.L.; Xi, Y.; Scholz, R.; López, A.; Lima, R.M.F.; Ferreira, C.M. Vibrational spectroscopic characterization of the phosphate mineral series eosphorite–childrenite–$(Mn,Fe)Al(PO_4)(OH)_2 \cdot (H_2O)$. *Vib. Spectr.* **2013**, *67*, 4–21.
35. Rondeau, B.; Fritsch, E.; Lefèvre, P.; Guiraud, M.; Fransolet, A.-M.; Lulzac, Y. A Raman investigation of the amblygonite-montebrasite series. *Can. Mineral.* **2006**, *44*, 1109–1117. [CrossRef]
36. Cerný, P.; Ercit, T.S. Mineralogy of niobium and tantalum: Crystal chemistry relationship, paragenetic aspects and their economic implications. In *Lanthanides, Tantalum and Niobium*; Moller, P., Černý, P., Saupé, F., Eds.; Springer: Berlin, Germany, 1989; pp. 27–29.
37. Marignac, C.; Cuney, M.; Cathelineau, M.; Lecomte, A.; Carocci, E.; Pinto, F. The Panasqueira rare metal granite suites and their involvement in the genesis of the world-class Panasqueira W–Sn–Cu vein deposit: A petrographic, mineralogical, and geochemical study. *Minerals* **2020**, *10*, 562. [CrossRef]
38. Debon, F.; Lefort, P. A cationic classification of common plutonic rocks and their magmatic associations: Principles, method, applications. *Bull. Minéralogie* **1988**, *111*, 493–510. [CrossRef]
39. Cerny, P. Geochemical and Petrogenetic Features of Mineralization in Rare Element Granitic Pegmatites in the Light of Current Research. *Appl. Geochem.* **1992**, *7*, 393–416.
40. Simmons, W.; Falster, A.; Webber, K.; Roda-Robles, E. Bulk composition of mt. Mica pegmatite, Maine, USA: Implications for the origin of an L.C.T. type pegmatite by anatexis. *Can. Mineral.* **2016**, *54*, 1053–1070. [CrossRef]
41. Deveaud, S.; Gumiaux, C.; Gloaguen, E.; Branquet, Y. Spatial statistical analysis applied to rare-element LCT-type pegmatite fields: An original approach to constrain faults-pegmatites-granites relationships. *J Geosci.* **2013**, *58*, 163–182. [CrossRef]
42. Roda-Robles, E.; Vieira, R.; Lima, A.; Errandonea-Martin, J.; Pesquera, A.; Cardoso-Fernandes, J.; Garate-Olave, I. Li-rich pegmatites and related peraluminous granites of the Fregeneda-Almendra field (Spain-Portugal): A case study of magmatic signature for Li enrichment. *Lithos* **2023**, *452–453*, 107195. [CrossRef]
43. Pichavant, M. Experimental Crystallization of the Beauvoir Granite as a Model for the Evolution of Variscan Rare Metal Magmas. *J. Petrol.* **2022**, *63*, egac120. [CrossRef]
44. London, D. Phosphorus in S-type magmas: The P_2O_5 content of feldspars from granites, pegmatites, and rhyolites. *Am. Mineral.* **1992**, *77*, 126–145.
45. London, D.; Morgan VI, G.B.; Babb, H.A.; Loomis, J.L. Behavior and effects of phosphorus in the system $Na_2O-K_2O-Al_2O_3-SiO_2-P_2O_5-H_2O$ at 200 MPa. *Contrib. Mineral. Petrol.* **1993**, *113*, 450–465. [CrossRef]
46. Bea, F.; Fershtater, G.B.; Corretge, L.G. The geochemistry of phosphorus in granite rocks and the effect of aluminium. *Lithos* **1992**, *29*, 43–56. [CrossRef]
47. Shand, S.J. *The Eruptive Rocks*, 2nd ed.; John and Wiley Sons: New York, NY, USA, 1943; 444p.
48. Frost, B.R.; Barnes, C.G.; Collins, W.J.; Arculus, R.J.; Ellis, D.J.; Frost, C.D. A Geochemical Classification for Granitic Rocks. *J. Petrol.* **2001**, *42*, 2033–2048. [CrossRef]
49. Villaseca, C.; Merino, E.; Oyarzun, R.; Orejana, D.; Pérez-Soba, C.; Chicharro, E. Contrasting chemical and isotopic signatures from Neoproterozoic metasedimentary rocks in the Central Iberian Zone (Spain) of pre-Variscan Europe: Implications for terrane analysis and Early Ordovician magmatic belts. *Precambrian Res.* **2014**, *245*, 131–145. [CrossRef]
50. Thomas, R.; Webster, J.D.; Rhede, D. Strong phosphorus enrichment in a pegmatite-forming melt. *Acta Universitatis Carolinae. Geologica* **1998**, *42*, 150–164.
51. Thomas, R.; Rhede, D.; Trumbull, R.B. Microthermometry of volatile-rich silicate melt inclusions in granitic rocks. *Z. Geol. Wiss.* **1996**, *24*, 505–526.
52. Webster, J.D.; Thomas, R.; Rhede, D.; Forster, H.J.; Seltmann, R. Melt inclusions in quartz from an evolved peraluminous pegmatite: Geochemical evidence for strong tin enrichment in fluorine-rich and phosphorus-rich residual liquids. *Geochim. Cosmochim. Acta* **1997**, *61*, 2589–2604. [CrossRef]
53. Harte, B.; Hunter, R.H.; Kinny, P.D. Melt geometry, movement and crystallization, in relation to mantle dykes, veins and metasomatism. *Trans. R. Soc. London* **1993**, *342*, 1–21.
54. London, D.; Morgan VI, G.B.; Wolf, M.B. Amblygonite-montebrasite solid solutions as monitors of fluorine in evolved granitic and pegmatitic melts. *Am. Mineral.* **2001**, *86*, 225–233. [CrossRef]

55. Frýda, J.; Breiter, K. Alkali feldspars as a main phosphorus reservoir in rare-metal granites: Three examples from the Bohemian Massif (Czech Republic). *Terra Nova* **1995**, *7*, 315–320. [CrossRef]
56. Kontak, D.J.; Martin, R.F.; Richard, L. Patterns of phosphorus enrichment in alkali feldspar, South Mountain Batholith, Nova Scotia, Canada. *Eur. J. Miner.* **1996**, *8*, 805–824. [CrossRef]
57. Cathelineau, M.; Boiron, M.-C.; Marignac, C.; Dour, M.; Dejean, M.; Carocci, E.; Truche, L.; Pinto, F. High pressure and temperatures during the early stages of tungsten deposition at Panasqueira revealed by fluid inclusions in topaz. *Ore Geol. Rev.* **2020**, *126*, 103741. [CrossRef]
58. Yakovenko, A. Estudo de Inclusões Fluidas Dos filões Pegmatíticos Litiníferos de Segura. Master's Thesis, Porto University, Porto, Portugal, 2021; 107p.
59. Yakovenko, A.; Guedes, A.; Boiron, M.C.; Cathelineau, M.; Martins, I.; Mateus, A. Fluid inclusion studies in quartz from the Li-rich pegmatite veins from Segura. In Proceedings of the Jornadas do ICT, Évora, Portugal, 10–11 February 2022; p. 43, ISBN 978-972-778-232-1.

Disclaimer/Publisher's Note: The statements, opinions and data contained in all publications are solely those of the individual author(s) and contributor(s) and not of MDPI and/or the editor(s). MDPI and/or the editor(s) disclaim responsibility for any injury to people or property resulting from any ideas, methods, instructions or products referred to in the content.

Article

Zircon U-Pb and Whole-Rock Geochemistry of the Aolunhua Mo-Associated Granitoid Intrusion, Inner Mongolia, NE China

Hao Li [1,2], Xuguang Li [1], Jiang Xin [3] and Yongqiang Yang [1,*]

[1] State Key Laboratory of Geological and Mineral Resources, China University of Geosciences, Beijing 100083, China; a2009020778a@163.com (H.L.); lix180520@gmail.com (X.L.)
[2] No. 5 Geological Party Limited Liability Company of Liaoning Province, Dashiqiao 115100, China
[3] Xinjiang Nonferrous Metal Industry (Group) Co., Ltd., Ürümqi 830000, China; xinjiang_111@163.com
* Correspondence: yangyonq@cugb.edu.cn

Abstract: The Aolunhua Mo deposit is a typical porphyry deposit, which is located in the middle southern section of the Da Hinggan Range metallogenic belt. Here, we report LA-ICP-MS zircon U-Pb age data from the Mo-associated granitoid, together with the element geochemistry of the zircons, discussing the source material of the ore-forming rock of the deposit. The zircon data constrain the crystallization age of the granite porphyry as 135.0 ± 1.0 Ma, correlating it with the widespread Yanshanian intermediate–felsic magmatic activity. The Th/U ratio of the zircon is greater than 0.1, with a significant positive Ce anomaly (Ce* = 1.72–188.71) and a negative Eu anomaly (Eu* = 0.05–0.57). The zircons show depleted LREE and enriched HREE patterns, as well as low La and Pr contents, suggesting crystallization from crust-derived magmas. Based on the geology of the ore deposit and the age data, in combination with the regional geodynamic evolution, we infer that the Aolunhua Mo deposit was formed near the peak stage of Sn poly-metallic metallogenesis in the Da Hinggan Range region at around 140 Ma, associated with a tectonic setting, characterized by the transition from compression to extension. Based on a comparison with the newly found Mo deposits along the banks of the Xilamulun River, we propose that the Tianshan–Linxi is an important Mo-metallogenic belt. It also suggests an increased likelihood for the occurrence of Mo along the north bank of the Xilamulun River.

Keywords: Mo-associated granitoid; metallogenic rock body; zircon trace elements; Aolunhua Mo deposit; Da Hinggan Range; NE China

Citation: Li, H.; Li, X.; Xin, J.; Yang, Y. Zircon U-Pb and Whole-Rock Geochemistry of the Aolunhua Mo-Associated Granitoid Intrusion, Inner Mongolia, NE China. *Minerals* **2024**, *14*, 226. https://doi.org/10.3390/min14030226

Academic Editors: Ignez de Pinho Guimarães and Jefferson Valdemiro De Lima

Received: 30 December 2023
Revised: 5 February 2024
Accepted: 19 February 2024
Published: 23 February 2024

Copyright: © 2024 by the authors. Licensee MDPI, Basel, Switzerland. This article is an open access article distributed under the terms and conditions of the Creative Commons Attribution (CC BY) license (https:// creativecommons.org/licenses/by/ 4.0/).

1. Introduction

Most of the important Cu, Mo, and Au deposits around the world are associated with porphyry systems [1–8]. A series of porphyry-type Mo deposits include some giant world-class deposits in the Central Asian metallogenic domain (CAMD), notably the Northeast China Mo-Cu metallogenic provinces [9]. In Northeast China, the Da Hinggan Range is an important poly-metallic metallogenic belt, hosting different types of hydrothermal deposits [10–18]. A number of ore deposits, such as the Dajin Cu-Ag-Sn-Pb-Zn mineral deposit, the Bairendaba Pb-Zn-Ag ore deposit, the Huanggangliang Fe-Sn ore deposit, and the Baiyinnur Pb-Zn ore deposit [11,12,18–20], have been discovered in this region, all of which are large or super-large ore deposits. Several previous investigations have addressed the regional metallogenic series, including the regional metallogenic characteristics, the kinematic background of metallogenesis, and the petrogenesis of the region [11–13,21–23]. The discovery of the Mo poly-metallic metallogenic belt on the south bank of the Xilamulun River has attracted wide attention [24]. However, there are relatively few studies of metallogenetic rule on the north bank of the Xilamulun River before the Aolunhua porphyry Mo deposit was discovered [25].

The present paper aims to study a rare porphyry type Mo deposit that carries abundant Mo metallogenetic elements, which is located along the north bank of the Xilamulun deep fracture belt. In the following, we present an analysis of mineralogical features, zircon U-Pb geochronology and trace element geochemistry, and the major and trace element geochemistry of the deposit-hosting granite porphyry. Based on these results, we further estimate the age of the ore formation, and interpret the geodynamic setting of the mineralization, both locally and regionally.

Spatially, the Mo-Cu ore deposits are clustered along margin fractures [26], such as the Xilamulun deep fracture belt [27]. The deposits along the south bank can be divided into three stages: a post-collisional orogenic stage at 250–220 Ma; a transitional geodynamic stage at 180–145 Ma; and finally, a stage associated with the large-scale thinning of the lithosphere at 140–120 Ma [28]. However, in the north bank, the ore-forming events occurred mainly around 140–130 Ma, such as the newly found Hashitu porphyry Mo deposit, with an ore-forming age of 148.8 ± 1.6 Ma [17,29], and the newly found Bianjia porphyry Mo-Sn ore deposit, with a quartz porphyry formation age of 140 Ma (see Figure 1). Combined with the geochronological data of the Aolunhua Mo-Cu ore deposit presented in this paper, together with those from the existing literature [30,31], the Tianshan–Linxi porphyry type Mo deposit can be considered as an important metallogenic belt, developed during the peak stage of metallogenesis under an extensional tectonic setting on the north bank of Xilamulun River, and may reinforce the interpretation of a large-scale thinning of the lithosphere during that time.

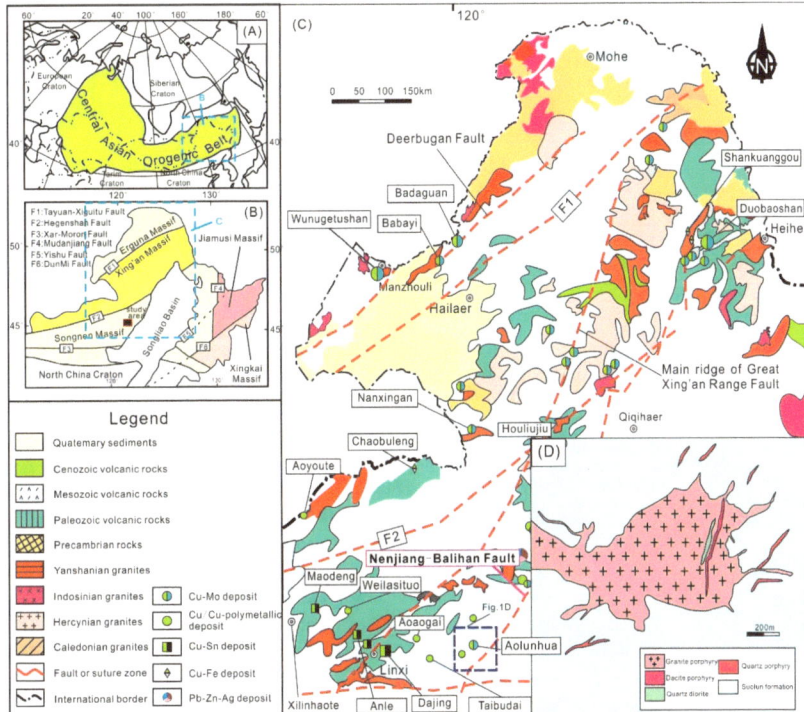

Figure 1. Simplified tectonic map of the southern Da Hinggan Range and its adjacent areas. Legend: 1—major fault; 2—boundary between countries; 3—fault numbers; 4—porphyry type ore deposits. ① Deep fracture of the north margin of the North China plate. ② Fault of the Xilamulun River. ③ Erenhot–Hegenshan deep fault. ④ Onor–Elunchun fault. ⑤ Derburgan fault. ⑥ Da Hinggan Range Major Fault. ⑦ Nenjiang fault. (**A**) Location of reginal tectonic of study area; (**B**) Location in Central Asian Orogenic Belt; (**C**) Geological map of the Da Hinggan Range; (**D**) Aolunhua area geological map.

2. Regional Geological Setting

The middle-south segment of the Da Hinggan Range is located within the collisional suture between the Siberian and North China plates [32] and is a mineral concentration district of the Cu-Sn-Mo poly-metallic metallogenic belt in North China (Figure 1).

The Aolunhua porphyry Cu-Mo deposit is located at the border of Alu-Kerqing Qi and Zarute Qi, with geographic coordinates of E 120°12′00″–120°15′30″; N 44°31′00″–44°34′00″. The Hing-Meng orogen, where the Aolunhua deposit is located, is the eastern elongation of the Central Asian Orogenic Belt, a composite orogen with an extensive history of accretion and terrane amalgamation that spans the Paleozoic and Mesozoic [11,33–37]. The major tectonic features of this region are fault structures. The near E-W trending Xilamulun River fault is a major boundary fault (Figure 1). Magmatism was common throughout the history of the region, including the Caledonian, Variscan, Indosinian, Yanshanian, and Cenozoic events. Among them, the Yanshanian magmatic events were the largest in scale, and were closely related to metallogenesis.

3. Geology of the Ore Deposit

The Aolunhua porphyry Mo deposit is located to the northwest of the Nenjiang faults (Figure 1). The strata around the Aolunhua Mo-ore deposit mainly belong to the Linxi group (P_{2l}) of the upper Permian and the Manketou Obo group ($J_2\ m$) of the upper Jurassic, as well as the Quaternary system (Figure 1D). The upper Permian Linxi group is mainly composed of sandstone and hornfels derived from igneous intrusions. The upper Jurassic Manketou Obo group is distributed in the northeast part of the ore area and shows a disconformable contact with the underlying Linxi group. In the ore area, faults and folds are weakly developed. However, the joints and fissures were filled with a hydrothermal solution, forming a stock work of quartz veins [38], which provided enough space to host the ore minerals of the Aolunhua ore deposit, where the strata were influenced by uplift intrusions. The joints developed in the granite porphyry are conspicuously different from those developed in the external contact zones. Due to the influence of the pre-metallogenic faulting and the pre-existing gneissic structure, the joints in the external contact zones are always steep, following the gneissic foliation, while, within the porphyry intrusion, there are networks of quartz veins with many different orientations.

The ore body is hosted within the inner and external contact zones of the Aolunhua granite porphyry intrusion. Mo and Cu are the two major metallogenic elements. The Mo industrial ore bodies enclosed in the porphyry rock body occupy the main part of the total reserve. It is especially enriched at the top of the rock body. The ore bodies are well preserved, occurring as stockworks. A preliminary investigation showed that there are two ore bodies, classified as the "upper" and "lower" bodies. The ore bodies in the external contact zone are mainly ore-bearing quartz veins, which are controlled by joints and fissures. The major mineralizations are molybdenitizations with the minor secondary oxidation enrichment of copper. The variations in Mo-ore grade are rather large.

In the ore body, the major ore minerals are molybdenite, chalcopyrite, pyrite, and arsenopyrite. The minor ore minerals are bornite, sphalerite, and galena. The vein minerals include quartz, feldspar, calcite, chlorite, and kaolinite. The textures of the ore are mainly granular, poikilitic, mosaic, porphyritic, and pseudomorphic. The structures of the ore are mainly impregnation, vein, stock work, and banded [30]. The wall rocks are characterized by potash alteration, silicification, propylitization, phyllite alteration, and argillation. Among the different types of alteration, potash alteration and silicification are most closely related to metallogenesis.

4. Sample Features and Laboratory Studies

4.1. Sample Features

The wall rocks hosting the ore body are generally granite porphyry. Some fine-grained porphyritic granodiorite and minor coarse-grained granites are also present. The transition between different rock types is gradual. Sample AL03 is a porphyritic granodiorite, collected from the quarry of the Aolunhua Mo-ore deposit. The rock is massive (Figure 2a), grayish-white in color, and porphyritic in texture. Under the microscope, the rock is mainly composed of plagioclase (35%–40%), K-feldspar (20%–25%), quartz (20%–25%), amphibole (2%–3%), and biotite (5%–7%). Minerals are homogeneously distributed with 15%–20% phenocrysts, showing a porphyritic texture (Figure 3b) with a granitic groundmass (Figure 2c). The phenocrysts are mainly composed of plagioclase, a few quartz, biotite, and amphibole phenocrysts. Most of the phenocrysts are euhedral with partial irregular fringes. The plagioclase phenocrysts are euhedral and platy, with prominent zoned structures (Figure 2b). The quartz phenocrysts are granular, and the biotite phenocrysts are idiomorphic flakes (Figure 2d), while the amphibole phenocrysts are euhedral and prismatic. The grain size of the phenocrysts ranges from 0.5 to 1.0 mm. In the groundmass, granular textures are common, with the plagioclase showing some euhedral to subhedral forms, but mostly anhedral forms. K-feldspars are anhedral, and quartz grains are anhedral granular, constituting the granitic texture of the rock (Figure 2c). Locally, K-feldspar and quartz intergrowths forming graphic textures can be found. The grain size of the groundmass minerals ranges from 0.2 to 1.0 mm. The biotites distributed in the groundmass form subhedral–anhedral plates, while amphiboles occur as anhedral short prisms.

Figure 2. Occurrence and mineral composition of granite porphyry in the Aolunhua deposit Mineral abbreviations: Qtz—quartz, Kfs—K feldspar, Pl—plagioclase, Bi—biotite. (**a**) Granite porphyry occurrence sample; (**b**) Photo under orthogonal polarization of granite showing Pl, Qtz; (**c**) Photo under orthogonal polarization of granite showing Kfs, Qtz; (**d**) Photo under orthogonal polarization of granite showing Bt, Pl, Qtz.

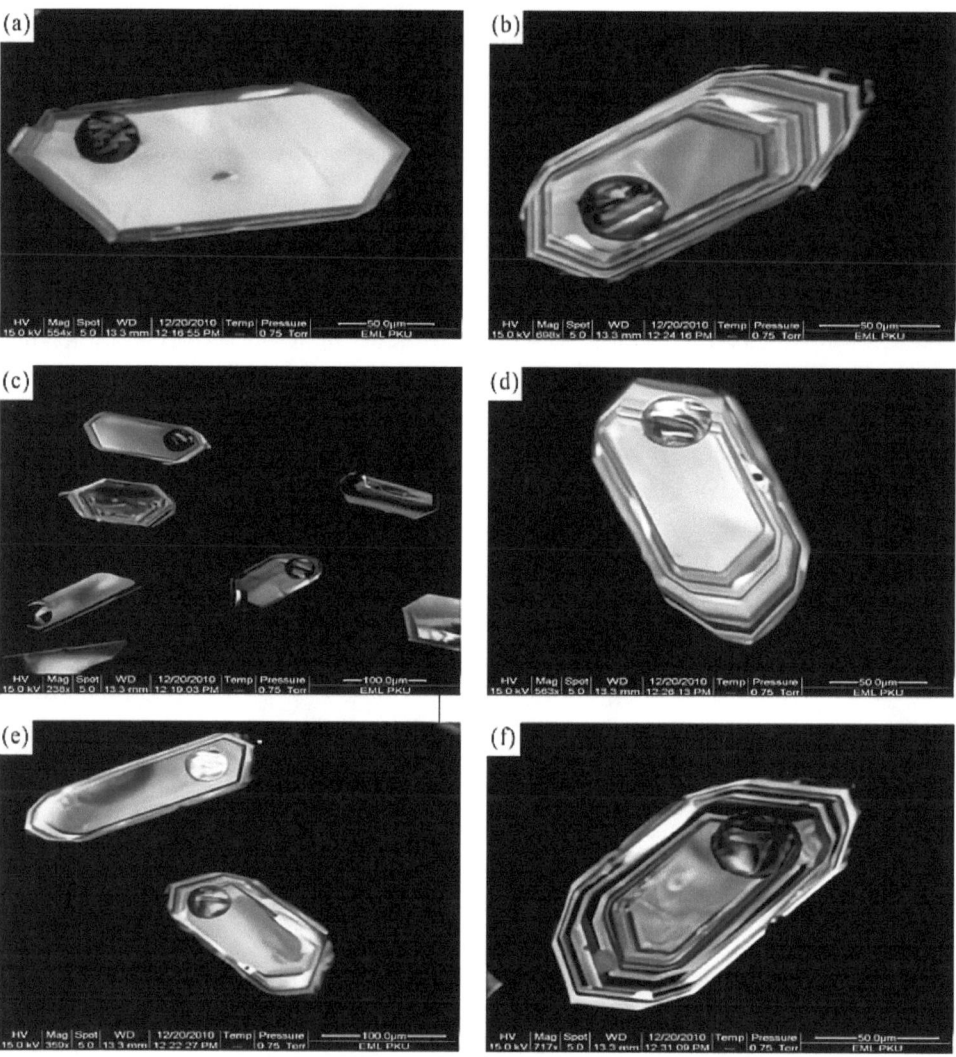

Figure 3. Cathodo-luminescence (CL) image of the representative zircons of the granite porphyries from the Aolunhua Mo deposit. (**a**) no band; (**b**) crystalline band; (**c**) no band; (**d**) oscillatory band; (**e**) no band; (**f**) oscillatory band.

4.2. Analytical Methods

4.2.1. Zircon U-Pb Geochronology and REE Elements Analyses

Twenty-seven granite samples were selected for zircon U-Pb dating. The zircon grains were separated through standard gravity and magnetic methods, followed by hand picking under the binocular microscope, all following the crushing of the rock sample. The zircon grains were mounted onto epoxy resin discs, and then polished to expose their internal texture. Before U-Pb dating, zircon grains were imaged under transmitted light, reflected light, and cathodoluminescence (CL) in order to allow for the evaluation of their internal textures. The most suitable grains were then selected for U-Pb analyses.

The U-Pb analyses were carried out with a laser ablation inductively coupled plasma mass-spectrometer (LA-ICP-MS), housed at the China University of Geosciences (Beijing,

China). The ICP-MS is made by the US Agilent company, and the mass spectrometer is of the 7500a type. The laser apparatus, of type number UP193SS, was made by the New Wave company in the USA. The laser wave length, laser spot diameter, and frequency were 193 nm, 36 μm, and 10 Hz, respectively, while the pre-ablation and ablation times were 5 s and 45 s, respectively. During the experimental process, He-gas was used as the carrier with a flow velocity of 0.8 L/min. The element integration time is 20 ms for U, Th, and Pb, 6 ms for Si and Zr, and 10 ms for the other elements. Raw data were processed using the GLITTER 4.41.1 program to calculate isotopic ratios and $^{207}Pb/^{206}Pb$, $^{206}Pb/^{238}U$, $^{207}Pb/^{235}U$ ages, respectively. The age calculation uses standard zircon TEM as the external standard for the correction of the isotope ratio. Standard zircon 91500 [39] and Qinghu were used as monitoring blind samples, the element contents are calculated by using the international standard NIST610 [39] as an external standard, and Si as an internal standard, NIST612 and NIST614 [39] are taken as monitoring blind samples. The correction of ^{204}Pb follows ref. [39]. Each analysis is reported at 1σ uncertainties, and isoplot 3.0 was used to calculate the U-Pb ages, as well as to make the Concordia plots [40].

4.2.2. Whole-Rock Geochemical Analyses

Fresh granitoid samples for elemental analyses were first trimmed and chipped, and then powdered in an agate mill to about 200 mesh for analyzing. The analyses of major elements were conducted on the basis of Rock Samples from the United States Geological Survey. The analyses of silicate petrochemistry were conducted by X-ray fluorescence spectrometry. The analysis apparatus is a Phillip X-ray fluorescence spectrometer PW 2014. The analytical precision and accuracy for most major elements measured are generally better than 5%. FeO was analyzed by the titration method, with a standard deviation less than 10 percent. The analyses of trace elements were conducted, according to the general rule of the ICP-MS method; the apparatus used was an HR-ICP-MS (element 1), made by Finigan MAT. Rhodium was used as an internal standard to monitor any signal drift during counting. The analytical error is generally less than 5% for trace elements. The analyses were conducted in the Analytical Center of the Geological Institute of the Ministry of Nuclear Industry. The analytical results are given in Table 1.

Table 1. The component of macro-elements (wt.%) and trace elements (10^{-6}) of granites from Aolunhua deposit.

Rock Types	Granite	Fine-Grained Granodiorite	Granite	Fine-Grained Granodiorite	Fine-Grained Granodiorite	Rock Types	Granite	Fine-Grained Granodiorite	Granite	Fine-Grained Granodiorite	Fine-Grained Granodiorite
Sample Numbers	AL-01	AL-03	AL-05	AL-03-1	AL-03-2	Sample Numbers	AL-01	AL-03	AL-05	AL-03-1	AL-03-2
Al_2O_3	12.87	13.30	12.93	10.64	12.21	Rb	116	97.1	125	113	115
SiO_2	73.03	71.04	70.00	77.05	74.41	Sr	466	431	233	409	453
CaO	1.55	2.24	2.72	1.08	1.54	Zr	66.4	72.8	63.0	56.1	65.8
K_2O	4.37	4.19	4.04	4.76	4.48	Nb	4.20	5.32	5.49	4.84	5.34
TiO_2	0.30	0.42	0.42	0.26	0.34	Hf	2.21	2.66	2.43	1.68	2.02
Fe_2O_3	2.30	2.95	3.23	1.05	1.47	Ta	0.402	0.438	0.416	0.361	0.401
MgO	0.61	0.81	0.78	0.51	0.69	W	25.1	3.10	15.8	12.4	11.4
Na_2O	3.92	4.37	3.04	2.98	3.56	Re	0.043	0.018	1.95	0.578	0.187
MnO	0.031	0.038	0.032	0.028	0.022	Tl	0.672	0.526	0.884	0.596	0.504
P_2O_5	0.18	0.18	0.16	0.09	0.12	Pb	37.3	229	36.5	21.0	9.74
FeO	0.90	2.10	1.90	0.65	1.20	Th	5.46	6.61	5.34	4.12	4.61
LOI	0.58	0.24	2.32	1.23	0.91	Bi	1.33	1.19	19.7	2.53	1.04
Total	100.641	101.878	101.572	100.328	100.952						
Li	9.93	13.2	21.7	8.47	9.01	U	0.990	7.15	1.58	2.28	1.73
Be	2.22	1.90	1.28	1.99	2.03	Dy	1.66	1.66	1.87	1.56	1.48

Table 1. Cont.

Rock Types	Granite	Fine-Grained Granodiorite	Granite	Fine-Grained Granodiorite	Fine-Grained Granodiorite	Rock Types	Granite	Fine-Grained Granodiorite	Granite	Fine-Grained Granodiorite	Fine-Grained Granodiorite
Sample Numbers	AL-01	AL-03	AL-05	AL-03-1	AL-03-2	Sample Numbers	AL-01	AL-03	AL-05	AL-03-1	AL-03-2
Sc	7.14	7.25	4.42	2.59	3.17	Ho	0.296	0.296	0.311	0.249	0.273
V	29.6	33.0	32.6	26.7	34.8	Er	0.816	0.880	0.855	0.685	0.762
Cr	4.19	4.56	5.89	8.36	8.17	Tm	0.135	0.129	0.135	0.118	0.136
Co	3.54	4.06	7.27	1.63	2.48	Yb	0.815	0.871	0.839	0.724	0.714
Ni	2.46	29.6	3.10	0.928	1.30	Lu	0.141	0.131	0.135	0.106	0.111
Cu	275	7069	2852	142	221	Y	9.00	8.73	9.16	8.40	8.31
Zn	114	5339	212	52.3	31.3	B	2.40	1.27	4.39	2.06	2.26
Ga	17.6	18.7	17.6	16.5	17.1						

5. Results

5.1. Zircon U-Pb Geochronology

The selected zircon grains are colorless, transparent-to-pale yellow, and euhedral, with a typical elongated prismatic shape (Figure 3). The Th/U ratios of the grains are greater than 0.1 (Table 2), indicating a magmatic origin [41]. Their magmatic origin is further supported by their typical magmatic oscillatory zoning under cathodoluminescence (CL) images (Figure 3). The lack of a core-mantle structure and deuteric alteration shells suggests the zircon crystals were originally crystallized from a common magma. Therefore, the age of the zircons could represent the timing of magma crystallization. A total of 27 zircon grains were analyzed from the granite porphyry sample, and the results are given in Table 2. The contents of Th and U are greatly variable, ranging from 18.46–467.96 ppm and 28.96 ppm–768.17 ppm, respectively. On the U-Pb concordia diagram (Figure 4a), all data are plotted along the concordia line or near to it, showing a high degree of concordance, without any loss or addition. The $^{206}Pb/^{238}U$-Pb age of the zircons varies from 131 Ma to 140 Ma (Figure 4a,b). The zircon ages range from ~140 Ma to 130 Ma, documenting the history from the magma emplacement to the crystallization. The large range of zircon ages that reflect the magma evolution history have also been reported in the Adamello Intrusive suite, N. Italy [42], and Acadian deformation and Devonian granites in northern England [43]. The analytical results accurately represent the crystallization age of the granite porphyry. Zircon U-Pb dating results indicate that the granite porphyries, which are closely related with the Aolunhua Mo deposit, are products of Yanshanian early Cretaceous magmatism.

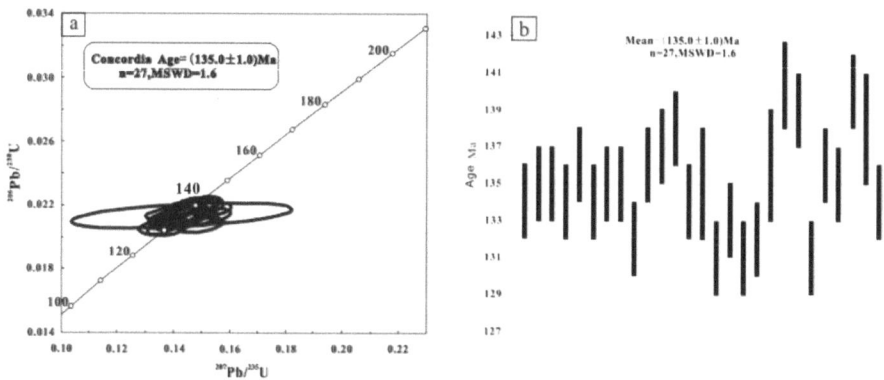

Figure 4. LA-ICP-MS U-Pb age concordia (**a**) and the weighted mean age histogram (**b**) of zircons from fine-grain porphyritic granodiorite (sample AL 03).

Table 2. The dating results of U-Pb isotopes for single zircon in sample AL03 of Aolunhua.

Sample Numbers	Contents/(μg/g)					Ratios									Age/Ma						Th/U
	^{206}Pb	^{207}Pb	^{208}Pb	^{232}Th	^{238}U	^{207}Pb/^{206}Pb	1σ	^{207}Pb/^{235}U	1σ	^{206}Pb/^{238}U	1σ	^{208}Pb/^{232}Th	1σ	^{207}Pb/^{206}Pb	1σ	^{206}Pb/^{238}U	1σ	^{207}Pb/^{235}U	1σ	^{208}Pb/^{232}Th	
AL03-01	29.49	1.620	1.831	149.73	346.53	0.04862	0.00332	0.14482	0.01016	0.02107	0.00037	0.00697	188	119	134	2	137	9	140	0.43
AL03-02	30.56	1.604	2.191	177.31	357.54	0.04873	0.00318	0.13898	0.00920	0.02116	0.00030	0.00705	81	116	135	2	132	8	142	0.50
AL03-03	35.75	1.92	2.640	211.88	418.38	0.04729	0.00145	0.14184	0.00416	0.02116	0.00039	0.00712	129	44	135	2	135	4	143	0.51
AL03-04	39.41	2.140	3.060	258.55	463.67	0.05230	0.00299	0.14286	0.00912	0.02105	0.00036	0.00677	158	105	134	2	136	8	136	0.56
AL03-05	24.90	1.339	1.292	106.04	289.12	0.04896	0.00201	0.14352	0.00604	0.02133	0.00035	0.00696	138	67	136	2	136	4	140	0.37
AL03-06	34.34	1.841	3.880	313.88	406.58	0.04883	0.00242	0.14033	0.00740	0.02093	0.00039	0.00705	130	84	134	2	133	6	142	0.77
AL03-07	30.08	1.616	2.086	164.92	351.59	0.04885	0.00297	0.14247	0.00908	0.0212	0.00040	0.00722	135	106	135	2	135	12	145	0.47
AL03-08	37.76	1.969	1.412	113.42	442.43	0.04878	0.00268	0.13795	0.00821	0.02115	0.00037	0.00711	64	95	135	2	131	6	143	0.26
AL03-09	47.40	2.790	4.440	364.70	569.44	0.04895	0.00205	0.14962	0.00630	0.02074	0.00033	0.00692	299	70	132	2	142	9	139	0.64
AL03-10	30.25	1.633	1.833	149.59	351.61	0.04971	0.00265	0.14399	0.00429	0.02132	0.00043	0.00700	146	45	136	2	137	10	141	0.43
AL03-11	18.91	1.028	1.393	86.78	190.36	0.04899	0.00147	0.14421	0.00753	0.02141	0.00036	0.00777	140	86	137	2	137	10	156	0.46
AL03-12	50.45	2.720	4.110	333.74	579.63	0.04858	0.00140	0.14541	0.00409	0.02158	0.00038	0.00704	141	42	138	2	138	5	142	0.57
AL03-13	32.05	1.724	2.079	169.46	377.09	0.04720	0.00206	0.14180	0.00656	0.02108	0.00039	0.00701	137	73	134	2	135	7	141	0.45
AL03-15	23.66	1.294	0.551	46.05	290.51	0.04734	0.00258	0.14554	0.01348	0.02123	0.00035	0.00702	181	140	135	2	138	7	141	0.16
AL03-16	45.95	2.470	4.860	415.08	555.05	0.04852	0.00332	0.13892	0.00788	0.02056	0.00042	0.00667	147	70	131	3	132	5	134	0.75
AL03-17	49.49	2.600	3.810	319.27	595.90	0.04863	0.00200	0.13985	0.00589	0.02088	0.00034	0.00656	128	127	133	2	133	14	132	0.53
AL03-18	62.73	3.280	5.610	467.96	768.17	0.04963	0.00514	0.13379	0.00998	0.02055	0.00031	0.00665	59	61	131	2	127	12	134	0.61
AL03-20	44.42	2.360	8.480	710.61	542.18	0.04938	0.00212	0.13466	0.00832	0.02063	0.00039	0.00677	66	96	132	2	128	9	136	0.76
AL03-21	2.49	0.133	0.226	18.46	28.96	0.04852	0.00514	0.14290	0.00624	0.02136	0.00038	0.00700	125	93	136	2	136	5	141	0.64
AL03-22	18.43	0.994	1.616	124.12	204.18	0.04859	0.00236	0.14766	0.01185	0.02202	0.00031	0.00733	130	132	140	2	140	6	148	0.61
AL03-23	29.89	1.637	3.410	249.50	340.75	0.04859	0.00146	0.14914	0.01102	0.02179	0.00030	0.00782	178	136	139	2	141	7	157	0.73
AL03-24	20.96	1.119	2.221	179.22	253.51	0.04888	0.00258	0.13845	0.01185	0.02060	0.00034	0.00696	135	143	131	2	132	7	140	0.71
AL03-26	36.54	1.960	2.850	232.71	425.82	0.04862	0.00143	0.14275	0.01561	0.02133	0.00037	0.00703	125	144	136	2	135	8	142	0.55
AL03-27	41.41	2.220	2.550	204.12	486.42	0.04873	0.00388	0.14181	0.00753	0.02116	0.00039	0.00718	128	136	135	2	135	4	145	0.42
AL03-28	30.19	1.620	2.122	166.71	343.07	0.04729	0.00332	0.14662	0.00906	0.02188	0.00036	0.00730	128	151	140	2	139	6	147	0.49
AL03-29	10.18	0.549	0.725	55.00	117.34	0.05230	0.00188	0.14542	0.01561	0.02157	0.00037	0.00757	142	127	138	2	138	7	152	0.47
AL03-30	47.22	2.520	4.700	399.45	560.76	0.04896	0.00332	0.13933	0.00125	0.02094	0.00030	0.00676	111	132	134	3	136	8	147	0.71

5.2. Zircon Trace Element Geochemistry

Using the method proposed by [6], after eliminating 3 of the granodiorite samples of zircon trace element data, 30 analytical points of trace element data are left for study. The contents of the trace elements in zircon from the Aolunhua granite porphyry are given in Table 3. Their distribution (Figure 5a) indicates that the zircons are enriched in large ion lithospheric elements of Th, Zr, and Hf, and have a strong negative anomaly of La, Nd, and Ti. The chondrite normalized REE distribution patterns of zircon (Figure 5b) show that all of the analytical results are depleted in LREE and enriched in HREE, and the Aolunhua granodiorite zircon samples show a prominent Ce positive anomaly (Ce* = 1.72–188.71) and prominent Eu negative anomaly (Eu* = 0.05–0.57). The zircons are high in ΣREE (299.02 × 10^{-6}–2548.68 × 10^{-6}) and have a high degree of variation, which are characteristic of a magmatic origin [44].

Table 3. Trace element analyses of zircons from fine-grain porphyritic granite of Aolunhua deposit (^{90}Zr, ^{178}Hf wt%, others 10^{-6}).

	AL03-1	AL03-2	AL03-3	AL03-4	AL03-5	AL03-6	AL03-7	AL03-8	AL03-9	AL03-10	AL03-11	AL03-12	AL03-13	AL03-14	AL03-15
^{49}Ti	3.03	2.87	3.32	2.42	2.15	2.60	7.32	2.49	2.82	2.74	10.91	2.94	2.73	6.95	3.63
^{89}Y	1100.39	1328.41	1134.30	1165.77	796.49	1541.33	1431.76	471.55	1038.42	1136.44	1020.77	1436.96	1058.42	1277.40	373.79
^{90}Zr	491,255.6	482,274.2	492,333.3	481,478.9	475,668.9	479,458.5	469,524.2	476,759.6	479,104.1	471,036.4	475,740.5	460,816.8	472,844.2	475,096.9	469,188.0
^{93}Nb	4.030	4.830	5.330	5.720	2.500	3.090	2.610	1.210	4.600	2.790	2.600	7.030	2.710	1.000	0.954
^{139}La	0.7780	0.0447	4.2000	0.1190	0.0333	0.0218	0.0469	0.0120	2.1600	0.2970	0.0320	0.1290	0.2950	0.9790	1.2500
^{140}Ce	26.44	33.11	43.62	31.86	19.09	48.33	16.23	13.42	41.95	21.66	19.30	38.28	23.80	7.37	5.90
^{141}Pr	0.1840	0.09970	1.5700	0.0800	0.03470	0.1540	0.2850	0.0460	0.6720	0.1740	0.0610	0.1020	0.1610	0.6280	0.5160
^{146}Nd	1.720	1.900	8.600	1.230	0.684	4.130	5.720	0.436	4.900	1.620	1.580	1.610	1.950	6.670	2.900
^{147}Sm	3.16	5.59	5.11	3.90	2.03	9.76	13.41	1.08	4.99	3.98	4.11	4.73	3.94	9.55	3.07
^{153}Eu	0.877	1.380	1.327	1.275	0.731	2.880	3.590	0.608	1.384	1.100	1.147	1.298	1.235	1.830	0.399
^{157}Gd	18.61	27.05	20.60	24.13	13.06	48.99	52.29	6.87	21.40	21.71	20.16	27.66	21.20	40.91	13.90
^{159}Tb	7.24	9.88	7.77	8.75	5.12	15.69	16.39	2.67	7.41	8.34	7.47	10.73	7.90	13.03	4.65
^{163}Dy	97.01	122.36	100.91	109.67	65.72	168.49	170.15	35.16	92.18	106.92	93.12	136.33	98.45	146.38	45.11
^{165}Ho	37.69	45.96	38.61	41.41	26.82	54.48	52.26	14.07	34.58	40.59	34.92	50.65	37.46	49.19	12.19
^{166}Er	172.27	203.10	176.73	184.69	124.45	214.97	194.73	69.62	161.81	180.72	155.47	218.66	164.39	194.64	46.24
^{166}Tm	45.13	51.88	45.97	46.79	33.33	51.10	43.99	20.24	41.67	45.44	40.38	55.47	42.11	45.06	10.90
^{172}Yb	535.18	617.11	551.00	547.79	422.97	581.25	475.64	283.62	512.19	528.58	477.09	637.11	492.08	485.27	128.23
^{175}Lu	100.05	114.43	102.37	101.96	84.01	105.86	78.24	65.12	101.36	95.48	88.44	117.67	90.63	87.80	23.77
^{178}Hf	9334.11	9017.03	10056.96	9951.02	9356.05	9261.37	8263.11	9917.20	9748.66	8998.79	7946.22	10054.13	9209.08	6019.05	8618.70
^{181}Ta	1.108	1.211	1.500	1.610	0.734	0.674	0.798	0.397	1.250	0.926	0.762	1.740	0.877	0.424	0.480
^{232}Th	149.73	177.31	211.88	258.55	106.04	313.88	164.92	113.42	364.70	149.59	86.78	333.74	169.46	67.59	46.05
^{238}U	346.53	357.54	418.38	463.67	289.12	406.58	351.59	442.43	569.44	351.61	190.36	579.63	377.09	139.77	290.51
Ce*	16.04	81.34	3.99	73.43	116.01	85.12	15.32	75.95	8.16	21.81	75.67	73.07	25.19	2.11	1.72
Eu*	0.28	0.29	0.35	0.31	0.38	0.33	0.37	0.52	0.35	0.29	0.32	0.27	0.33	0.24	0.16
	AL03-16	AL03-17	AL03-18	AL03-19	AL03-20	AL03-21	AL03-22	AL03-23	AL03-24	AL03-25	AL03-26	AL03-27	AL03-28	AL03-29	AL03-30
^{49}Ti	3.72	3.83	3.32	12.42	7.05	11.01	11.40	2.63	2.03	2.33	3.88	3.57	2.90	6.13	3.92
^{89}Y	1420.71	1651.60	1985.24	808.60	3286.52	721.15	1569.56	934.91	570.98	1084.91	1359.21	1516.20	980.03	973.57	1396.37
^{90}Zr	465,131.3	460,473.9	462,432.5	467,410.5	463,055.8	476,714.8	469,992.5	466,770.8	463,436.1	462,163.6	478,752.2	468,543.7	482,700.4	486,641.5	48,9760.7
^{93}Nb	5.250	8.460	9.950	0.677	4.260	0.831	3.100	3.740	2.240	4.290	5.550	7.800	3.510	2.630	5.980
^{139}La	1.4900	0.2700	0.1170	1.1300	0.0760	0.0174	0.5410	0.0522	0.0720	0.0111	0.0390	0.0228	1.3700	0.0195	0.8830
^{140}Ce	41.10	41.18	55.61	9.51	67.41	5.99	22.16	30.18	23.38	12.34	39.94	35.97	32.66	12.05	41.58
^{141}Pr	0.3190	0.1450	0.1100	0.6380	0.8260	0.0820	0.2850	0.0380	0.0361	0.0633	0.0670	0.0460	0.3700	0.0670	0.2730
^{146}Nd	3.24	1.97	2.30	5.68	15.06	1.40	3.24	1.18	0.799	1.32	1.56	0.97	2.26	1.51	2.84
^{147}Sm	6.10	5.47	7.57	7.83	30.49	3.42	7.15	3.21	2.30	3.87	4.86	3.90	3.04	4.46	5.26
^{153}Eu	1.920	1.610	2.360	2.800	8.160	1.590	2.600	1.269	0.882	0.199	1.630	1.205	1.124	0.953	1.590

Table 3. *Cont.*

	AL03-16	AL03-17	AL03-18	AL03-19	AL03-20	AL03-21	AL03-22	AL03-23	AL03-24	AL03-25	AL03-26	AL03-27	AL03-28	AL03-29	AL03-30
^{157}Gd	32.30	34.50	43.97	29.60	115.02	15.54	36.74	17.36	13.01	22.44	28.33	23.40	17.49	21.28	28.99
^{159}Tb	11.38	12.89	16.55	8.64	35.21	5.64	13.34	6.40	4.44	8.56	10.37	9.96	6.69	8.03	10.59
^{163}Dy	138.56	163.86	202.52	90.63	371.49	68.59	157.20	82.55	53.64	110.33	129.74	132.80	85.17	96.29	130.07
^{165}Ho	50.40	59.37	73.14	30.23	118.31	25.66	56.19	31.52	20.04	41.38	47.63	51.92	32.94	35.80	48.71
^{166}Er	216.57	254.55	303.76	119.21	449.70	113.19	233.53	142.86	88.65	176.82	208.86	240.48	151.89	155.34	215.35
^{166}Tm	54.05	63.27	73.96	28.19	99.14	28.36	56.42	37.61	22.57	42.27	51.69	62.62	41.24	37.72	54.59
^{172}Yb	644.15	708.51	817.84	313.02	1065.86	327.56	622.93	465.45	281.64	468.54	602.31	747.83	509.66	426.10	637.30
^{175}Lu	118.93	121.39	136.34	57.87	171.93	61.99	111.33	91.36	53.03	81.36	107.05	139.08	103.82	77.49	119.44
^{178}Hf	9204.88	9583.04	9201.58	6589.12	7582.62	6711.38	7167.53	9346.73	9878.15	9320.66	9328.82	9310.72	9502.13	8372.76	9474.74
^{181}Ta	1.160	2.350	2.550	0.150	0.975	0.187	0.736	0.845	0.684	1.500	1.550	1.950	0.795	0.673	1.490
^{232}Th	415.08	319.27	467.96	35.43	710.61	18.46	124.12	249.5	179.22	107.43	232.71	204.12	166.71	55.00	399.45
^{238}U	555.05	595.90	768.17	27.25	542.18	28.96	204.18	340.75	253.51	279.21	425.82	486.42	343.07	117.34	560.76
Ce*	13.52	48.08	91.94	2.57	22.53	19.33	13.06	150.50	106.55	52.26	140.08	188.71	10.66	46.35	19.84
Eu*	0.34	0.28	0.31	0.50	0.37	0.57	0.40	0.42	0.39	0.05	0.34	0.30	0.37	0.25	0.32

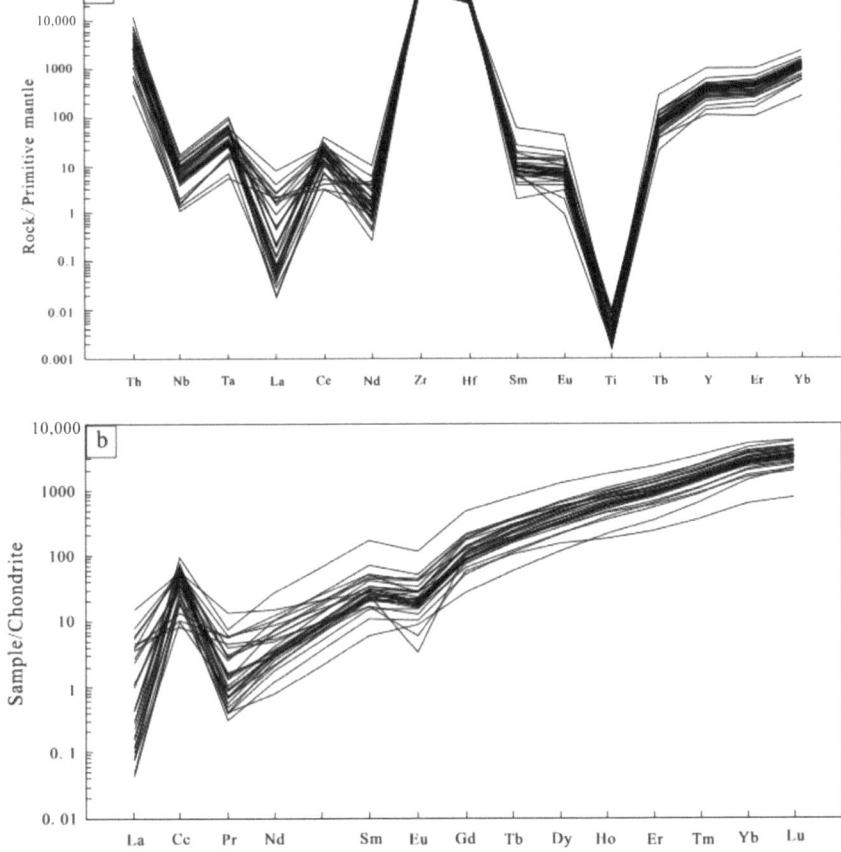

Figure 5. Primitive mantle normalized trace element spider diagrams (**a**) and the chondrite normalized REE patterns (**b**) for zircon grains from the Aolunhua granodiorite.

5.3. Whole-Rock Geochemistry

5.3.1. Major Elements

The Aolunhua granite porphyry samples (AL01, AL03, AL05, AL03-01, AL03-02) have SiO_2 contents ranging between 70.00%–77.05%, Al_2O_3 contents of 10.64%–13.30%, K_2O contents of 4.04%–4.76%, and Na_2O contents of 2.98%–4.37%, indicating that they are acidic and K-rich. The Al saturation index (A/CNK) ranges between 0.838–0.916. On the diagram of A/CNK-A/NK, the Aolunhua samples display a meta-aluminous series, similar to the characteristics of I-type granites. On the R1-R2 diagram, these samples plot in the field of syn–orogenic granites (Figure 6b).

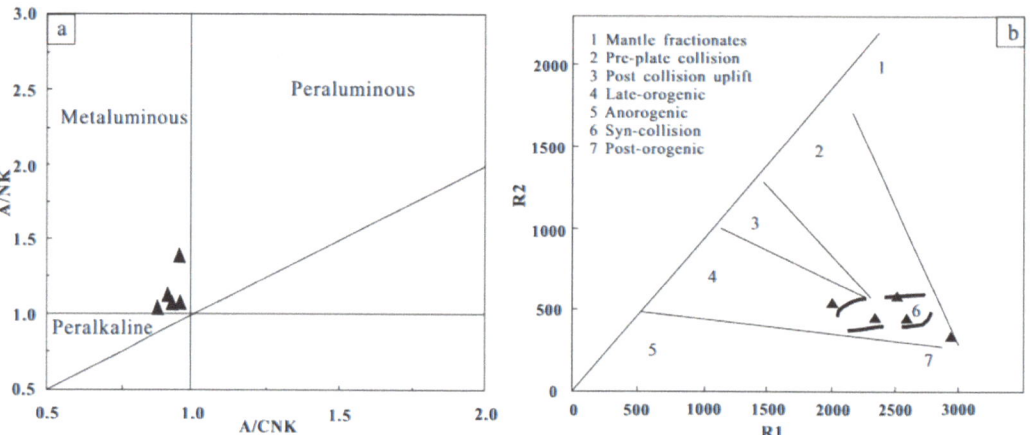

Figure 6. The A/CNK-A/NK (**a**) and R1-R2 (**b**) diagrams for rock samples from the Aolunhua deposit. In (**b**): 1—mantle differentiation; 2—pre-collisional; 3—post collisional uplift; 4—late orogenic; 5—non-orogenic; 6—syn-collisional; 7—post orogenic.

5.3.2. Trace Elements

The trace element content of the Aolunhua granite porphyries (AL01, AL03, AL05, AL03-01, AL 03-02) are given in Table 1. The prominent characteristics of the samples are their enrichment in Rb, Th, etc., with Rb variations in the range of 97.1–125 ppm, and Th variations in the range of 4.12–6.61 ppm. The Hf, Y, and Yb concentrations are low. The Hf content ranges from 1.68–2.66 ppm. Y varies from 8.31 ppm to 9.16 ppm, whereas Yb varies from 0.714 ppm to 0.871 ppm.

6. Discussion

6.1. The Chronology of Petrogenesis

The accurate determination of the chronology of petrogenesis is important in understanding the genesis of an ore deposit. The zircon isotope system is a reliable method to obtain the age of petrogenesis. Therefore, the results are quite reliable. The present study has found the zircon U-Pb LA-ICP-MS crystallization age to be 135.0 ± 1.0 Ma. As viewed from the perspective of metallogenesis, the ore deposit is directly hosted by the contact zone between the granite porphyry body and its wall rocks, which is closely related in space with the granite porphyry. Therefore, the study of the characteristics of the ore deposit and the metallogenic geochronology suggest that the Aolunhua ore deposit is the product of the Yanshanian intermediate–felsic magmatic hydrothermal activity.

The crystallization age of the Aolunhua ore-bearing granite porphyry, i.e., 135.0 ± 1.0 Ma, as presented in this paper, is nearly identical to the zircon U-Pb SHRIMP age of 134 ± 4 Ma obtained by [30], which suggests that the magmatism originated at about 130 Ma. In addition, the Re-Os isochron age of 131.2 ± 1.9 Ma was obtained for the Aolunhua porphyry Mo deposit [45]. In the Banlashan Mo deposit, 50 km SW of the Aolunhua Mo deposit, the crystalline age of the granite porphyry is dated at 131.1 ± 1.8 Ma [27]. Additionally, the Yangchang Mo deposit provides an isochron age of 138.5 ± 4.5 Ma for two groups of molybdenite [46]. The ages are also close to the petrogenic age of the Aolunhua Mo deposit. It is clear that the deposits are products of the intensive Cretaceous tectono-magmatic activities, which also confirms that the large scale magmatism-fluid-metallogenic events developed in Da Hinggan Range, and even in East China, happened at around 130 Ma [46,47].

Besides the geotectonic situation of the Aolunhua Mo-deposit at the north of the Xilamulun River deep fracture belt, the discovery of the Banlashan and Yangchang Mo-deposit further confirms that the southern section of the Da Hinggan Range not only hosts large scale, mainly Cu, Sn, Pb, and Zn ore deposits, but is also an excellent prospect for further exploration of mainly Mo-metallogenic element deposits in this region.

6.2. The Tracer Significance of Trace Elements in Zircon

The trace element composition in zircon is important for developing a proper understanding of the petrogenesis of zircon and its host rocks. The contents of the trace elements in zircon from the Aolunhua granite porphyry are given in Table 3. The hydrothermal zircons are believed to precipitate from aqueous fluids, in most cases, at relatively low temperatures, rather than from magmas. The Th content of the zircon is 18.96 ppm–710.61 ppm, and the U content is 27.25 ppm–768.7 ppm; thus, the relevant Th/U ratios range from 0.6–0.76 (larger than 0.1), indicating a magmatic origin [44,46]. There are local testing sites higher in U content. Chakoumakos et al. [47] studied zircons from Sri Lanka and found that the metamict domains in zircon had a very high U-content. The average U content is 3000 ppm. After studying different zircon samples from the Adirondack terrain, Valley et al. [48] found that the highly magnetic, high metamictized zircons had higher U contents than those of low magnetism and weak metamictization. All research results show that the metamict domain in zircon is high in U content.

The REE distribution patterns of zircon from Aolunhua granite porphyry are shown in Figure 5b. The ΣREE values are high and the variation is large, ranging from 299.02×10^{-6} to 2548.08×10^{-6}. All of the analytical spots show a prominent Ce positive anomaly and an Eu negative anomaly (Figure 5b), with a large variation range: Ce* = 1.72–188.71, Eu* = 0.05–0.57 [49,50], which are within the range of crust-derived zircons [51]. Unlike other REEs that have only +3 valency, Ce and Eu commonly have two oxidation states in terrestrial magmas, and zircon more preferentially incorporates the oxidized cations Ce^{4+} (0.97 Å) and Eu^{3+} (1.07 Å) into the Zr^{4+} (0.84 Å) site of its structure than the reduced Ce^{3+} (1.14 Å) and Eu^{2+} (1.25 Å) [49]. Thus, high Ce^{4+}/Ce^{3+} and Eu/Eu* ratios usually reflect the high oxygen fugacity (fO_2) of the parental magmas.

In addition, all of the testing spots show the characteristics of enrichment in HREE (Figure 5b). According to [44], if the range of variation of the LREE in recrystallized zircon is clearly larger than that of the HREE, it can be interpreted as the result of the greater instability of LREE in the zircon. It is obvious that during recrystallization, LREE in zircons are easier to drive out of the zircon lattice, leading to a decrease in LREE contents. The low contents of La and Pr and the Ce positive anomaly are characteristic of REE in crust-derived magmatic zircons.

6.3. The Tectonic Setting of the Formation of Ore-Bearing Intrusive Body

The major ore-hosting igneous rock in Aolunhua is fine-grained porphyritic granodiorite, and chronological studies show that the porphyritic granodiorite zircon dating yields the age of 135.0 ± 1.0 Ma. The petrochemical data indicate that the Aolunhua Mo-deposit porphyritic granodiorites are meta-aluminous. On the R1-R2 discrimination diagram (Figure 6b), they plot into the field of syn-collisional granites. However, for the plots on the (Yb+Nb)-Rb diagram (Figure 7a), the Aolunhua ore-bearing igneous rocks plot on the boundary between the syn-collisional (Syn-COLG) and the volcanic arc granite (VAG) fields. The plots on the Y-Nb diagram (Figure 7b) are within the field of the volcanic arc and the syn-collisional granites (VAG+syn-COLG), while the plots in Y-Sr/Y diagram (Figure 7c) show that they are adakaitic in character.

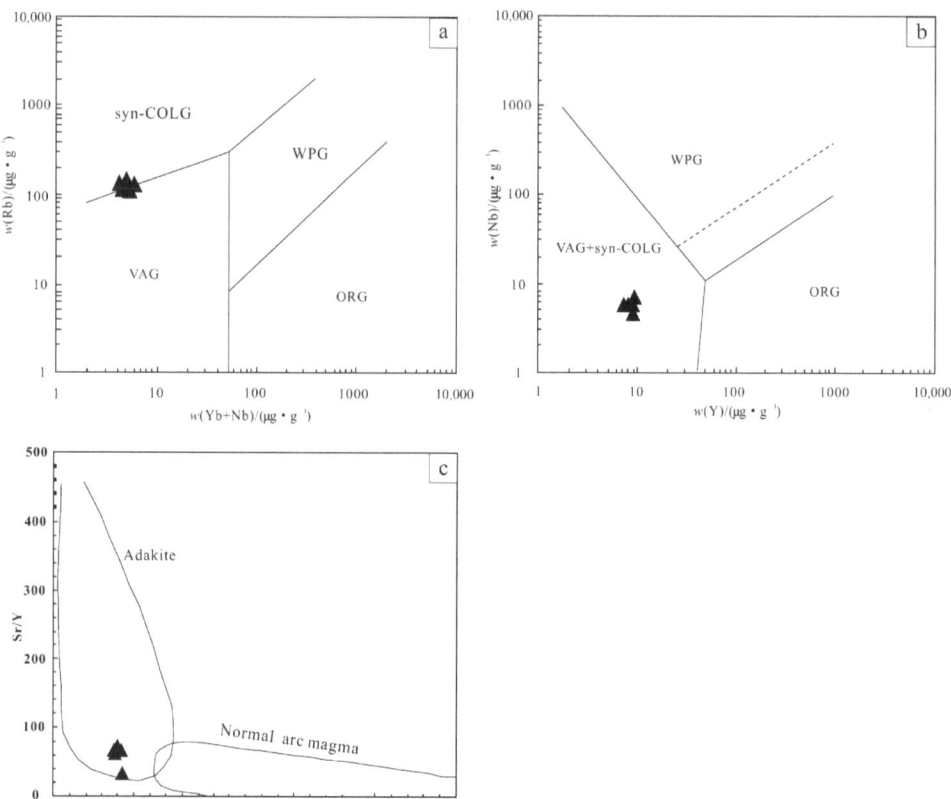

Figure 7. The (Yb+Nb)-Rb (**a**), Y-Nb (**b**) and Y-Sr/Y (**c**) diagrams for rock samples from the Aolunhua deposit.

Previous studies have shown that the southern segment of the Da Hinggan Range experienced partial subduction and continent–continent collision in the Paleo-Asian Ocean, between the North China and Siberian Paleo continents, which formed the Hing-Meng orogenic belt [52]. In recent years, more studies have confirmed that the collision between the Siberian and North China plates likely happened in the mid-late Permian period and continued to the mid Triassic period. By the late Jurassic period (150 Ma), the western segment of the Mongol–Okhotsk ocean had closed, and the continental collision ended [53]. For the Da Hinggan Range in northeastern China, the NNW subduction of the Izanagi plate beneath the Eurasian plate triggered intensive magmatism and a mineralization event in this region, which is the most important tectonic activity in the Mesozoic [28].

After the conclusion of the Da Hinggan Range orogeny, the crust of the southern segment of the Da Hinggan Range gradually experienced a transition to an extensional setting. On the R1-R2 discrimination diagram, the granite porphyries of the Aolunhua Mo deposit mainly plot into the syn-collision field, whereas on the Y-Sr/Y diagram, they plot into the field of adakaite, indicating that the petrogenesis occurred under rather high pressures. Combining this knowledge with the petrogenic age of 135.0 ± 1.0 Ma obtained in this study, these results suggest that the tectonic setting of the magmatic rocks in the ore area was in transition between a compression-orogenic and extensional back-arc regime.

Moreover, Wang et al. [54] proposed that block margins and suture zones are preferred settings for large scale metallogenesis. The main metallogenic pulse formed in response to the magmatic activity and minerogenesis, during which an extensional back-arc regime was developed in the Xilamulun area, following accretionary orogenesis and the thickening of the continental crust.

In fact, at this regional, the large-scale porphyry-type Mo mineralization was dominated by a tectonic-magmatic event: slab rollback, accompanied by related lower crust delamination, asthenospheric upwelling, and lithospheric thinning in eastern China during the Cretaceous era (140–90 Ma). Although old crustal material may have been involved in the genesis of Mo-forming magmas, the partial melting of the juvenile lower crust provided significant contributions. In addition, other factors, including high magma oxygen fugacity, the efficient exsolution of metal-bearing fluids from magmas, and the boiling or immiscibility of fluids, may have played a fundamental role in the Mo enrichment and subsequent mineralization [9]. The Mo minera lizations of the Aolunhua deposit, the Banlashan deposit, and the Yangchang deposit occurred within or near the Cretaceous intrusive stocks. The geochronology of the ore-forming and intrusive rocks is consistent. Therefore, there is a close spatial and temporal relationship between the Mo mineralization and Cretaceous intrusive stocks in DHMP. The Mo mineralization took place in Cretaceous regional volcanic-magmatism. It is consistent with the fastigium of lithospheric thinning in North China.

7. Conclusions

(1) The zircon U-Pb LA ICP-MS age of 135.0 ± 1.0 Ma for the granite porphyry of the Aolunhua Mo deposit of Inner Mongolia indicates that the granite porphyry is the product of early Cretaceous magmatic activity.

(2) The Th/U ratio of the zircon from the ore-bearing igneous body of the ore deposit is greater than 0.1, with a prominent Ce positive anomaly (Ce* = 1.72–188.71) and an Eu negative anomaly (Eu* = 0.05–0.57), indicating typical magmatic zircon, with depletion in LREE and enrichment in HREE, and low La and Pr contents. The positive Ce anomaly reveals the characteristics of a crust-derived magmatic zircon.

(3) The Aolunhua ore-deposit is formed within the Cu-Mo metallogenic belt at the northern flank of the Xilamulun River deep fracture, which constitutes the Linxi–Tianshan Cu-Mo ore deposit belt and was formed during the peak stage of metallogeny at 140 Ma, caused by the magmatic activity, developed during the transitional stage between compression-orogeny and back-arc extension.

Author Contributions: H.L.: Conceptualization, Methodology, Validation, Formal analysis, Investigation, Data curation, Writing—original draft. X.L.: Conceptualization, Validation, Formal analysis,. J.X.: Methodology, Validation, Formal analysis. Y.Y.: Conceptualization, Methodology, Validation, Formal analysis, Investigation, Data curation, Writing—original draft, Supervision, Funding acquisition. All authors have read and agreed to the published version of the manuscript.

Funding: The study is supported by the Natural Science Foundation (Grant No. 41272110), China Nuclear Uranium Co., Ltd. in joint with East China University of Technology (Grant No. 2023NRE-LH-06), and the special program from Geological Survey of China (No. 12120113089400, 12120114076801).

Data Availability Statement: Data are contained within the article.

Acknowledgments: We are thankful to Pirajno Franco for his valuable help with the revision of manuscript. The CL image is conducted by Chen Li of the Physics Institute of Peking University, and the field work was assisted by the geologists of the Aolunhua Mo deposit, to whom the authors are deeply indebted.

Conflicts of Interest: Hao Li is an employee of No. 5 Geological Party Limited Liability Company of Liaoning Province. Jiang Xin is an employee of Xinjiang Nonferrous Metal Industry Group Co., Ltd. The paper reflects the views of the scientists and not the companies.

References

1. Berzina, A.N.; Berzina, A.P.; Gimon, V.O. The Aksug Porphyry Cu–Mo Ore-Magmatic System (Northeastern Tuva): Sources and Formation of Ore-Bearing Magma. *Russ. Geol. Geophys.* **2021**, *62*, 445–459. [CrossRef]
2. Hou, X.G.; Sun, D.Y.; Gou, J.; Yang, D.G. The origin of variable-$\delta^{18}O$ zircons in Jurassic and Cretaceous Mo-bearing granitoids in the eastern Xing-Meng Orogenic Belt, Northeast China. *Int. Geol. Rev.* **2019**, *61*, 129–149. [CrossRef]
3. Patrick, B.R.; Marco, T.E. The Bingham Canyon Porphyry Cu–Mo–Au deposits. I. Sequence of intrusion, vein formation, and sulfide deposition. *Econ. Geol.* **2010**, *105*, 43–68.
4. Pirajno, F.; Santosh, M. Rifting, intraplate magmatism, mineral systems and mantle dynamics in central-east Eurasia: An overview. *Ore Geol. Rev.* **2014**, *63*, 265–295. [CrossRef]
5. Richards, J.P. Tectono-Magmatic Precursors for Porphyry Cu-(Mo- Au) Deposit Formation. *Econ. Geol.* **2003**, *98*, 1515–1533. [CrossRef]
6. Wang, L.L.; Zhang, D.H.; Tian, L. Silicate melt inclusions in the Qiushuwan granitoids, northern Qinling belt, China: Implications for the formation of a porphyry Cu–Mo deposit as a reduced magmatic system. *Comptes Rendus Geosci.* **2014**, *346*, 190–199. [CrossRef]
7. Wu, H.Y.; Zhang, L.C.; Gao, J.; Zhang, M.; Zhu, M.T.; Xiang, P. U-Pb geochronology, isotope systematics, and geochemical characteristics of the Triassic Dasuji porphyry Mo deposit, Inner Mongolia, North China: Implications for tectonic evolution and constraints on the origin of ore related granitoids. *J. Asian Earth Sci.* **2018**, *165*, 132–144. [CrossRef]
8. Wyman, D.A.; Cassidy, K.F.; Hollongs, P. Orogenic gold and the mineral systems approach: Resolving fact, fiction and fantasy. *Ore Geol. Rev.* **2016**, *78*, 322–335. [CrossRef]
9. Gao, J.; Qin, K.Z.; Zhou, M.F.; Zaw, K. Large-scale porphyry-type mineralization in the Central Asian Metallogenic Domain: Geodynamic background, magmatism, fluid activity and metallogenesis. *J. Asian Earth Sci.* **2018**, *165*, 1–6. [CrossRef]
10. Chen, Y.J.; Zhan, C.; Li, N.; Yang, Y.F.; Deng, K. Geology of the Mo Deposits in Northeast China. *J. Jilin Univ. Earth Sci. Ed.* **2012**, *42*, 1223–1268, (In Chinese with English Abstract).
11. Liu, J.M.; Zhang, R.; Zhang, Q.Z. The regional metallogeny of Da Hinggan Ling, China. *Earth Sci. Front.* **2004**, *11*, 269–277, (In Chinese with English Abstract)
12. Mao, J.W.; Wang, Y.T.; Zhang, Z.H.; Yu, J.J.; Niu, B.G. Geodynamic settings of Mesozoic large-scale mineralization in North China and adjacent areas. *Sci. China D* **2003**, *46*, 838–851. [CrossRef]
13. Mao, J.W.; Xie, G.Q.; Zhang, Z.H.; Li, X.F.; Wang, Y.T.; Zhang, C.Q.; Li, Y.F. Mesozoic large-scale metallogenic pluses in North China and corresponding geodynamic settings. *Acta Petrol.* **2005**, *21*, 169–188, (In Chinese with English Abstract)
14. Wu, F.Y.; Sun, D.Y.; Ge, W.C.; Zhang, Y.B.; Grant, M.L.; Wilde, S.A.; Jahn, B.M. Geochronology of the Phanerozoic granitoids in northeastern China. *J. Asian Earth Sci.* **2011**, *41*, 1–30. [CrossRef]
15. Wu, G.; Li, X.Z.; Xu, L.Q.; Wang, G.R.; Liu, J.; Zhang, T.; Quan, Z.X.; Wu, H.; Li, T.G.; Zeng, Q.T.; et al. Age, geochemistry, and Sr–Nd–Hf–Pb isotopes of the Caosiyao porphyry Mo deposit in Inner Mongolia, China. *Ore Geol. Rev.* **2017**, *81*, 706–727. [CrossRef]
16. Zeng, Q.D.; Liu, J.M.; Zhang, Z.L.; Chen, W.J.; Zhang, W.Q. Geology geochronology of the Xilamulun molybdenum metallogenic belt in eastern Inner Mongolia, China. *Int. J. Earth Sci.* **2011**, *100*, 1791–1809. [CrossRef]
17. Zhai, D.G.; Liu, J.J.; Wang, J.P.; Yang, Y.Q.; Zhang, H.Y.; Wang, X.L.; Zhang, Q.B.; Wang, G.W.; Liu, Z.J. Zircon U-Pb and molybdenite Re-Os geochronology, and whole-rock geochemistry of the Hashitu molybdenum deposit and host granitoids, Inner Mongolia, NE China. *J. Asian Earth Sci.* **2014**, *79*, 144–160. [CrossRef]
18. Zhou, Z.H.; Liu, H.W.; Chang, G.X.; Lu, L.S.; Li, T.; Yang, Y.J.; Zhang, R.J.; Ji, X.H. Mineralogical characteristics of skarns in the Huanggangliang Sn-Fe deposit of Inner Mongolia and their metallogenic indicating significance. *Acta Petrol. Sin.* **2011**, *30*, 97–112, (In Chinese with English Abstract)
19. Liu, J.J.; Xing, Y.L.; Wang, J.P. Discovery of Falkmanite from the Bairendaba Superlarge Ag-Pb-Zn Polymetallic Deposit, Inner Mongolia and Its Origin Significance. *J. Jilin Univ. Earth Sci. Ed.* **2010**, *40*, 565–572, (In Chinese with English Abstract)
20. Zhao, Y.M.; Zhang, D.Q. *Metallogeny and Prospective Evaluation of Copper-Polymetallic Deposits in the Da Hinggan Mountains and Its Adjacent Regions*; Seismological Press: Beijing, China, 1997; pp. 83–106, (In Chinese with English Abstract).
21. Chen, Z.G.; Zhang, L.C.; Wan, B.; Wu, H.Y.; Cleven, N. Geochronology and geochemistry of the Wunugetushan porphyry Cu–Mo deposit in NE China, and their geological significance. *Ore Geol. Rev.* **2011**, *43*, 92–105. [CrossRef]
22. Xue, H.M.; Guo, L.J.; Hou, Z.Q. SHRIMP zircon U-Pb ages of the middle Neopaleozoic unmetamorphosed magmatic rocks in the southwestern slope of the Da Hinggan Mountains, Inner Mongolia. *Acta Petrol. Sin.* **2010**, *29*, 811–823, (In Chinese with English Abstract).
23. Zhang, S.T.; Zhao, P.D. Porphyry ore deposits: Important study subjects of nontraditional mineral resources. *Earth Sci. J. China Univ. Geosci.* **2011**, *36*, 247–254, (In Chinese with English Abstract)
24. Sun, X.G.; Liu, J.M.; Qin, F. The new progress on polymetallic studies in Daxinganling—The revelation of the southern bank Xilamulun river Molybdenum polymetallic belt. *Min. Mag.* **2008**, *17*, 75–78, (In Chinese with English Abstract)
25. Xia, X.H.; Zhao, Y.H.; Yuan, J.Z.; Yao, M.C.; Ye, F. Geology and mineralizing regularity of poly-metal sulfide deposits in Linxi-Tianshan district of Inner Mongolia. *Geol. Chem. Miner.* **2002**, *24*, 198–206, (In Chinese with English Abstract)

26. Chen, X.L.; Huang, W.T.; Chen, L.; Zou, S.H.; Zhang, J.; Li, K.X.; Liang, H.Y. Controlling factors of different Late Cretaceous granitoid-related mineralization between western margin of the Yangtze Block and the neighbor Yidun arc. *Ore Geol. Rev.* **2021**, *139*, 104554. [CrossRef]
27. Zeng, Q.D.; Liu, J.M. Zircon SHRIMP U-Pb dating and geological significance of the granite porphyry from Banlashan porphyry molybdenum deposit in Xilamulun molybdenum metallogenic Belt. *J. Jilin Univ. Earth Sci. Ed.* **2010**, *40*, 828–834, (In Chinese with English Abstract)
28. Zhang, J.F.; Quan, H.; Wu, G. Tectonic setting of Mesozoic volcanic rocks in Northern China. *J. Precious Met. Geol.* **2000**, *9*, 33–38, (In Chinese with English Abstract)
29. Zhang, K.; Nie, F.J.; Hou, W.R.; Li, C.; Liu, Y. Re-Os isotopic age dating of molybdenite separates from Hashitu Mo deposit in Linxi County of Inner Mongolia and its geological significance. *Miner. Deposita* **2012**, *31*, 129–138, (In Chinese with English Abstract)
30. Ma, X.H.; Chen, B.; Lai, Y. Petrogenesis and mineralization chronology study on the Aolunhua porphyry Mo deposit, Inner Mongolia, and its geological implications. *Acta Petrol.* **2009**, *5*, 2939–2950, (In Chinese with English Abstract)
31. Shu, Q.H.; Jiang, L.; Lai, Y.; Lu, Y.H. Geochronology and fluid inclusion study of the Aolunhua Porphyry Cu-Mo Deposit in Arhorqin Area, Inner Mongolia. *Acta Petrol.* **2009**, *25*, 2601–2614, (In Chinese with English Abstract)
32. Geology and Mineral Resource Bureau of Inner Mongolia Autonomous Region. *Geological Magazine*; Geological Publishing House: Beijing, China, 1991; pp. 13–26.
33. Chen, B.; Jahn, B.M.; Tian, W. Evolution of the Solonker suture zone: Constraints from zircon U-Pb ages, Hf isotopic ratios and whole-rock Nd-Sr isotope compositions of subduction and collision-related magmas and forearc sediments. *J. Asian Earth Sci.* **2009**, *34*, 245–257. [CrossRef]
34. Li, J.Y.; Gao, L.M.; Sun, G.H.; Wang, Y.B. Shuangjingzi middle Triassic syn-collisional crust-derived granite in the east Inner Mongolia and its contraint on the timing of collision between Siberian and Sino-Korean paleo-plates. *Acta Petrol.* **2007**, *23*, 565–582.
35. Ren, J.S.; Chen, Y.Y.; Niu, B.G. *The Crust Tectonic Evolution and Metallogenic at Eastern China and Adjacent Area*; Science Press: Beijing, China, 1992; pp. 1–205, (In Chinese with English Abstract)
36. Nie, F.J.; Zhang, W.Y.; Du, A.D. Re-Os isotopic dating on molybdenite separates from the Xiaodonggou porphyry Mo deposit, Hexigten Qi, Inner Mongolia. *Acta Petrol.* **2007**, *81*, 898–905, (In Chinese with English Abstract)
37. Xiao, W.J.; Windley, B.F.; Hao, J.; Zhai, M.G. Accretion leading to collision and the Permian Solonker suture, Inner Mongolia, China: Termination of the central Asian orogenic belt. *Tectonics* **2003**, *22*, 1069–1089. [CrossRef]
38. Xu, Q.; Fu, S.X.; Yuan, J.M. Geological characteristics and prospecting marks of the Aolunhua porphyry Mo-Cu deposit, Inner Mongolia. *Geol. Explor.* **2010**, *46*, 1019–1028, (In Chinese with English Abstract)
39. Anderson, T. Correction of common lead in U-Pb analyses that do not report ^{204}Pb. *Chem. Geol.* **2002**, *192*, 59–79. [CrossRef]
40. Black, L.P.; Kamo, S.L.; Allenc, M. TEMORA 1: A new zircon standard or Phanerozoic U-Pb geochronology. *Chem. Geol.* **2003**, *200*, 155–170. [CrossRef]
41. Belousova, E.A.; Griffin, W.L.; Pearson, N.J. Trace element composition and catholuminescence properties of southern African kimberlitic zircons. *Miner. Mag.* **1998**, *62*, 355–366. [CrossRef]
42. Schaltegger, U.; Nowak, A.; Ulianov, A.; Fisher, C.M.; Gerdes, A.; Spikings, R.; Whitehouse, M.J.; Bindeman, I.; Hanchar, J.M.; Duff, J. Zircon Petrochronology and 40Ar/39Ar Thermochronology of the Adamello Intrusive Suite, N. Italy: Monitoring the Growth and Decay of an Incrementally Assembled Magmatic System. *J. Petrol.* **2019**, *60*, 701–722. [CrossRef]
43. Nigel, H. Woodcock; N. Jack Soper and Andrew, J. Miles. Age of the Acadian deformation and Devonian granites in northern England: A review. *Proc. Yorks. Geol. Soc.* **2019**, *62*, 238–253.
44. Hoskin, P.W.O.; Schaltegger, U. The composition of zircon and igneous and metamorphic petrogenesis. *Rev. Mineral. Geochem.* **2003**, *53*, 27–62. [CrossRef]
45. Zeng, Q.D.; Liu, J.M.; Zhang, Z.L. Re-Os Geochronology of Porphyry Molybdenum Deposit in South Segment of Da Hinggan Mountains, Northeast China. *J. Earth Sci.* **2010**, *21*, 392–401. [CrossRef]
46. Corfu, F.; Hanchar, J.M.; Hoskin, P.W.O.; Kinny, P. Altas of zircon textures. *Rev. Mineral. Geochem.* **2003**, *53*, 469–500. [CrossRef]
47. Chakoumakos, B.C.; Murakami, T.; Lumpkin, G.R. Alpha-decay induced fracturing in zircon: The transition from the crystalline to the metarniet statc. *Science* **1987**, *235*, 1556–1559. [CrossRef]
48. Valley, J.W.; Chiarenzelli, J.R.; MeLelland, J.M. Oxygen isotope geochemistry of zircon. *Earth Planet. Sci. Lett.* **1994**, *126*, 187–206. [CrossRef]
49. McDonough, W.F.; Sun, S. The composition of the Earth. *Chem. Geol.* **1995**, *120*, 223–253. [CrossRef]
50. Shu, Q.; Chang, Z.; Lai, Y.; Hu, X.; Wu, H.; Zhang, Y.; Wang, P.; Zhai, D.; Zhang, C. Zircon trace elements and magma fertility: Insights from porphyry (-skarn) Mo deposits in NE China. *Miner. Depos.* **2019**, *54*, 645–656. [CrossRef]
51. Li, X.; Liang, X.; Sun, M. Geochronology and geochemistry of single-grain zircons: Simultaneous in-situ analysis of U-Pb age and trace elements by LAM-ICP-MS. *Eur. J. Miner.* **2000**, *12*, 1015–1024. [CrossRef]
52. Zhang, Y.B.; Sun, S.H.; Mao, Q. Mesozoic O-type adakitic volcanic rocks and its petrogenesis, paleo-tectonic dynamic and mineralization significance of the eastern side of southern Da Hinggan, China. *Acta Petrol.* **2006**, *22*, 2289–2304, (In Chinese with English Abstract)

53. Van der Voo, R.; van Hinsbergen, D.J.J.; Domeier, M.; Spakman, W.; Torsvik, T.H. Latest Jurassic–earliest Cretaceous closure of the Mongol-Okhotsk Ocean: A paleomagnetic and seismological-tomographic analysis. In *Late Jurassic Margin of Laurasia—A Record of Faulting Accommodating Plate Rotation*; Anderson, T.H., Didenko, A.N., Johnson, C.L., Khanchuk, A.I., MacDonald, J.H., Jr., Eds.; GeoScience World: McLean, VA, USA, 2015; Volume 513, pp. 1–18. [CrossRef]
54. Wang, J.B.; Xu, X. Post-collisional tectonic evolution and metallogenesis in Northern Xinjiang, China. *Acta Petrol.* **2006**, *80*, 23–31, (In Chinese with English Abstract)

Disclaimer/Publisher's Note: The statements, opinions and data contained in all publications are solely those of the individual author(s) and contributor(s) and not of MDPI and/or the editor(s). MDPI and/or the editor(s) disclaim responsibility for any injury to people or property resulting from any ideas, methods, instructions or products referred to in the content.

Article

U-Pb Geochronology, Geochemistry and Geological Significance of the Yongfeng Composite Granitic Pluton in Southern Jiangxi Province

Yunbiao Zhao [1], Fan Huang [1,*], Denghong Wang [1], Na Wei [2], Chenhui Zhao [1] and Ze Liu [1]

1. MNR Key Laboratory of Metallogeny and Mineral Assessment, Institute of Mineral Resources, Chinese Academy of Geological Sciences, Beijing 100037, China; zybcags@163.com (Y.Z.)
2. Department of Transportation Engineering, Hebei University of Water Resources and Electric Engineering, Cangzhou 061001, China
* Correspondence: hfhymn@163.com

Abstract: The Yongfeng composite granitic pluton, located in the southern section of the Nanling area, is composed of the Yongfeng and Longshi biotite monzonitic granites. In order to reveal the genesis of this composite granitic pluton and its relationship with mineralization, this study conducted zircon U-Pb dating, whole-rock major and trace element analysis, and biotite electron probe analysis. The results show that the Yongfeng composite granitic pluton is rich in silicon and alkali, weakly peraluminous, and poor in calcium and iron. It shows the enrichment of light rare earth elements and a significant fractionation of light and heavy rare earth elements. It also shows the enrichment of large ion lithophile elements and depletion of Ba, K, P, Eu, and Ti relative to the primitive mantle. The contents of TFe_2O_3, MgO, CaO, TiO_2, and P_2O_5 are low and decrease with increasing SiO_2 content. The Yongfeng composite granitic pluton does not contain alkaline dark minerals. Its average zircon saturation temperature is 776 °C, average TFe_2O_3/MgO is 4.81, and average Zr + Nb + Ce + Y is 280.6 ppm, which correspond to a highly fractionated I-type granite. The Yongfeng and Longshi granites were respectively formed at 152.0 ± 1.0 Ma–151.3 ± 1.1 Ma and 148.9 ± 1.2 Ma. They were formed in the extensional tectonic setting during the post-orogenic stage, under the control of the breakup or retreat of the backplate after the subduction of the Pacific Plate into the Nanling hinterland. The magmatic system of the Yongfeng composite granitic pluton is characterized by high fractionation, high content of F, high temperature, and low oxygen fugacity, which is conducive to mineralization of Sn, Mo, and fluorite.

Keywords: Yongfeng composite granitic pluton; highly fractionated granite; zircon U-Pb dating; granite petrogenesis; tectonic setting; southern Jiangxi Province

Citation: Zhao, Y.; Huang, F.; Wang, D.; Wei, N.; Zhao, C.; Liu, Z. U-Pb Geochronology, Geochemistry and Geological Significance of the Yongfeng Composite Granitic Pluton in Southern Jiangxi Province. *Minerals* **2023**, *13*, 1457. https://doi.org/10.3390/min13111457

Academic Editor: Clemente Recio

Received: 19 August 2023
Revised: 24 October 2023
Accepted: 24 October 2023
Published: 20 November 2023

Copyright: © 2023 by the authors. Licensee MDPI, Basel, Switzerland. This article is an open access article distributed under the terms and conditions of the Creative Commons Attribution (CC BY) license (https://creativecommons.org/licenses/by/4.0/).

1. Introduction

The Nanling region is an important mineral deposit area in China because it features a wide distribution of granites that are rich in non-ferrous and rare metal mineral resources, such as W, Sn, Li, Be, Nb, Ta, and U. In this region, mineralization is spatially related to the widespread intrusion of highly fractionated granites [1]. After undergoing high crystallization differentiation, granitic magma becomes enriched in volatile components, REEs, and rare metals in the residual melt-hydrothermal fluid system, providing the appropriate conditions for the formation of metal deposits [2–4].

The Yongfeng composite granitic pluton is located in Xingguo County, southern Jiangxi Province, east section of the Nanling Range. It consists of Yongfeng medium–coarse grained porphyritic biotite monzonitic granite and Longshi fine grained biotite monzonitic granite, exhibiting characteristics of highly fractionated granite. Multiple molybdenum deposits have been discovered at the contacting zone between the Yongfeng composite granitic pluton and its host rocks. The largest deposit is the Leigongzheng molybdenum deposit [5]. In addition, a large-scale fluorite deposit (Longping fluorite deposit) has been found in the contact zone outside the Longshi granite [6]. Previous studies have generally believed that the Leigongzheng molybdenum deposit and the Longping fluorite deposit are closely related to the adjacent Yongfeng composite granitic pluton [7,8]. However, only a few studies have investigated the Yongfeng composite granitic pluton, and only Yang et al. [9] have suggested it as an A-type granite. Nevertheless, it exhibits distinct characteristics that differentiate it from typical A-type granites, and its genetic type remains to be further determined. In order to reveal the genesis of this composite granitic pluton and its relationship with mineralization, this study systematically conducted petrographic analysis, zircon U-Pb dating, whole-rock major and trace element analysis, and biotite electron probe analysis.

2. Geological Setting and Petrography

Southern Jiangxi province is located in the eastern part of the Nanling tectono-magmatic zone [10] (Figure 1b). The stratigraphy in the area comprises Pre-Carboniferous, Lower Paleozoic, Upper Paleozoic, Mesozoic, and Cenozoic age rocks (Figure 1a). Regional metamorphism generally occurred in the Carboniferous and Lower Paleozoic, forming epimetamorphic rocks, which are covered by Upper Paleozoic shallow marine carbonate and siliciclastic sedimentary rocks. Mesozoic volcanoclastic rocks and terrigenous red-bed sandstones are presented in faulted basins. The lithology of the Cenozoic is composed of loose mud and sand [11]. Southern Jiangxi province is mostly controlled by faults in EW, NNE, and SWW trends [12]. Granites are widespread in the region and mainly occur in Yanshanian [13] (Figure 1a).

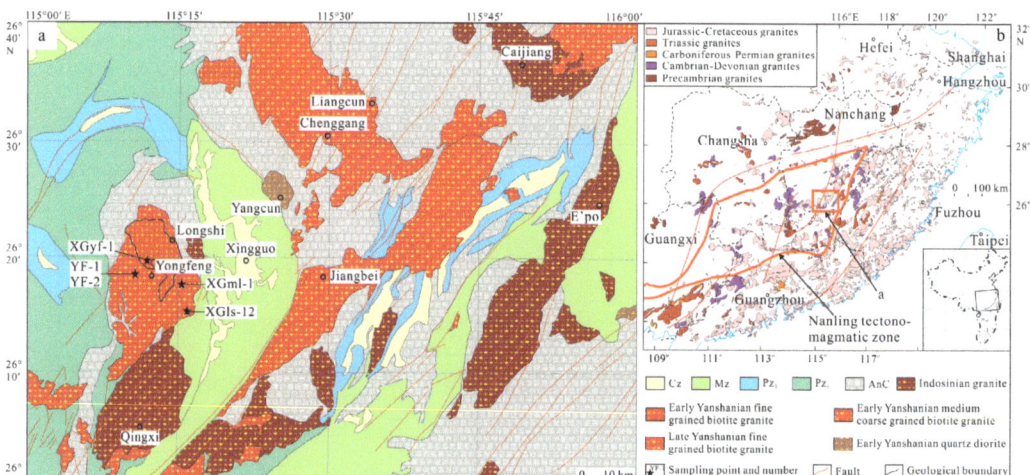

Figure 1. (a) Regional geological sketch of the Yongfeng composite granite pluton (modified according to [14]); (b) Distribution map of granite in South China (modified according to [15]).

The Yongfeng composite granitic pluton consists of Yongfeng, medium- to coarse-grained biotite porphyritic monzonite granite and Longshi fine-grained biotite monzonite granite, with an outcrop area of 350 km^2. The northern, eastern, and western parts of the composite granitic pluton are in contact with epimetamorphic rocks of the Laohutang formation and Bali formation, and the southern part intrudes the Qingxi granite (Figure 1a). The Yongfeng granite occurs along the margins of the composite granitic pluton, predominantly composed of reddish or grayish white medium–coarse grained porphyritic biotite monzonitic granite (Figure 2), occasionally with potash feldspar phenocrysts. The main minerals are alkali feldspar, plagioclase, and quartz, with a small amount of biotite and muscovite, and the accessory minerals are zircon, apatite, monazite, and rutile. Alkaline feldspar mainly consists of microcline and perthite, with a semi-idiomorphic plate shape. The length ranges from 0.2 mm to 8.0 mm, with a content of 35 vol%. Tartan twinning and striped structure can be observed (Figure 3a,b). Plagioclase occurs in the form of semi-idiomorphic plate, and polysynthetic twin can be observed, with a length of 0.2–7.0 mm and a content of 32 vol%. The alkali feldspar and plagioclase have undergone strong sericitization alteration (Figure 3a,b). Quartz has a xenomorphic-granular shape, with a diameter of 0.2–5.0 mm and a content of 30 vol%. In the hand specimen, biotite appears as a clump-like aggregate, and under the microscope, it appears scaly, with a length of 0.2–10 mm and a content of 2 vol%. Muscovite also appears scaly (Figure 3a,d), with a length of 0.2–0.6 mm and a content of 1 vol%. The biotite and muscovite have undergone strong chloritization alteration. The Longshi granite is located in the middle of the composite granitic pluton and intrudes the Yongfeng granite. The edges of the granite are heavily weathered. The lithology of the Longshi granite is similar to that of the Yongfeng granite, with only differences in mineral grain size, content, and the phenocrysts content. The lithology of the Longshi pluton is fine-grained biotite monzonitic granite, occasionally with feldspar phenocrysts. The main minerals include potassium feldspar (~20 vol%), plagioclase feldspar (~30 vol%), and quartz (~40 vol%), followed by biotite (~8 vol%) and muscovite (~2 vol%). Potassium feldspar occurs in semi-idiomorphic plate form with grain sizes of 0.2–1.5 mm, with occasional occurrences of striped feldspar and microcline. Plagioclase feldspar occurs in semi-idiomorphic plate form with grain sizes of 0.2–2.0 mm, showing the development of polysynthetic twins. Quartz occurs in granular form with grain sizes of 0.1–2.5 mm. Biotite occurs in scaly form, with a length of 0.1–1 mm. Muscovite occurs in scaly shape, with a length of 0.1–0.6 mm. The accessory minerals mainly include apatite, xenotime, zircon, and a small amount of metallic minerals. The alteration mainly includes K-feldsparization, silicification, sericitization, chloritization, epidotization, and carbonatization (Figure 3d).

Figure 2. Field photos of the Yongfeng composite granitic pluton. (**a**) reddish medium–coarse grained porphyritic biotite monzonitic granite; (**b**) grayish white medium–coarse grained porphyritic biotite monzonitic granite.

Figure 3. *Cont.*

Figure 3. Typical petrographic photos of the Yongfeng composite granitic pluton (**a,b,d**) are orthogonal-polarization micrographs and (**c**) is a single-polarization micrograph. (Bit-biotite, Cal-calcite, Mic-microcline, Pth-striped feldspar, Ms-muscovite, Qtz-quartz, Pl-plagioclase, Srt-sericite).

3. Sample Information and Analysis Methods

The sample information is provided in Supplementary Table S1 and Figure 1a. All samples were subjected to zircon U-Pb dating. Samples YF-1 and YF-2 were analyzed for whole-rock major and trace elements, zircon trace elements, and biotite composition.

3.1. Zircon U-Pb Dating and Trace Elements

The zircon grains were separated from five samples (YF-1, YF-2, XGml-1, XGls-12, XGyf-1) using heavy liquid and magnetic techniques. Representative zircons were hand-picked and mounted in epoxy resin and then polished and coated with carbon. The internal morphology was examined using cathodoluminescence (CL) prior to U-Pb analyses. Zircon U-Pb dating and trace element analysis were completed in the MC-ICP-MS laboratory of the Institute of Mineral Resources, Chinese Academy of Geological Sciences, using a Bruker M90 ICP-MS equipped with a RESOlution S-155 193 nm laser. The $^{207}Pb/^{206}Pb$ and $^{206}Pb/^{238}U$ ratios were calculated using the ICP-MS Data Cal 8.0 program and corrected using zircon GJ-1 as external calibration. These correction factors were then applied to each sample to correct for both instrumental mass bias and depth-dependent elemental and isotopic fractionation. Common Pb content was evaluated using the method described by Andersen [16]. The concordia diagrams were plotted using ISOPLOT (version 3.0). The errors quoted in the tables and figures are at the 1σ level. Instrument operation and data processing methods are described in Hou et al. [17]. Zircon trace elements were quantified using SRM610 as the external standard.

3.2. Major and Trace Element Analysis

Whole-rock samples were trimmed to remove weathered surfaces, cleaned with deionized water, crushed, and then powdered through a 200-mesh screen using a tungsten carbide ball mill. Major elements were analyzed using an X-ray fluorescence (XRF) spectrometer (PW4400) at the National Geological Experimental Testing Center. The detection method is based on GB/T 14506.28-2010. Trace elements were determined using an inductively coupled plasma mass spectrometer (PE300D) at the National Geological Experimental Testing Center. The detection method is based on GB/T 14506.30-2010. The error of the analysis results is less than 5%.

3.3. Electron Probe Microanalysis (EPMA)

Biotite without alteration or with weak alteration was selected for compositional analysis. Biotite composition analysis was performed on a JXA-8230 electron probe analyzer, which is housed at the MLR Key Laboratory of Metallogeny and Mineral Assessment, Institute of Mineral Resources, Chinese Academy of Geological Sciences. The working conditions were as follows: 15 kV voltage, 20 nA current, 5 μm beam spot. Fe^{3+} and Fe^{2+} were obtained according to the calculation method of Lin et al. [18]. The structural formula of biotite was calculated on the basis of 22 oxygen atoms.

4. Results

4.1. Zircon Trace Element Geochemistry

The results of trace elements of zircons from samples YF-1 and YF-2 are shown in Supplementary Table S2. The content of total rare-earth elements (ΣREE) in zircons from sample YF-1 was 518.25–1679.63 ppm, with an average of 921.04 ppm. The content of light rare-earth elements (LREEs) was 6.53–102.18 ppm, with an average of 27.36 ppm. The content of heavy rare-earth elements (HREEs) was 504.49–577.45 ppm, with an average of 893.69 ppm. The LREE/HREE ratio was 0.01–0.07, with an average of 0.03, indicating HREE enrichment and LREE depletion. Chondrite-normalized REE patterns of zircons invariably showed a left-leaning trend (Figure 4). Eu presented a significant negative anomaly (Eu/Eu* = 0.04–0.61, mean of 0.16), whereas Ce presented a significant positive anomaly (Ce/Ce* = 1.10–593.83, mean of 73.65). The content of ΣREE in zircons from sample YF-2 was 508.96–1434.22 ppm, with an average of 991.67 ppm. The content of LREEs was 9.72–234.52 ppm, with an average of 37.80 ppm. The content of HREEs was 499.05–1414.01 ppm, with an average of 953.87 ppm. The LREE/HREE ratio was 0.01–0.32, with an average of 0.04. The zircons in sample YF-2 were also enriched in HREE and depleted in LREE. Chondrite-normalized REE patterns of zircons from sample YF-2 also invariably showed a left-leaning trend (Figure 4). Eu presented a significant negative anomaly (Eu/Eu* = 0.04–0.43, mean of 0.13) and Ce presented a significant positive anomaly (Ce/Ce* = 1.30–147.88, average of 18.97). The chondrite-normalized REE patterns of zircons from the Yongfeng complex pluton are similar to the chondrite-normalized REE patterns of typical magmatic zircons (Figure 4), indicating that the zircons are magmatic zircon.

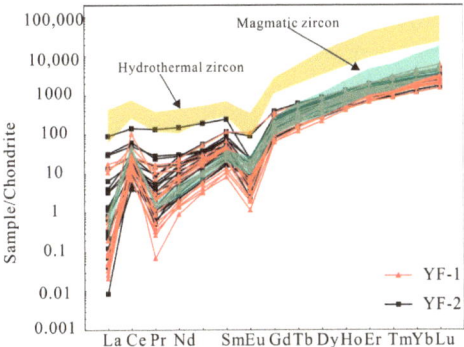

Figure 4. Chondrite-normalized REE patterns for the zircons from the Yongfeng composite granitic pluton (REE data for magmatic and hydrothermal zircons from [19]; the chondrite normalization values are from [20]).

4.2. Zircon U-Pb Age

Zircon cathodoluminescence (CL) images of five samples are shown in Figure 5 and U-Pb dating results are shown in Supplementary Table S3 and Figure 6.

The zircons from sample YF-1 were generally euhedral, measuring up to 80–170 μm, with a length/width ratio of 2:1. Most zircons were colorless or light brown and transparent to subtransparent. They exhibited clear oscillatory zoning in CL images and some of them had inherited cores. The Th content in the zircons was 90.15–666.90 ppm, U content was 146.74–2316.61 ppm, and Th/U ratio was 0.08–1.67, with an average of 0.59. The Th/U ratio of sample point 8 was 0.09, but there was a distinct oscillatory zoning, indicating that it should also be the product of magmatic crystallization. A total of 20 analytical data points were extracted from sample YF-1, and one of the data points was discarded because the concordance was below 90%. Six zircons had older apparent ages, and the other 13 zircons showed relatively consistent apparent age, with a weighted mean age of 151.5 ± 1.2 Ma (MSWD = 1.18, n = 13, 1σ) (Figure 6a).

Figure 5. The cathodoluminescence (CL) images of zircons from the Yongfeng composite granitic pluton (the white circle is the analysis location).

The zircons from sample YF-2 were also generally euhedral, measuring up to 80–160 μm with a length/width ratio of 2:1. Most zircons were colorless or light brown and transparent to subtransparent. They exhibited clear oscillatory zoning in CL images and some of them had inherited cores. The Th content in the zircons was 77.49–426.99 ppm, U content was 115.16–874.41 ppm, and Th/U ratio was 0.25–1.05, with an average of 0.59, which are typical of magmatic zircon [21]. A total of 21 analytical data points were obtained, and three of those data points were discarded because the concordance was below 90%. Six zircons showed older apparent ages, and the other 12 zircons showed relatively consistent apparent ages, with a weighted mean age of (151.3 ± 1.1) Ma (MSWD = 0.51, n = 12, 1σ) (Figure 6b).

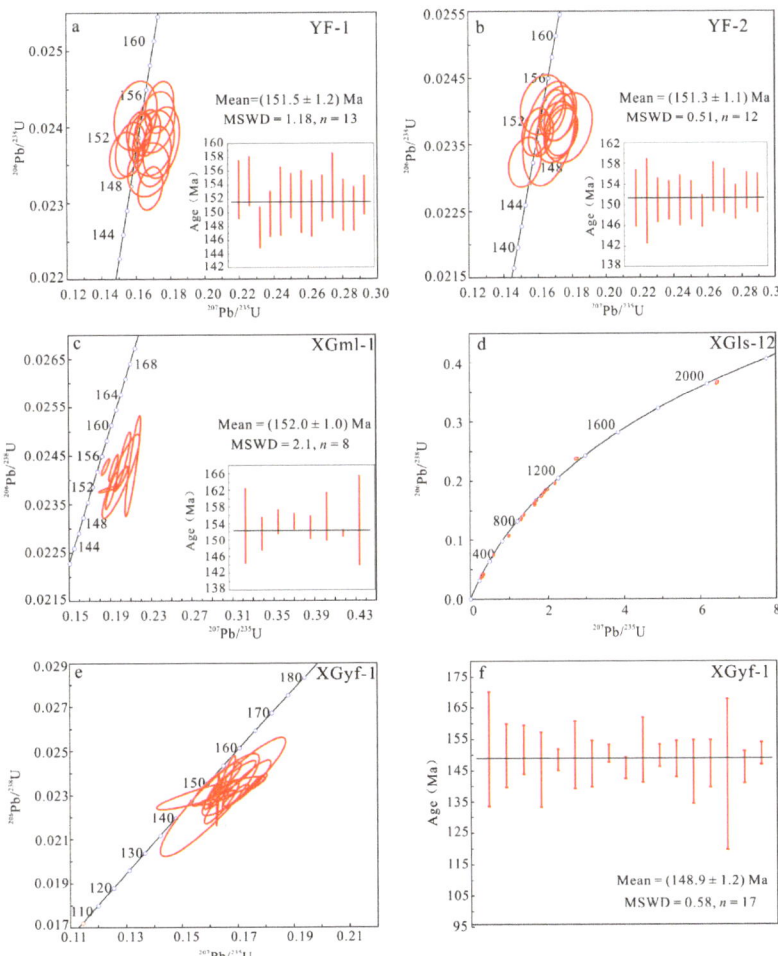

Figure 6. The zircon U-Pb age concordance diagram and weighted average age diagram of the Yongfeng composite granitic pluton (**a**) YF-1, (**b**) YF-2, (**c**) XGml-1; The zircon U-Pb age concordance diagram of (**d**) XGls-12 and (**e**) XGyf-1; (**f**) The weighted average age diagram of XGyf-1.

The zircons from sample XGml-1 were generally long columnar and xenomorphic-subhedral, measuring up to 100–210 μm with a length/width ratio of 2:1–4:1. The internal morphology of the zircons was complex, with clear oscillatory zoning in some, and unclear or no oscillatory zoning in others. A few zircons had inherited cores, while oscillatory zoning was invisible. Some zircons were black, indicating a high content of U. The Th content of the zircons was 29.32–1061.16 ppm, U content was 52.46–1475.35 ppm, and Th/U ratio was 0.02–1.73, with an average of 0.76. The Th/U ratio of sample point 9 was 0.02, and its oscillatory zoning was not prominent, suggesting that this zircon was a metamorphic zircon [22]. A total of 20 analytical data points were obtained from sample XGml-1, and four of the data points were discarded because their concordance was below 90%. Eight zircons showed older apparent ages, and the other eight zircons showed relatively consistent apparent ages, with a weighted mean age of 152.0 ± 1.0 Ma (MSWD = 2.1, n = 8, 1σ) (Figure 6c).

The zircons from sample XGls-12 were morphologically diverse and xenomorphic-subhedral, measuring up to 50–200 μm with a length/width ratio of 2:1. The internal morphology of the zircons was complex, with clear oscillatory zoning in some, but unclear or no oscillatory zoning in others. A few zircons had inherited cores, while oscillatory zoning was invisible. Some zircons were black in color, indicating a high U content. The Th content in zircons was 9.51–2478.66 ppm, U content was 133.76–2759.93 ppm, and Th/U ratio was 0.05–1.32, with an average of 0.55. The Th/U ratios of sample points 1 and 2 were less than 0.1, and their oscillatory zonings were not prominent, indicating they were metamorphic zircon [22]. A total of 20 analytical data points were obtained from sample XGls-12, and two of the data points were discarded because their concordance was below 90%. The weighted mean age could not be determined because the apparent ages were highly scattered (Figure 6d).

The zircons from sample XGyf-1 were short columnar, long columnar, and round, measuring up to 50–230 μm with a length/width ratio of 2:1–5:1. The internal morphology of the zircons is complex, with clear oscillatory zoning in some, but unclear or no oscillatory zoning in others. A zircon had an inherited core. The Th content in zircons was 47.01–6213.76 ppm, U content was 116.37–7576.37 ppm, and the Th/U ratio was 0.07–1.37, with an average of 0.58. The Th/U ratio of point 6 was 0.07, and its oscillatory zoning was not prominent, indicating it was metamorphic zircon [22]. A total of 20 analytical data points were obtained, and three of the data points were discarded because their concordance was below 90%. The data points were distributed on or near the concordance curve, with a weighted mean age of (148.9 ± 1.2) Ma (MSWD = 0.58, n = 17, 1σ) (Figure 6f).

4.3. Element Geochemistry of Granite

The results of major elements, trace elements, and rare-earth elements for samples YF-1 and YF-2 are presented in Supplementary Table S4.

As shown in Supplementary Table S4, the major elemental composition of the Yongfeng composite granitic pluton can be characterized as follows: (1) High SiO_2. The contents of SiO_2 in YF-1 and YF-2 were 72.48%–73.28% (average of 72.87%) and 72.41%–73.21% (average of 72.76%), respectively. The average content of SiO_2 in the Longshi granite was 74.30% [8]. (2) High K_2O. The contents of K_2O in YF-1 and YF-2 were 4.99%–5.53% (average of 5.35%) and 5.25%–5.53% (average of 5.35%), respectively. The K_2O/Na_2O ratios in YF-1 and YF-2 were 1.45–1.70 (average of 1.57) and 1.68–1.79 (average of 1.72), respectively. The content of K_2O in the Longshi granite was 5.15%–6.06%, with an average of 5.60%. The K_2O/Na_2O ratio was 1.71–2.13, with an average of 2.00. In the SiO_2-K_2O diagram (Figure 7c), most samples of the Yongfeng composite granitic pluton are distributed within the range of shoshonite. (3) High alkali. The contents of (Na_2O + K_2O) in YF-1 and YF-2 were 8.39%–9.14% (average of 8.76%) and 8.37%–8.62% (average of 8.48%), respectively. The alkali aluminum index (AKI) values of YF-1 and YF-2 were 0.80–0.91 (average of 0.86) and 0.79–0.80 (average of 0.80), respectively. The content of (Na_2O + K_2O) in the Longshi granite was 8.09%–8.96%, with an average of 8.40%. Its AKI was 0.76–0.87, with an average value of 0.80. In the ANOR-Q′ diagram (Figure 7a), the samples of the Yongfeng composite granitic pluton were distributed within the range of alkaline-granite and syenogranite. In the SiO_2 − (Na_2O + K_2O − CaO) diagram (Figure 7b), the samples of the Yongfeng composite granitic pluton were distributed within the range of alkaline-calcic and calc-alkalic. (4) Slightly peraluminous. The A/CNK values of YF-1 and YF-2 were 0.95–1.04 (average of 1.00) and 1.06–1.07 (average of 1.06), respectively. The A/CNK value of the Longshi granite was slightly higher, ranging from 1.05 to 1.25, with an average of 1.20. In the A/CNK-A/NK diagram (Figure 7d), the samples of the Yongfeng composite granitic pluton were located in the ranges of metaluminous and peraluminous. (5) High degree of fractionation. The differentiation index (DI = quartz + orthoclase + albite + nepheline + leucite + kalsilite) values of YF-1 and YF-2 were 88.40–91.80 (average of 90.39) and 88.55–89.38 (average of 89.13), respectively. The average DI of Longshi granite was 93.43. The contents of TFe_2O_3, MgO, CaO, TiO_2, and P_2O_5 were low in both the Yongfeng granite and Longshi granite, and gradually decreased with increasing SiO_2 content (Figure 8). These results indicate that the magma experienced a high degree of fractionation. In addition, the sample points of the Longshi granite were close to the end of evolution line (Figure 8), indicating a higher degree of crystallization differentiation compared to the Yongfeng granite.

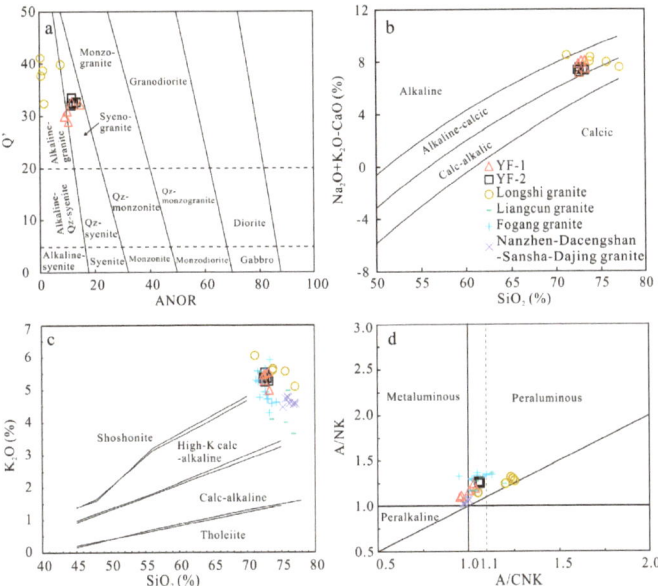

Figure 7. ANOR−Q' diagram (**a**), SiO$_2$ − (Na$_2$O + K$_2$O − CaO) diagram (**b**), SiO$_2$ − K$_2$O diagram (**c**), and A/CNK-A/NK diagram (**d**) of the Yongfeng composite granitic pluton. The data of the Longshi granite, Liangcun granite, Fogang granite, and Nanzhen-Daliangshan-Sansha-Dajing granite are respectively cited from [9,23–25].

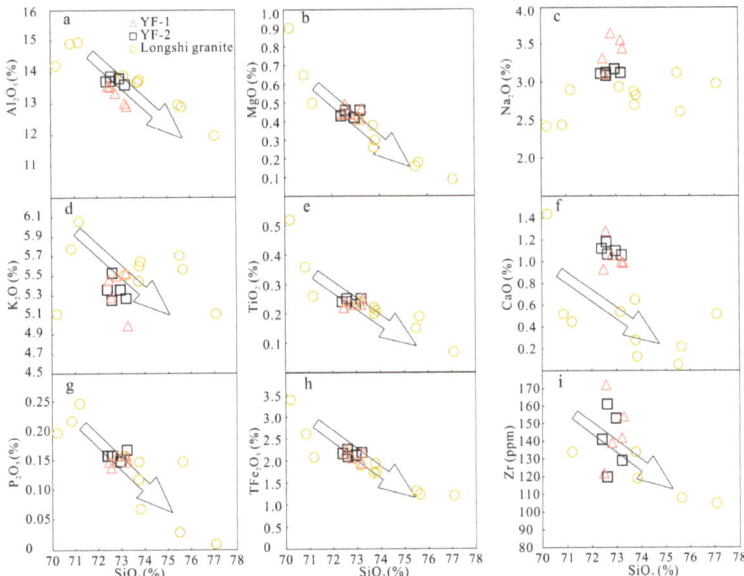

Figure 8. SiO$_2$ vs. Al$_2$O$_3$ (**a**), MgO (**b**), Na$_2$O (**c**), K$_2$O (**d**), TiO$_2$ (**e**), CaO (**f**), P$_2$O$_5$ (**g**), TFe$_2$O$_3$ (**h**), and Zr (**i**) variation diagrams of the Yongfeng composite granitic pluton. The data of the Longshi granite is cited from [9].

The ΣREE, LREE, and HREE contents of YF-1 were 186.85–211.23 ppm (average of 199.49 ppm), 170.42–188.95 ppm (average of 181.27 ppm), and 16.43–22.28 ppm (average value of 18.23 ppm), respectively. The LREE/HREE ratio was 8.48–10.98, with an average of 10.03. The $(La/Yb)_N$ value was 10.33–15.55, with an average of 13.97. The fractionation of LREEs and HREEs was distinct, showing a rightward inclination in the chondrite-normalized REE diagram (Figure 9a). Meanwhile, Eu presented a significant negative anomaly (Eu/Eu* = 0.25–0.28, average of 0.27). The primitive mantle-normalized trace element spider diagram (Figure 9b) shows the enrichment of large ion lithophile elements (LILEs) such as Rb, Th, U, and Pb, and depletion of Ba, K, P, Eu, and Ti. The ΣREE, LREE, and HREE contents of YF-2 were 199.72–227.30 ppm (average of 209.40 ppm), 182.56–203.42 ppm (average of 190.55 ppm), and 17.16–23.88 ppm (average of 18.84 ppm), respectively. The LREE/HREE ratio was 8.52–11.00, with an average of 10.23. The $(La/Yb)_N$ value was 10.55–16.28, with an average of 14.21. The chondrite-normalized REE diagram also shows a rightward trend (Figure 9a). Eu presented a significant negative anomaly (Eu/Eu* = 0.23–0.27, average of 0.26). The primitive mantle-normalized trace element spider diagram (Figure 9b) also shows the enrichment of LILEs such as Rb, Th, U, and Pb, and the depletion of Ba, K, P, Eu, and Ti. In the chondrite-normalized REE diagram (Figure 9a), the distribution curves of the Longshi granite were generally flat, with less fractionation of LREE and HREE, and stronger negative Eu anomaly, indicating a higher degree of differentiation.

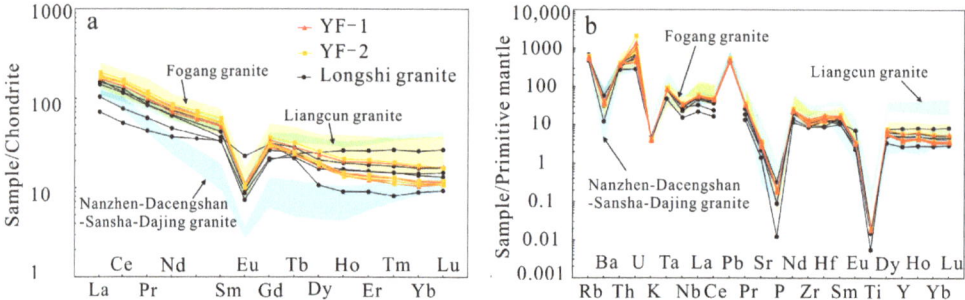

Figure 9. Chondrite-normalized REE patterns (**a**) and primitive-mantle-normalized trace element spider diagram (**b**) of the Yongfeng composite granitic pluton (chondrite and primitive mantle normalizing values from [20]. The data of the Longshi granite, Liangcun granite, Fogang granite, and Nan-zhen-Daliangshan-Sansha-Dajing granite are respectively cited from [9,23–25].

In general, the Yongfeng granite and the Longshi granite exhibit similar geochemical characteristics, suggesting that they are products of the same magmatic system. The Longshi granite shows a higher degree of crystallization differentiation than the Yongfeng granite. Overall, the element geochemical characteristics of the Yongfeng composite granitic pluton are similar to those of highly fractionated I-type granites, such as the Liangcun granite, Fogang granite, and Nanzhen-Dacengshan-Sansha-Dajing granites (Figures 7 and 9).

4.4. Element Geochemistry of Biotite

The SiO_2, FeO, TiO_2, and MgO contents of biotite were 34.05%–35.16% (average of 34.63%), 21.43%–23.91% (average of 22.63%), 1.97%–3.44% (average of 2.68%), and 4.91%–5.93% (average of 5.24%), respectively, indicating that biotite is enriched in Fe and Ti. The Fe/(Fe + Mg) and Mg/(Mg + Fe) ratios of biotite were 0.68–0.72 (average of 0.71) and 0.28–0.32 (average of 0.29), indicating that biotite is enriched in Fe but depleted in Mg. The classification diagram of biotite (Figure 10b) shows that biotite belongs to ferruginous biotite. In the $10 \times TiO_2 - FeO - MgO$ diagram (Figure 10a), most sample points are located in the area of primary magmatic biotites.

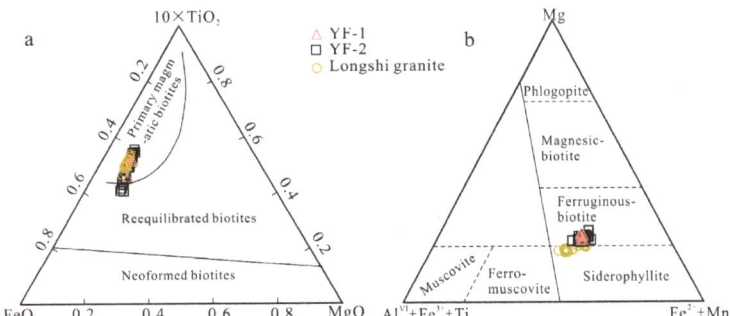

Figure 10. (a) $(10 \times TiO_2) - FeO - MgO$ diagram, base map according to [26]; (b) $Mg - (Al^{VI} + Fe^{3+} + Ti) - (Fe^{2+} + Mn)$ diagram, base map according to [27] of the biotite from the Yongfeng composite granitic pluton. The data of the Longshi granite is cited from [8].

5. Discussion

5.1. Rock-Forming Age

Liu and Li [28] reported that the zircon U-Pb ages of the Yongfeng granite are (160.0 ± 1.1) Ma and (155 ± 2.2) Ma. Qiu [29] reported that the U-Pb zircon age of the Longshi granite is (142.9 ± 0.8) Ma. Yang et al. [9] reported that the zircon U-Pb ages of the Yongfeng granite are (155.8 ± 2.0) Ma and (154.7 ± 0.75) Ma, while those of the Longshi granite are (154.0 ± 2.2) Ma and (156.9 ± 1.8) Ma. In this study, the zircon U-Pb ages of the Yongfeng granite and Longshi granite were determined to be 152.0 ± 1.0 Ma–151.3 ± 1.1 Ma and 148.9 ± 1.2 Ma, respectively. Regionally, the rock formation ages of the Liangcun, Jiangbei, and Donggu granites are 158–147 Ma, 156–153 Ma, and 152 ± 2 Ma, respectively [30–33], indicating the occurrence of large-scale magmatism during the late Jurassic in southern Jiangxi province. The above evidence proves that the Yongfeng composite granitic pluton formed in the late Jurassic.

Inherited zircons were observed in samples (Figure 5) and they could be divided into four groups according to their ages: (1) 164–239 Ma. This group of inherited zircons may indicate the presence of Mesozoic magmatism in the study area. For example, the rock-forming age of the Qingxi granite is 229.3 ± 0.8 Ma [34]. (2) 275–462 Ma. This group of inherited zircons may indicate the presence of Paleozoic magmatism in the study area. For example, the rock-forming ages of the Hanfang, Dabu, and Danqian granites are 438.0 ± 1.7 Ma, 434.1 ± 2.0 Ma, and 427.3 ± 1.8 Ma, respectively [23]. (3) 656–990 Ma. Magmatism in the Cathaysian block was relatively weak during the Neoproterozoic and mainly distributed in the northern Wuyi area, which is located in the east of the Cathaysian block [35]. The Yongfeng composite granitic pluton is located in the southern part of the Wuyi area, where the magmatism is weak during Neoproterozoic, while the northern part of the Wuyi area has strong magmatism during Neoproterozoic [35]. This group of inherited zircons may indicate that some rocks from northern Wuyi area entered magma. (4) 1035–2661 Ma. The age of the basement of the Cathaysian block is mainly early middle Proterozoic [36]. The presence of inherited zircons from this period suggests that early middle Proterozoic basement rocks from the Cathaysian block may have been mixed with magma.

5.2. Rock-Forming Physicochemical Conditions

5.2.1. Temperature

The zircon saturation temperature (T_{Zr}) and zircon Ti temperature (T_{Ti}) can be used to estimate the crystallization temperature of granitic magma [37–39]. In this study, the T_{Zr} of samples YF-1 and YF-2 were calculated using the method of Miller et al. [38]. The T_{Zr} of sample YF-1 was 763–792 °C, with an average of 774 °C. The T_{Zr} of sample YF-2 was 764–789 °C, with an average of 777 °C (Supplementary Table S4).

The presence of alkali-rich melt and inherited zircon can lead to overestimation of the calculated zircon saturation temperature [38,40]. The samples of the Yongfeng granite are alkali-rich and have inherited zircons. Therefore, the crystallization temperature was lower than the zircon saturation temperature of the Yongfeng granite. In addition, the T_{Ti} of samples YF-1 and YF-2 was calculated using the method of Ferry and Watson [37]. Except for the low T_{Ti} (631 °C, 679 °C) and high T_{Ti} (907 °C), the normal T_{Ti} ranged from 707 °C to 868 °C, with an average of 773 °C (Supplementary Table S2), which is consistent with T_{Zr}.

5.2.2. Oxygen Fugacity

Oxygen fugacity is one of the crucial factors influencing metal mineralization, as it controls the migration and enrichment of metallic elements. Trail et al. [41] proposed a method for calculating the absolute oxygen fugacity. However, natural zircon has very low La and Pr content and often contains LREE-rich mineral inclusions. This will lead to inaccurate calculations of the absolute oxygen fugacity [42]. These interferences can be avoided by using the Geo-fO_2 software developed by Li et al. [42]. In this study, the oxygen fugacity of magma was calculated using this software. The results are listed in Supplementary Table S2. Because the ages of some zircons are much older than the rock formation age of the Yongfeng granite, the calculation results may not represent the true oxygen fugacity. After removing these points, the Ce^{4+}/Ce^{3+} ratio was 1.93–121.23, with an average of 21.49, which is much lower than the Ce^{4+}/Ce^{3+} value of zircon in the porphyry Cu, Mo deposit metallogenic granite (>300, [43,44]), but similar to the Ce^{4+}/Ce^{3+} value of zircon in the Sn deposit metallogenic granite [45]. The absolute oxygen fugacity ($\lg f (O_2)$) ranged from −24.47 to −10.23, with an average value of −16.64. On the whole, the range of oxygen fugacity presented significant variations with generally low values.

In addition, the absolute oxygen fugacity of magma can be estimated according to the empirical formula of biotite: $\lg f (O_2) = 10.9 - 27,000/T$ [46]. Using this formula, this study calculated the absolute oxygen fugacity of magma was −20.56--−17.48, with an average of −18.80 (Supplementary Table S5), which also indicates low oxygen fugacity.

5.3. Genetic Type

The determination of the genetic type of granite is an important and basic problem in the study of granite [47]. Previous researchers have proposed several criteria from multiple perspectives [48–53]. However, for granite that has experienced strong evolution, as its mineral and chemical compositions are close to hypoeutectic granite, established indicators are usually inapplicable for accurately determining the genesis type [48,54].

Yang et al. [9] considered the Yongfeng composite granitic pluton to be an A-type granite, but the geochemical characteristics clearly differ from those of A-type granites. For example, the TFe_2O_3/MgO ratio < 10, Zr + Nb + Ce + Y content < 350 ppm and the zircon saturation temperature was lower than 800 °C. Samples YF-1 and YF-2 also exhibit mineralogical and geochemical characteristics distinctly different from those of A-type granites: (1) They do not contain alkaline dark minerals. (2) The Yongfeng composite granitic pluton is depleted in high field strength elements (HFSEs) such as Zr, Nb, Y, REE, and Ga. (3) The TFe_2O_3/MgO is low. The TFe_2O_3/MgO of YF-1 and YF-2 were 4.63–4.93 (4.73 on average) and 4.74–5.05 (4.89 on average), respectively, which are smaller than that of typical A-type granites (>10, [52]). (4) The Zr + Nb + Ce + Y value is low. The Zr + Nb + Ce + Y values of YF-1 and YF-2 were 248.60–319.10 ppm (average of 279.92 ppm) and 256.70–297.30 ppm (average of 281.28 ppm), which are smaller than that of typical A-type granites (>350 ppm, [52]). On the (Zr + Nb + Ce + Y) − (TFe_2O_3/MgO) diagram (Figure 11a) and (Zr + Nb + Ce + Y) − ($K_2O + Na_2O$)/CaO diagram (Figure 11b), wherein the vast majority of sample points are located in the range of fractionated felsic granite. (5) The crystallization temperature of magma is low. The average zircon saturation temperatures of YF-1 and YF-2 were 774 °C and 777 °C, respectively. The crystallization temperature is lower than that of A-type granite (>800 °C, [55]), but close to that of highly fractionated I-type granite (764 °C, [55]).

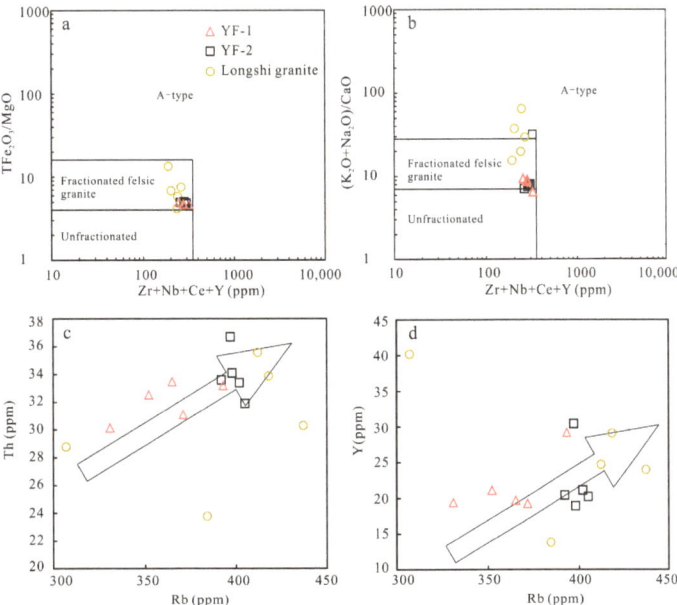

Figure 11. (**a**) (Zr + Nb + Ce + Y) − (TFe$_2$O$_3$/MgO) diagram, base map according to [52]; (**b**) (Zr + Nb + Ce + Y) − (K$_2$O + Na$_2$O)/CaO diagram, base map according to [52]; (**c**) Rb-Th diagram; (**d**) Rb-Y diagram of the Yongfeng composite granitic pluton. The data of Longshi granite is cited from [9].

The aluminum saturation index (A/CNK) of the Yongfeng composite granitic pluton varied widely. Some samples reached the level of peraluminousgranite, similar to S-type granite formed by the partial melting of metasedimentary rocks [56]. Mineralogically, the Yongfeng composite granitic pluton does not contain aluminum-rich minerals, except for a small amount of muscovite. Furthermore, all samples showed decreases in P$_2$O$_5$ with increasing SiO$_2$ (Figure 8g), and increases in Th and Y with increasing Rb (Figure 11c,d). This feature is consistent with I-type granite, while opposite to S-type granite [54]. The Yongfeng composite granitic pluton has high SiO$_2$, high DI, and significant negative Eu anomaly. These characteristics indicate that the Yongfeng composite granitic pluton has experienced a high degree of fractionation. Wu et al. [47] believe that the most feasible method for determining the genetic type of highly fractionated granite is to compare the genetic type of contemporaneous paragenetic granite. The petrological and geochemical characteristics of the Yongfeng composite granitic pluton are similar to those of highly fractionated I-type granites that formed in the Nanling area during the late Jurassic, such as the Liangcun granite, Fogang granite, and Nanzhen-Dacengshan-Sansha-Dajing granites [23–25]. In conclusion, the Yongfeng composite granitic pluton is a highly fractionated I-type granite, rather than an A-type granite as previously believed.

5.4. Petrogenesis

The εHf(t) value of the Yongfeng composite granitic pluton ranged from −27.71 to −9.92. The corresponding two stage model age was 1.83–2.93 Ga, which is within the Paleoproterozoic [9]. The zircon U-Pb age was homogeneous whereas the εHf(t) value varied widely, indicating that magma came from different sources [57]. Although the εHf(t) value of the Yongfeng composite granitic pluton was negative, it varied widely. This evidence indicates that the magmatic source is mainly composed of Paleoproterozoic basement rocks, and there may be a small amount of mantle magma. Regionally, the εHf(t) value of the adjacent Liangcun granite ranged from −12.96 to −7.4, and the corresponding two-stage model age was 1.67–2.0 Ga. Its magmatic source is estimated to contain 25% mantle magma [23]. The εHf(t) and δ^{18}O values of the zircons from the biotite granite in the Longyuanba composite granitic pluton indicated that the magmatic source is dominated by crust-derived sediments, with a small proportion of mantle magma [58]. No petrographic evidence of crust-mantle mixing was found in the Yongfeng composite granitic pluton. Intermediate-basic intrusive rocks and alkaline rock-syenite in the Nanling area occurred during late Jurassic, such as the Chencun diorite, Wushi diorite-hornblende gabbro, Nankunshan alkaline granite, Ejinao alkaline syenite, Qinghu quartz monzonite, Luorong-Mashan granitoid complex, Huashan-Guposhan granite, and syenite-granitoid in southern Jiangxi [59–63]. Moreover, mafic mineral inclusions are found in the Fogang, Guposhan, and Dadongshan granites [64,65]. This evidence indicates that crust-mantle mixing was widespread in the Nanling area during the late Jurassic.

The Rb/Sr and Rb/Nb ratios of the Yongfeng composite granitic pluton are significantly higher than those of the global upper crust (0.32 and 4.5, respectively [66]), indicating the possible occurrence of continental crust rocks with high maturity in the magmatic source. In the (Rb/Sr) − (Rb/Ba) diagram (Figure 12a), a positive correlation was observed between Rb/Sr and Rb/Ba. Except for one sample point being located in the region of clay-poor sources, other sample points were located in the region of clay-rich sources. This indicates that the magmatic source corresponds to the partial melting of argillaceous rocks. It is enriched in LREEs and features a significant negative Eu anomaly, indicating that its magma originates from the melting of crustal rocks. In the (Al$_2$O$_3$ + TFeO + MgO + TiO$_2$) − Al$_2$O$_3$/(TFeO + MgO + TiO$_2$) diagram (Figure 12b), most sample points of the Yongfeng composite granitic pluton are located in the area corresponding to the partial melting of amphibolite and mafic argillaceous rocks. This indicates the presence of magma derived from the partial melting of basaltic rocks and metamorphosed argillaceous rocks [67]. Therefore, the magmatic source of the Yongfeng composite granitic pluton is suggested to be the mixing of basaltic magma and ancient metamorphic argillaceous magma.

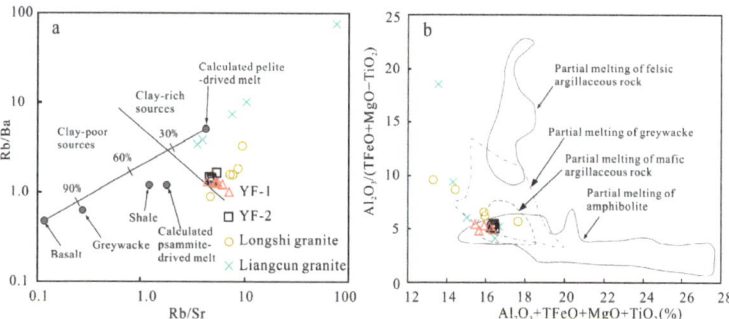

Figure 12. (Rb/Sr) − (Rb/Ba) diagram. (**a**) base map according to [68]) and (Al$_2$O$_3$ + TFeO + MgO + TiO$_2$)-Al$_2$O$_3$/(TFeO + MgO + TiO$_2$); (**b**) base map according to [69]) of the Yongfeng composite granitic pluton. The data of the Longshi granite and Liangcun granite are respectively cited from [9,23].

As mentioned earlier, the Yongfeng composite granitic pluton has the following characteristics: high SiO$_2$ and DI; low TFe$_2$O$_3$, MgO, CaO, TiO$_2$, and P$_2$O$_5$, which gradually decrease with increasing SiO$_2$; depletion of trace elements such as Ba, P, Ti, and Nb; and significant negative Eu anomaly. These features indicate that the Yongfeng composite granitic pluton experienced a high degree of crystallization differentiation. The strong depletion of Ba, Eu, and enrichment of Rb are mainly caused by the crystallization differentiation of K-feldspar and plagioclase. The variation of REEs is controlled by the crystallization differentiation of zircon and monazite. The depletion of Ti is caused by the crystallization differentiation of Ti-rich minerals, such as ilmenite and rutile. In addition, the distribution coefficients of Nb and Ta in rutile are high [70]. Therefore, the crystallization differentiation of rutile is an important reason for the Nb depletion in the Yongfeng composite granitic pluton. From the Yongfeng granite to the Longshi granite, the SiO_2 content, DI, Rb/Sr, and Rb/Ba values increased, whereas the contents of CaO, MgO, and TFe_2O_3 decreased, and the depletion of Sr, Ti, P, and Eu intensified; nevertheless, their Hf isotope compositions were similar. These features indicate that the Yongfeng granite and Longshi granite formed through the same magmatic system under different degrees of crystallization differentiation. The degree of crystallization differentiation of the Longshi granite is higher than that of the Yongfeng granite.

5.5. Tectonic Implication

The early Yanshanian granites in the Nanling area are distributed in EW and NE directions. Ling et al. [71] believe that the formation of the early Yanshanian granite with the EW distribution may be related to the revival and extension of the existing EW distribution of the Indosinian structure under the influence of the subduction of the Pacific Plate, while the formation of the early Yanshanian granite with the NE distribution is mainly related to the intracontinental fault extension caused by the subduction of the Pacific Plate. In addition, alkaline granites, bimodal volcanic rocks, alkaline basalt, and basic-acid complexes developed in the Nanling area during the early Yanshanian [60,72–74]. The above evidence indicates that the Nanling area has been in an extension-thinning tectonic setting since the Middle Jurassic.

The composition of trace elements in igneous rocks varies significantly under different tectonic settings. Therefore, the tectonic background of igneous rock can be restored according to differences in trace element compositions [75]. In the Yb-Ta diagram (Figure 13a), the sample points are located in the range of syn-collision granite and within plate granite. In the (Yb + Nb) − Rb and (Rb/30) − Hf − (Ta × 3) diagrams (Figure 13b,c), the sample points are located in the range of syn-collision granite and post-collision granite. In the $SiO_2 - lg[CaO/(K_2O + Na_2O)]$ diagram (Figure 13d), the sample points are located in the area of extrusion and extension. Based on the geochronology, geochemistry, and tectonic setting discrimination diagrams, the Yongfeng composite granitic pluton is suggested to have formed under the extensional setting during the post-collision stage, which was controlled by the breakup or retreat of the backplate after the subduction of the Pacific Plate into the Nanling hinterland [1].

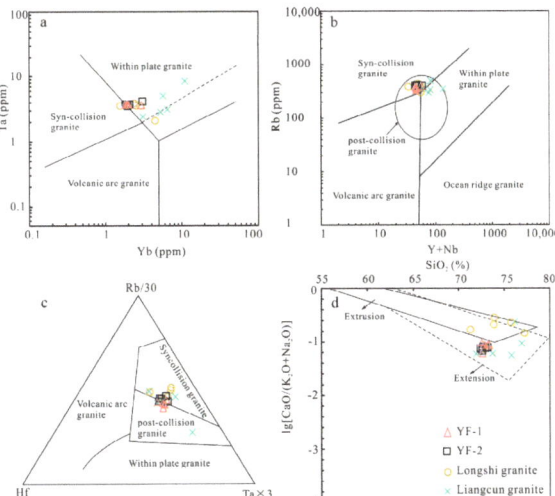

Figure 13. Tectonic setting discrimination diagrams of the Yongfeng composite granitic pluton and Liangcun granite (**a–d**), with base map according to [75–78], respectively. The data of the Longshi granite and Liangcun granite are respectively cited from [9,23]).

5.6. Relationship between Magmatism and Mineralization

The Leigongzhang molybdenum deposit is a medium-sized quartz vein type molybdenum deposit, which is different from the porphyry molybdenum deposit and quartz vein molybdenum polymetal deposit in the Nanling area. Previous studies suggest that it has a close spatiotemporal relationship with the Yongfeng composite granitic pluton [7,79]. Mineral deposits closely associated with granite are influenced by factors such as magmatic source, oxygen fugacity, degree of fractionation, and volatile content [80,81]. During the intrusive process, high oxygen fugacity magmas not only assimilate sulfides, but also suppress the separation of metals as magma sulfide melts (such as MoS_2), thereby retaining them in large quantities in the exsolved liquid phase [82] which is favorable for mineralization. However, the oxygen fugacity of the Yongfeng granite is low compared to that of porphyry type molybdenum mineralization granite. Oxygen fugacity is not the only factor controlling the formation of the Leigongzhang molybdenum deposit. Granite related to Mo deposits has undergone a high degree of crystallization differentiation and has a high content of F [80,83]. A high content of F in magma can promote the enrichment of metal elements such as Sn, Mo, and W during magma intrusion and fractionation processes. For example, the magmatic systems of the Zhuxiling Mo-W deposit and the Shizhuyuan W-Sn-Mo-Bi polymetallic deposit have a high content of F [84,85]. The biotite in Longshi granite has a high content of F (>1%) [8], which is similar to biotite in granite related to fluorite mineralization in the region [86]. Furthermore, a large fluorite deposit is developed in the outer contact zone of Longshi granite. This evidence indicates that the Longshi granite has a high content of F, which is conducive to molybdenum mineralization.

The Sn deposit is related to the crystallization differentiation of reducing magma [87]. Most Sn deposits in the South China region are associated with low oxygen fugacity granite, such as the Guposhan granite and the Qitianling granite [45]. Additionally, the temperature of magma also plays a role in controlling Sn enrichment. Stemprok [88] notes that higher temperatures during magma crystallization lead to higher SnO_2 content. The magmatic systems of the Gejiu Sn deposit and the Xitian W-Sn deposit also exhibit high temperature characteristics [89,90]. The magmatic system of the Yongfeng composite granitic pluton has the characteristics of high temperature, low oxygen fugacity, and high content of F, similar to the magmatic system of a typical Sn deposit. In addition, the Yongfeng composite granitic pluton shows high Sn anomaly on the Sn element anomaly map [91]. In summary, the Yongfeng composite granitic pluton has significant Sn, Mo, and fluorite mineralization potential.

6. Conclusions

(1) The Yongfeng composite granitic pluton is rich in silicon, potassium, and alkali, weakly peraluminous, poor in calcium and iron, and has a high content of $\sum REE$, with the enrichment of LREEs and significant fractionation of LREEs and HREEs. It is also enriched in LILEs such as Rb, Th, U, and Pb, and strongly depleted in elements such as Ba, K, P, Eu, and Ti, showing a clear negative Eu anomaly.

(2) The Yongfeng composite granitic pluton does not contain alkaline dark minerals, and has low contents of TFe_2O_3, MgO, CaO, TiO_2, and P_2O_5, which decrease with increasing SiO_2 content. The negative Eu anomaly is significant and the Rb/Sr ratio is high. This evidence indicates the crystallization differentiation of minerals such as plagioclase, potassium feldspar, biotite, zircon, monazite, and ilmenite/rutile during magma evolution. The average zircon saturation temperature is 776 °C, the average TFe_2O_3/MgO ratio is 4.81, and the average Zr + Nb + Ce + Y content is 280.6 ppm, indicating that it is a highly fractionated I-type granite.

(3) The crystallization age of the Yongfeng granite and the Longshi granite are 152.0 ± 1.0 Ma–151.3 ± 1.1 Ma and 148.9 ± 1.2 Ma, respectively. They are products of large-scale magmatic activity in the Nanling region during the Late Jurassic. Further, they formed in the extensional tectonic setting during the post-orogenic stage,

under the control of the breakup or retreat of the backplate after the subduction of the Pacific Plate into the Nanling hinterland.

(4) The magmatic system of the Yongfeng composite granitic pluton is characterized by high fractionation, high F content, high temperature, and low oxygen fugacity, which is conducive to the large-scale mineralization of Sn, Mo, and fluorite.

Supplementary Materials: The following supporting information can be downloaded at: https://www.mdpi.com/article/10.3390/min13111457/s1, Table S1: List of samples information; Table S2: Results of trace elements in zircons from samples YF-1 and YF-2 (ppm); Table S3: LA-ICP-MS U-Pb zircon data of the Yongfeng composite pluton; Table S4: Major (wt%), trace, and rare earth (ppm) element data of the Yongfeng granite; Table S5: The chemical composition of biotite from the Yongfeng granite [92].

Author Contributions: Conceptualization: Y.Z., F.H. and D.W.; Data curation, Formal analysis, Methodology, Writing-original draft: Y.Z.; Supervision, Project administration: F.H.; Funding acquisition: F.H. and D.W.; Investigation: D.W., N.W., C.Z. and Z.L.; Resources: N.W. All authors have read and agreed to the published version of the manuscript.

Funding: This paper is financed by the China Geological Survey's projects (DD20221695; DD20190379) and the National Natural Science Foundation of China (42172097).

Data Availability Statement: The data that support the findings of this study are available from the corresponding author.

Conflicts of Interest: The authors declare no conflict of interest.

References

1. Wu, F.Y.; Guo, C.L.; Hu, F.Y.; Liu, X.C.; Zhao, J.X.; Li, X.F.; Qin, K.Z. Petrogenesis of the highly fractionated granites and their mineralizations in Nanling Range, South China. *Acta Petrol. Sin.* **2023**, *39*, 1–36. (In Chinese with English Abstract) [CrossRef]
2. Webster, J.D. Exsolution of magmatic volatile phases from Cl-enriched mineralizing granitic magmas and implications for ore metal transport. *Geochim. Cosmochim. Acta* **1997**, *61*, 1017–1029. [CrossRef]
3. Jahn, B.M.; Wu, F.Y.; Capdevila, R.; Martineau, F.; Zhao, Z.H.; Wang, Y.X. Highly evolved juvenile granites with tetrad REE patterns: The Woduhe and Baerzhe granites from the Great Xing'an Mountains in NE China. *Lithos* **2001**, *59*, 171–198. [CrossRef]
4. Wu, F.Y.; Sun, D.Y.; Jahn, B.M.; Wilde, S.A. A Jurassic garnet-bearing granitic pluton from NE China showing tetrad REE patterns. *J. Asian Earth Sci.* **2004**, *23*, 731–744. [CrossRef]
5. Tong, R.F.; Wu, Y.R.; Xing, Q.M. Geological Report on Detailed Survey of Mo Deposte in Leigongzhang, Xingguo County, Jiangxi Province. The National Geological Archive Center. 2011. Available online: http://dc.ngac.org.cn/museumDetails?id=10.35080%2Fn01.c.114677 (accessed on 1 July 2011). (In Chinese with English Abstract)
6. Zhuang, X.G.; Cai, H.J.; Feng, W.D. Verification Report on Resource Reserves in the Longping Fluorite Deposite in Xingguo County, Jiangxi Province. 2010. The National Geological Archive Center. Available online: http://dc.ngac.org.cn/museumDetails?id=10.35080%2Fn01.c.158108 (accessed on 28 June 2010). (In Chinese with English Abstract)
7. Tong, R.F.; Yang, R.D. Molybdenum mineralization characteristics of Leigongzhang molybdenum deposit and prospecting targets, Jiangxi. *Miner. Explor.* **2012**, *3*, 755–760. (In Chinese with English Abstract)
8. Yang, S.W.; Feng, C.Y.; Lou, F.S.; Zhang, F.R.; He, B.; Cao, Y.B. A preliminary study on metallogenic age and genesis of Longping fluorite deposit in Southern Jiangxi Province: Evidence from Sm-Nd isochron dating of fluorite and electron probe of biotite. *Geol. J. China Univ.* **2019**, *25*, 341–351. (In Chinese with English Abstract) [CrossRef]
9. Yang, S.W.; Lou, F.S.; Zhang, F.R.; Wu, Z.C.; Feng, C.Y. Late Jurassic aluminum A-type granite belt in southern Jiangxi and its geological significance. *Geol. Sci. Technol. Inf.* **2019**, *38*, 12–29. (In Chinese with English Abstract) [CrossRef]
10. Xu, Z.G.; Chen, Y.C.; Wang, D.H.; Chen, Z.H. *Classification Scheme of China Metallogenic Zone*; Geological Publishing House: Beijing, China, 2008. (In Chinese with English Abstract)
11. Wang, D.Z.; Shu, L.S. Late Mesozoic basin and range tectonics and related magmatism in Southeast China. *Geosci. Front.* **2012**, *3*, 109–124. [CrossRef]
12. Li, J.W.; Zhou, M.F.; Li, X.F.; Fu, Z.R.; Li, Z.J. The Hunan-Jiangxi strike-slip fault system in southern China: Southern termination of the Tan-Lu fault. *J. Geodyn.* **2001**, *32*, 333–354. [CrossRef]
13. Wang, D.H.; Wang, Y.; Qin, J.H. *Geology of Mineral Resources in China·Volume of Nanlling*; Geological Publishing House: Beijing, China, 2022; pp. 82–84. (In Chinese with English Abstract)
14. Wang, Y.; Wang, D.H.; Qin, J.H. *Geology of Mineral Resources in China·Metallogenic Regularity Map of the Nanling Metallogeny Belt (1:800000)*; Geological Publishing House: Beijing, China, 2022. (In Chinese with English Abstract)
15. Zhao, Z.; Wang, D.H.; Bagas, L.; Chen, Z.Y. Geochemical and REE mineralogical characteristics of the Zhaibei Granite in Jiangxi Province, southern China, and a model for the genesis of ion-adsorption REE deposits. *Ore Geol. Rev.* **2022**, *140*, 104579. [CrossRef]

16. Andersen, T. Correction of common lead in U-Pb analyses that do not report 204Pb. *Chem. Geol.* **2002**, *192*, 59–79. [CrossRef]
17. Hou, K.J.; Li, Y.H.; Tian, R.Y. In situ U-Pb zircon dating using laser ablation-multi ion counting-ICP-MS. *Miner. Depos.* **2009**, *28*, 481–492. (In Chinese with English Abstract) [CrossRef]
18. Lin, W.W.; Peng, L.J. The estimation of Fe^{3+} and Fe^{2+} contents in amphibole and biotite from EMPA data. *J. Chang. Univ. Earth Sci.* **1994**, *24*, 155–162. (In Chinese with English Abstract)
19. Hoskin, P.W.O. Trace-element composition of hydrothermal zircon and the alteration of Hadean zircon from the Jack Hills, Australia. *Geochim. Cosmochim. Acta* **2005**, *69*, 637–648. [CrossRef]
20. Sun, S.S.; McDonough, W.F. Chemical and isotopic systematics of oceanic basalts: Implications or mantle composition and processes. *Geol. Soc. Lond. Spec. Publ.* **1989**, *42*, 313–345. [CrossRef]
21. Rubatto, D.; Gebauer, D. Use of cathodoluminescence for U-Pb zircon dating by IOM Microprobe: Some examples from the western Alps. In *Cathodoluminescence in Geoscience*; Pagel, M., Barbin, V., Blanc, P., Ohnenstetter, D., Eds.; Springer: Berlin/Heidelberg, Germany, 2000; pp. 373–400.
22. Wu, Y.B.; Zheng, Y.F. Genesis of zircon and its constraints on interpretation of U-Pb age. *Chin. Sci. Bull.* **2004**, *49*, 1554–1569. (In Chinese with English Abstract) [CrossRef]
23. Wang, L.L. Geochemistry and Petrogenesis of Early Paleozoic-Mesozoic Granites in Ganzhou, Jiangxi Province, South China Block. Ph.D. Thesis, China University of Geosciences (Beijing), Beijing, China, 2015. (In Chinese with English Abstract)
24. Li, X.H.; Li, Z.X.; Li, W.X.; Liu, Y.; Yuan, C.; Wei, G.J.; Qi, C.S. U-Pb zircon, geochemical and Sr-Nd-Hf isotopic constraints on age and origin of Jurassic I- and A-type granites from central Guangdong, SE China: A major igneous event in response to foundering of a subducted flat-slab? *Lithos* **2007**, *96*, 186–204. [CrossRef]
25. Qiu, J.S.; Xiao, E.; Hu, J.; Xu, X.S.; Jiang, S.Y.; Li, Z. Petrogenesis of highly fractionated I-type granites in the coastal area of northeastern Fujian Province: Constraints from zircon U-Pb geochronology, geochemistry and Nd-Hf isotopes. *Acta Petrol. Sin.* **2008**, *24*, 2468–2484. (In Chinese with English Abstract)
26. Nachit, H.; Ibhi, A.; Abia, E.H.; Ben Ohoud, M. Discrimination between primary magmatic biotites, equilibrated biotites and neoformed biotites. *Comptes Rendus Geosci.* **2005**, *337*, 1415–1420. [CrossRef]
27. Foster, M.D. *Interpretation of the Composition of Trioctahedral Micas*; United States Geological Survey Professional Paper 354B; United States Geological Survey: Reston, VA, USA, 1960; pp. 11–49.
28. Liu, B.X.; Li, Y.M. Chief progress in 1:50000 regional geological survey during the 9th five-year-plan in Jiangxi. *Jiangxi Geol.* **2001**, *15*, 225–229. (In Chinese with English Abstract)
29. Qiu, Y.L. Description of 1/50000 Geologic Map of Xingguo County Sheet (G-50-63-B). The National Geological Archive Center. 1989. Available online: http://dc.ngac.org.cn/museumDetails?id=10.35080%2Fn01.c.77463 (accessed on 22 April 2021). (In Chinese with English Abstract)
30. Cui, Y.Y. Geochronology, Geochemistry and Petrogenesis of the Granitoids in the Sanming-Ganzhou Area, South China. Master's Thesis, China University of Geosciences (Beijing), Beijing, China, 2014. (In Chinese with English Abstract)
31. Feng, C.Y.; Zeng, Z.L.; Qu, W.J.; Liu, J.S.; Li, H.P. A geochronological study of granite and related mineralization of the Zhangjiadi molybdenite-tungsten deposit in Xingguo County, southern Jiangxi Province, China, and its geological significance. *Acta Petrol. Sin.* **2015**, *31*, 709–724. (In Chinese with English Abstract)
32. Shu, X.J.; Chen, Z.H.; Zhu, Y.H.; Liao, S.B.; Zhou, B.W.; Li, G.Z.; Zhao, X.C.; Liu, S.; Chen, L.C. Genesis of Donggu highly fractionated granites, southern Jiangxi, and its geological significance. *Geol. Rev.* **2018**, *64*, 108–126. (In Chinese with English Abstract) [CrossRef]
33. Wang, W.P.; Chen, Y.C.; Wang, D.H.; Chen, Z.Y. Zircon LA-ICP-MS U-Pb dating and petrogeochemistry of the Liangcun granites and their petrogenesis, south Jiangxi. *Geotecton. Metallog.* **2014**, *38*, 347–358. (In Chinese with English Abstract)
34. Yu, Y.; Chen, Z.Y.; Chen, Z.H.; Hou, K.J.; Zhao, Z.; Xu, J.X.; Zhang, J.J.; Zeng, Z.L. Zircon U-Pb dating and mineralization prospective of the Triassic Qingxi pluton in southern Jiangxi Province. *Geotecton. Metallog.* **2012**, *36*, 413–421. (In Chinese with English Abstract) [CrossRef]
35. Yang, S.W.; Lou, F.S.; Zhang, F.R.; Zhou, C.H.; Xia, M.; Ling, L.H.; Feng, Z.H. Detrital zircon U-Pb geochronology of the Xunwu Formation in Cathaysia Block and its geological significance. *J. East China Univ. Technol. (Nat. Sci.)* **2022**, *45*, 207–222. (In Chinese with English Abstract)
36. Shen, W.Z.; Zhu, J.C.; Liu, C.S.; Xu, S.J.; Lin, H.F. Sm-Nd isotopic study of basement metamorphic rocks in South China and its constraint on material sources of granitoids. *Acta Petrol. Sin.* **1993**, *9*, 115–124. (In Chinese with English Abstract)
37. Ferry, J.M.; Watson, E.B. New thermodynamic models and revised calibrations for the Ti-in-zircon and Zr-in-rutile thermometers. *Contrib. Mineral. Petrol.* **2007**, *154*, 429–437. [CrossRef]
38. Miller, C.F.; McDowell, S.M.; Mapes, R.W. Hot and cold granites? Implications of zircon saturation temperatures and preservation of inheritance. *Geology* **2003**, *31*, 529–532. [CrossRef]
39. Watson, E.B.; Harrison, T.M. Zircon thermometer reveals minimum melting conditions on earliest earth. *Science* **2005**, *308*, 841–844. [CrossRef]
40. Watson, E.B.; Harrison, T.M. Zircon saturation revisited: Temperature and composition effects in a variety of crustal magma types. *Earth Planet Sci. Lett.* **1983**, *64*, 295–304. [CrossRef]
41. Trail, D.; Watson, E.B.; Tailby, N.D. The oxidation state of Hadean magmas and implications for early Earth's atmosphere. *Nature* **2011**, *480*, 79–82. [CrossRef]

42. Li, W.K.; Cheng, Y.Q.; Yang, Z.M. Geo-fO2: Integrated software for analysis of magmatic oxygen fugacity. *Geochem. Geophys. Geosystems* **2019**, *20*, 2542–2555. [CrossRef]
43. Ballard, J.R.; Palin, M.J.; Campbell, I.H. Relative oxidation states of magmas inferred from Ce(IV)/Ce(III) in zircon: Application to porphyry copper deposits of northern Chile. *Contrib. Mineral. Petrol.* **2002**, *144*, 347–364. [CrossRef]
44. Liang, H.Y.; Campbell, I.H.; Allen, C.; Sun, W.D.; Liu, C.Q.; Yu, H.X.; Xie, Y.W.; Zhang, Y.Q. Zircon Ce^{4+}/Ce^{3+} ratios and ages for Yulong ore-bearing porphyries in eastern Tibet. *Miner. Depos.* **2006**, *41*, 152–159. [CrossRef]
45. Sun, Z.L. Geochronology and oxygen fugacity of Mesozoic granites in Nanling area of South China. *J. Earth Sci. Environ.* **2014**, *36*, 141–151. (In Chinese with English Abstract)
46. David, R.W.; Hans, P.E. Stability of biotite: Experiment, theory, and application. *Am. Mineral.* **1965**, *50*, 228–1272.
47. Wu, F.Y.; Li, X.H.; Yang, J.H.; Zheng, Y.F. Discussions on the petrogenesis of granites. *Acta Petrol. Sin.* **2007**, *23*, 1217–1238. (In Chinese with English Abstract)
48. Chappell, B.W. Aluminium saturation in I- and S-type granites and the characterization of fractionated haplogranites. *Lithos* **1999**, *46*, 535–551. [CrossRef]
49. Dan, W.; Li, X.H.; Wang, Q.; Wang, X.C.; Liu, Y.; Wyman, D.A. Paleoproterozoic S-type granites in the Helanshan Complex, Khondalite Belt, North China Craton: Implications for rapid sediment recycling during slab break-off. *Precambrian Res.* **2014**, *254*, 59–72. [CrossRef]
50. Kemp, A.I.S.; Hawkesworth, C.J.; Foster, G.L.; Paterson, B.A.; Woodhead, J.D.; Hergt, J.M.; Gray, C.M.; Whitehouse, M.J. Magmatic and crustal differentiation history of granitic rocks from Hf-O isotopes in zircon. *Science* **2007**, *315*, 980–983. [CrossRef]
51. Wolf, M.B.; London, D. Apatite dissolution into peraluminous haplogranitic melts: An experimental study of solubilities and mechanisms. *Geochim. Cosmochim. Acta* **1994**, *58*, 4127–4245. [CrossRef]
52. Whalen, J.B.; Currie, K.L.; Chappell, B.W. A-type granites: Geochemical characteristics, discrimination and petrogenesis. *Contrib. Mineral. Petrol.* **1987**, *95*, 407–419. [CrossRef]
53. Zhou, Z.M.; Ma, C.Q.; Wang, L.X.; Chen, S.G.; Xie, C.F.; Li, Y.; Liu, W. A source-depleted Early Jurassic granitic pluton from South China: Implication to the Mesozoic juvenile accretion of the South China crust. *Lithos* **2018**, *300–301*, 278–290. [CrossRef]
54. Chappell, B.W.; White, A.J.R. I- and S-type granites in the Lachlan Fold Belt. *Trans. R. Soc. Edinb. Earth Sci.* **1992**, *83*, 1–26. [CrossRef]
55. King, P.L.; White, A.J.R.; Chappell, B.W.; Allen, C.M. Characterization and origin of aluminous A-type granites from the Lachlan Fold Belt, Southeastern Australia. *J. Petrol.* **1997**, *38*, 371–391. [CrossRef]
56. Clemens, J.D. S-type granitic magmas-pertogenetic issues, model and evidence. *Earth-Sci. Rev.* **2003**, *61*, 1–18. [CrossRef]
57. Griffin, W.L.; Wang, X.; Jackson, S.E.; Pearson, N.J.; O'Reilly, S.Y.; Xu, X.S.; Zhou, X.M. Zircon chemistry and magma genesis, SE China: In-situ analysis of Hf isotopes, Tonglu and Pingtan Igneous complexes. *Lithos* **2002**, *61*, 237–269. [CrossRef]
58. Tao, J.; Li, W.; Li, X.; Cen, T. Petrogenesis of early Yanshanian highly evolved granites in the Longyuanba area, southern Jiangxi Province: Evidence from zircon U-Pb dating, Hf-O isotope and whole-rock geochemistry. *Sci. Sin. Terrae* **2013**, *43*, 760–778. (In Chinese with English Abstract) [CrossRef]
59. Chen, P.R.; Zhou, X.M.; Zhang, W.L.; Li, H.M.; Fan, C.F.; Sun, T.; Chen, W.F.; Zhang, M. Genesis and significance of the Early Yanshanian syenite-granite complex in the eastern section of Nanling. *Sci. China (Ser. D)* **2004**, *48*, 912–924. (In Chinese with English Abstract) [CrossRef]
60. Guo, X.S.; Chen, J.F.; Zhang, X.; Tang, J.F.; Xie, Z.; Zhou, T.X.; Liu, Y.L. Nd isotopic ratios of K-enriched magmatic complexes from southeastern Guangxi Province: Implications for upwelling of the mantle in southeastern China during the Mesozoic. *Acta Petrol. Sin.* **2001**, *17*, 19–27. (In Chinese with English Abstract)
61. Liu, C.S.; Chen, X.M.; Wang, R.C.; Hu, H. Origin of Nankunshan aluminous A-type granite, Longkou County, Guangdong Province. *Acta Petrol. Mineral.* **2003**, *22*, 1–10. (In Chinese with English Abstract) [CrossRef]
62. Liu, Y.S.; Hu, Z.S.; Zong, K.Q.; Gao, C.G.; Gao, S.; Xu, J.; Chen, H.H. Reappraisement and refinement of zircon U-Pb isotope and trace element analyses by LA-ICP-MS. *Chin. Sci. Bull.* **2010**, *55*, 1535–1546. [CrossRef]
63. Xu, X.S.; Lu, W.M.; He, Z.Y. The age and genesis of Fogang granitic batholith and Wushi diorite-hornblende gabbro. *Sci. China (D)* **2007**, *37*, 27–38. (In Chinese with English Abstract) [CrossRef]
64. Li, X.H.; Li, W.X.; Wang, X.C.; Li, Q.L.; Liu, Y.; Tang, G.Q. Role of mantle-derived magma in genesis of Early Yanshanian granites in the Nanling Range, South China: In situ zircon Hf-O isotopic constraints. *Sci. China (Ser. D)* **2009**, *52*, 1262–1278. (In Chinese with English Abstract) [CrossRef]
65. Zhao, K.D.; Jiang, S.Y.; Zhu, J.C.; Li, L.; Dai, B.Z.; Jiang, Y.H.; Ling, H.F. Hf isotopic composition of zircons from the Huashan-Guposhan intrusive complex and their mafic enclaves in northeastern Guangxi: Implication for petrogenesis. *Chin. Sci. Bull.* **2009**, *54*, 3716–3725. (In Chinese with English Abstract) [CrossRef]
66. Taylor, S.R.; McLennan, S.M. *The Continental Crust: Its Composition and Evolution*; Blackwell Scientific Publisher: Hoboken, NJ, USA, 1985.
67. Altherr, R.; Holl, A.; Hegner, E.; Langer, C.; Kreuzer, H. High-potassium, calc-alkaline I-type plutonism in the European Variscides: Northern Vosges (France) and northern Schwarzwald (Germany). *Lithos* **2000**, *50*, 51–73. [CrossRef]
68. Patino-Douce, A.E.; Harris, N. Experimental constraints on Himalayan anatexis. *J. Petrol.* **1998**, *39*, 689–710. [CrossRef]

69. Patino-Douce, A.E. What do Experiments Tell Us about the Relative Contributions of Crust and Mantle to the Origin of Granitic Magmas; Castro, A., Fernandez, C., Vigneresse, J.L., Eds.; Geological Society, London, Special Publication: London, UK, 1999; pp. 55–75. [CrossRef]
70. Foley, S.F.; Barth, M.G.; Jenner, G.A. Rutile/Melt partition coefficients for trace elements and an assessment of the influence of rutile on the trace element characteristics of subduction zone magmas. Geochim. Cosmochim. Acta 2000, 64, 933–938. [CrossRef]
71. Ling, H.F.; Shen, W.Z.; Sun, T.; Jiang, S.Y.; Jiang, Y.H.; Ni, P.; Gao, J.F.; Huang, G.L.; Ye, H.M.; Tan, Z.Z. Genesis and source characteristics of 22 Yanshanian granites in Guangdong Province: Study of element and Nd-Sr isotopes. Acta Petrol. Sin. 2006, 22, 2687–2703. (In Chinese with English Abstract)
72. Chen, P.R.; Zhang, B.T.; Kong, X.G.; Cai, B.C.; Ling, H.F.; Ni, Q.S. Geochemical characteristics and tectonic implication of Zhaibei A-type granitic intrusives in south Jiangxi Province. Acta Petrol. Sin. 1998, 14, 289–298. (In Chinese with English Abstract)
73. Chung, S.L.; Cheng, H.; Jahn, B.M.; O'Reilly, S.Y.; Zhu, B.Q. Major and trace element, and Sr-Nd isotope constraints on the origin of Paleogene volcanism in South China prior to the South China sea opening. Lithos 1997, 40, 203–220. [CrossRef]
74. Zhao, Z.H.; Bao, Z.W.; Zhang, B.Y. Geochemical characteristics of Mesozoic basanitoids in south Hunan Province. Sci. China (Ser. D) 1998, 28, 7–14. (In Chinese with English Abstract) [CrossRef]
75. Pearce, J.A.; Harris, N.B.W.; Tindle, A.G. Trace element discrimination diagrams for the tectonic interpretation of granitic rocks. J. Petrol. 1984, 25, 956–983. [CrossRef]
76. Pearce, J.A. Sources and settings of granitic rocks. Episodes 1996, 19, 120–125. [CrossRef]
77. Harris, N.B.W.; Pearce, J.A.; Tindle, A.G. Geochemical characteristics of Collision-Zone Magmatism. Geol. Soc. Lond. Spec. Publ. 1986, 19, 67–81. [CrossRef]
78. Brown, G.C. Calc-alkaline intrusive rocks: Their diversity, evolution, and relation to volcanic arcs. In Andesites-Orogenic Andesites and Related Rocks; Thorpe, R.S., Ed.; Wiley: Hoboken, NJ, USA, 1982; pp. 437–464.
79. Zhao, Y.B.; Huang, F.; Chi, L.; Wang, Y.; Tong, R.F. Re-Os dating of molybdenites from Leigongzhang molybdenum deposit of southern Jiangxi Province and its geological significance. Miner. Depos. 2023, 42, 66–76. (In Chinese with English Abstract)
80. Lehmann, B.; Ishihara, S.; Michel, H.; Miller, J.; Rapela, C.W.; Sanchez, A.; Tistl, M.; Winkelmann, L. The Bolivian tin province and regional tin distribution in the central Andes: A reassessment. Econ. Geol. 1990, 85, 1044–1058. [CrossRef]
81. Sun, W.D.; Liang, H.Y.; Ling, M.X.; Zhan, M.Z.; Ding, X.; Zhang, H.; Yang, X.Y.; Li, Y.L.; Ireland, T.R.; Wei, Q.R.; et al. The link between reduced porphyry copper deposits and oxidized magmas. Geochim. Cosmochim. Acta 2013, 103, 263–275. [CrossRef]
82. Sun, W.D.; Huang, R.F.; Li, H.; Hu, Y.B.; Zhang, C.C.; Sun, S.J.; Zhang, L.P.; Ding, X.; Li, C.Y.; Zartman, R.E.; et al. Porphyry deposits and oxidized magmas. Ore Geol. Rev. 2015, 65, 97–131. [CrossRef]
83. Candela, P.A.; Holland, H.D. A mass transfer model for copper and molybdenum in magmatic hydrothermal systems: The origin of porphyry-type ore deposits. Econ. Geol. 1986, 81, 1–19. [CrossRef]
84. Chen, Y.X.; Li, H.; Sun, W.D.; Ireland, T.; Tian, X.F.; Hu, Y.B.; Yang, W.B.; Chen, C.; Xu, D.R. Generation of Late Mesozoic Qianlishan A_2-type granite in Nanling Range, South China: Implications for Shizhuyuan W-Sn mineralization and tectonic evolution. Lithos 2016, 266–267, 435–452. [CrossRef]
85. Zhang, Z.; Duan, X.X.; Chen, B.; Wang, Z.Q.; Sun, K.K.; Yan, X. Implications of biotite geochemical characteristics for difference of ore-related magmatic system between Wushan copper deposit and Zhuxiling tungsten deposit. Acta Petrol. Mineral. 2019, 38, 673–692. (In Chinese with English Abstract)
86. Fang, G.C.; Wang, D.H.; Chen, Z.Y.; Chen, Z.H.; Zhao, Z.; Guo, N.X. Metallogenetic specialization ofthe fluorite-bearing granites in the northern part of the eastern Nanling region. Geotecton. Metallog. 2014, 38, 312–324. (In Chinese with English Abstract)
87. Blevin, P.L.; Chappell, B.W. The role of magmatic sources, oxidation states and fractionation in determining the granite metallogeny of eastern Australia. In Earth and Environmental Science Transactions of the Royal Society of Edinburgh; Brown, P.E., Chappell, B.W., Eds.; Geological Society of America: Boulder, CO, USA, 1992; pp. 305–316.
88. Stemprok, M. Solubility of tin, tungsten and molybdenum oxides in felsic magmas. Miner. Depos. 1990, 25, 205–212. [CrossRef]
89. Zhang, Y.C.; Chen, S.Y.; Zhao, J.N.; Zhao, Y.H.; Li, J. Compositional characteristics and petrogentic and metallogenic significance of biotite in granite in Gejiu tin polymetallic ore concentration area, Yunnan Province. Mineral. Petrol. 2020, 40, 76–88. (In Chinese with English Abstract)
90. Zhou, Y.; Liang, X.Q.; Cai, Y.F.; Fu, W. Petrogenesis and mineralization of Xitian tin-tungsten polymetallic deposit: Constraints from mineral chemistry of biotite from Xitian A-type granite, eastern Hunan Province. Earth Sci. 2017, 42, 1647–1657. (In Chinese with English Abstract)
91. Yang, R.D.; Tong, R.F.; Shao, W.J.; Zeng, Y. Potential for exploring large-sized and high grade deposits in Xingguo vortex structure area, southern Jiangxi. Miner. Explor. 2013, 4, 121–130. (In Chinese with English Abstract) [CrossRef]
92. Henry, D.J.; Guidotti, C.V.; Thomson, J.A. The Ti-saturation surface for low-to-medium pressure metapelitic biotites: Implications for geothermometry and Ti-substitution mechanisms. Am. Mineral. 2005, 90, 316–328. [CrossRef]

Disclaimer/Publisher's Note: The statements, opinions and data contained in all publications are solely those of the individual author(s) and contributor(s) and not of MDPI and/or the editor(s). MDPI and/or the editor(s) disclaim responsibility for any injury to people or property resulting from any ideas, methods, instructions or products referred to in the content.

Article

Integration of Whole-Rock Geochemistry and Mineral Chemistry Data for the Petrogenesis of A-Type Ring Complex from Gebel El Bakriyah Area, Egypt

Ahmed A. Abd El-Fatah [1,*], Adel A. Surour [1,2], Mokhles K. Azer [3] and Ahmed A. Madani [1]

1. Department of Geology, Faculty of Science, Cairo University, Giza 12613, Egypt; adelsurour@cu.edu.eg (A.A.S.); aamadani@sci.cu.edu.eg (A.A.M.)
2. Department of Geological Sciences, Faculty of Science, Galala University, New Galala City 43511, Egypt
3. Geological Sciences Department, National Research Centre, Giza 12622, Egypt; mk.abdel-malak@nrc.sci.eg
* Correspondence: aabdeldayiem@sci.cu.edu.eg

Abstract: El Bakriyah Ring Complex (BRC) is a prominent Neoproterozoic post-collisional granite suite in the southern part of the Central Eastern Desert of Egypt. The BRC bears critical materials (F, B, Nb, and Ta) in appreciable amounts either in the form of rare-metals dissemination or in the form of fluorite and barite vein mineralization. The complex consists of inner syenogranite and outer alkali feldspar granite that have been emplaced in a Pan-African assemblage made up of granitic country rocks (granodiorite and monzogranite), in addition to post-collisional fresh gabbro as a part of the Arabian-Nubian Shield (ANS) in northeast Africa. Granites of the BRC are characterized by enrichment in silica, alkalis, Rb, Y, Ga, Nb, Ta, Th, and U and depletion in Sr, Ba, and Ti. Geochemical characterization of the BRC indicates that the magma is a crustal melt, which originated from the partial melting of metasedimentary sources. Concentrations of rare-earth elements (REEs) differ in magnitude from the ring complex and its granitic country rocks but they have similar patterns, which are sub-parallel and show LREEs enrichment compared to HREEs. The presence of a negative Eu anomaly in these rocks is related to plagioclase fractionation. The abundance of fluorine (F) in the different granite varieties plays an important role in the existence of a tetrad influence on the behavior of REEs (TE1, 3 = up to 1.15). Geochemical parameters suggest the crystallization of the BRC granite varieties by fractional crystallization and limited assimilation. Mn-columbite and Mn-tantalite are the most abundant rare-metals dissemination in the BRC granite varieties. We present combined field, mineralogical and geochemical data that are in favor of magma originating from a metasedimentary source for the BRC with typical characteristics of A-type granites. Our geodynamic model suggests that the Gebel El Bakriyah area witnessed the Neoproterozoic post-collisional stage of the ANS during its late phase of formation. This stage was characterized by the emplacement of fresh gabbros followed by the syenogranite and alkali-feldspar granite of the BRC into an arc-related assemblage (granodiorite and monzogranite). It is believed that the mantle-derived magma was interplated and then moved upward in the extensional environment to a shallower level in the crust owing to events of lithospheric delamination. This presumably accelerated the processes of partial melting and differentiation of the metasedimentary dominated source (Tonian-Cryogenian) to produce the A-type granites building up the BRC (Ediacaran).

Keywords: El Bakriyah ring complex; A-type granites; fractional crystallization; metasedimentary source; Ediacaran post-collisional magmatism

Citation: Abd El-Fatah, A.A.; Surour, A.A.; Azer, M.K.; Madani, A.A. Integration of Whole-Rock Geochemistry and Mineral Chemistry Data for the Petrogenesis of A-Type Ring Complex from Gebel El Bakriyah Area, Egypt. *Minerals* 2023, 12, 1273. https://doi.org/10.3390/min13101273

Academic Editors: Ignez de Pinho Guimarães and Jefferson Valdemiro De Lima

Received: 23 August 2023
Revised: 23 September 2023
Accepted: 27 September 2023
Published: 29 September 2023

Copyright: © 2023 by the authors. Licensee MDPI, Basel, Switzerland. This article is an open access article distributed under the terms and conditions of the Creative Commons Attribution (CC BY) license (https://creativecommons.org/licenses/by/4.0/).

1. Introduction

The Eastern Desert of Egypt, as a part of the Neoproterozoic Arabian-Nubian Shield (ANS), is occupied by voluminous masses of granitic rocks that formed from different melt compositions and emplaced in different tectonic regimes (e.g., [1–3]). In the Gebel El Bakriyah area, which is located in the central part of the Eastern Desert (CED), prominent granitic masses are present that have field and laboratory characteristics, which can be integrated with other rock units such as gabbros to build up a geodynamic model for better understanding of the ANS evolution. The term "granitic rocks" is used here to denote two contrasting rock types with diversity in mineral composition, chemical composition, and tectonic setting. They include "proper granites" and "granitoids" such as quartz-diorite, granodiorite, and tonalite. They together constitute about 60% of plutonic rocks of the Nubian Shield in the Eastern Desert and can be classified and distinguished according to mineral and chemical compositions, magma type, tectonic setting, and age (e.g., [2,4,5]). In a recent review, [3] classified the Egyptian granitic rocks (mainly in the Eastern Desert and the Sinai Peninsula) based on the tectonic regime into four types namely, subduction-related, collisional, calc-alkaline post-orogenic, and alkaline post-tectonic. The subduction-related category (around 700–660 Ma, [6]) is represented by granitoids that are tholeiitic to calc-alkaline and is categorized as the older granitoids [6]. On the other hand, collision-related granites (around 610–590 Ma, [7]) are medium- to high-K calc-alkaline and are represented by younger granites (monzo- to syenogranite) that have been formed by the lower continental crust melting (e.g., [6,8]. Calc-alkaline post-orogenic granite (between 590 and 540 Ma, [3]) is characterized by a high-K calc-alkaline to shoshonite nature and comprises A_2-type younger granites, which originated from a melting process of lower crustal rocks in an extension setting during the transitional stage between orogenic and anorogenic regimes (e.g., [6,9]). The last phase of younger granites (A_1-type) (around 540 Ma, [7]) is the alkaline post-tectonic granite formed by the differentiation of alkali-basaltic magmas (e.g., [9]).

Globally, A-type granites are classified into anorogenic and post-orogenic granites [10], and they represent the youngest phase of felsic magmatism in the Eastern Desert of Egypt and the entire ANS [2,3,11]. For the last three decades, A-type granites in the Eastern Desert attracted considerable attention because of their economic potentialities and occasional enrichment of rare metals (e.g., Nb, Ta, Sn, W, Be, and B), rare-earth elements (REEs) and radioactive materials [4,12]. For example, [13] reported that these mineralizations are in intimate association with accessory minerals like beryl, columbite, tantalite, fergusonite, thorite, cassiterite, wolframite, and monazite. In comparison with I-type granites, A-type granites are deficient in Sr, Ba, Eu, Ca, and large ion lithophile elements (LILE) with abundance in silica, K, Rb, Y, REEs, Zr, and high-field strength elements (HFSE) [14].

A-type granites show vast geographic distribution in the Eastern Desert and extend in age from 630 Ma to 550 Ma [12]. From the petrological point of view, it is believed that these granites have been developed by partial melting of a variety of pre-existing crustal protoliths or by fractionation of mantle-derived mafic magmas (e.g., [15,16]). In the context of our present study, we investigate the Gebel El-Bakriyah younger granite pluton in the central Eastern Desert (CED) of Egypt to have a detailed and comprehensive account of the petrogenetic processes leading to the formation of this prominent A-type granite ring complex. Previously, country rocks of the El Bakriyah Ring Complex (BRC) were investigated by [17,18], whereas additional information about the ring complex is lacking. Therefore, we integrate field and laboratory studies to build up a reasonable geotectonic model in terms of the geodynamic evolution with respect to the main phases of magmatism and tectonism in the northwestern part of the ANS in the CED. Also, we aim to focus on possible rare-metal enrichment in the BRC, which was known in the past for its fluorite and barite mineralizations [19]. We use combined whole-rock geochemistry and mineral chemistry data for mineral characterization and emphasizing magma source and tectonic setting.

2. Geological Setting and Field Observation

The evolution of the basement rocks in the ANS during the Neoproterozoic rocks (850–590 Ma) involved three major stages: (1) pre-collisional, (2) collisional, and (3) post-collisional [2,3]. Ophiolites and metamorphosed volcano-sedimentary successions were formed during the pre-collisional stage (850–700 Ma). The collisional stage (670–630 Ma) witnessed the emplacement of calc-alkaline magmatic rocks represented by arc-related gabbros and granitoids and their volcanics. The BRC in the CED, which is the main target of the present study, belongs to the third stage (the post-collisional) that is characterized by alkaline magmatism including A-type granites [20,21]. Felsic and mafic magmatism in the CED are abundant and include fresh gabbros side by side with the A-type granites [22]. Association of fresh gabbro and granite is known in the Gebel El Bakriyah area (Figure 1b) and other localities in the CED, e.g., at Wadi El-Faliq and Gebel Atud area e.g., [23,24]. In addition, the CED has post-collisional felsic intrusions at Abu Dabab, Wadi El-Igla, Gebel Mueilha, Homret Waggat, Gebel El-Ineigi e.g., [9,12,13].

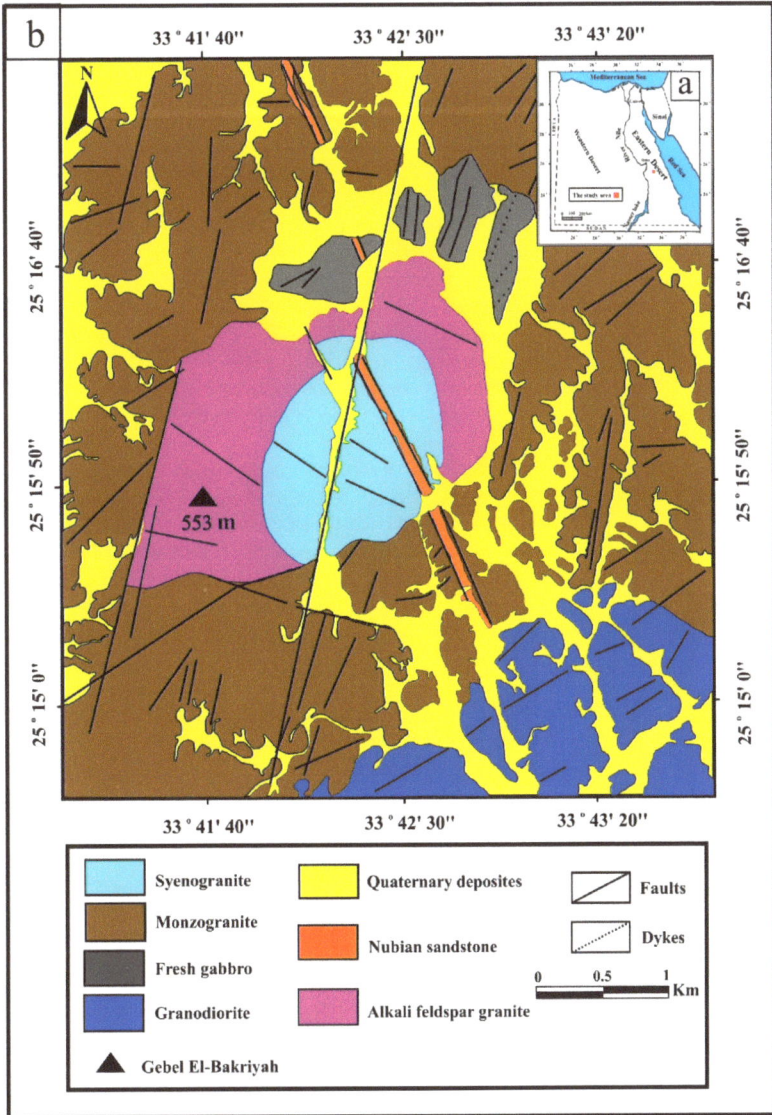

Figure 1. (**a**) Location map of the study area. (**b**) Geological map of the Gebel El Bakriyah area was recently constructed by the present authors [25].

Gebel El Bakriyah area (Figure 1a,b) is located about 15 km north of the El-Barramiya gold mine near the Idfu-Marsa Alam asphaltic road that connects the Red Sea with the Nile Valley. The area is bound by longitudes from 33°41′ to 33°43′ E and by latitudes from 25°15′ to 25°17′ N (Figure 1a,b). In general, the area is characterized by moderate relief, and the only exceptional peak is El Bakriyah Mountain, which has an altitude of ~550 m above the mean sea level. According to [16], Precambrian lithologies in the Gebel El Bakriyah area comprise two main granitic rock types: (1) older granitoids (diorite-granodiorite), and (2) younger granites, mostly syenogranite and much lesser granodiorite. All of them have been emplaced in folded and metamorphosed sequences of metagabbros (Figure 1b). Recently, [23] investigated the fresh post-collisional gabbros and distinguished them into three varieties represented by troctolite, olivine gabbro, and altered hornblende gabbro. Ref. Abd El-Fatah et al. [25] used the remote sensing technique to construct a new geological map of the Gebel El-Bakriyah area (Figure 1b). Using the processed satellite images, it was possible to confirm the rock units and varieties that have been identified in the field and assorted on a mineralogical/petrographical basis. Chronologically, the rocks are arranged from oldest to youngest as follows: granodiorite, fresh gabbro, monzogranite, the BRC (syenogranite and alkali feldspar granite), Nubian sandstone and Quaternary deposits [17,18]. The area is dissected by faults of different ages and types trending N-S, NW-SE, and NE-SW [17]. The only available age (520–506 Ma) for the complex was presented by [26] based on Rb-Sr isochron. In the field, it is easy to distinguish lithological contacts between the core of the BRC (syenogranite) from its rim, which is made up of alkali feldspar granite (Figure 2a). In some instances, and due to intense faulting, contact between the two granite varieties of the BRC is distinctly sharp, i.e., structural. Alkali feldspar granite (Figure 2b) is yellowish to reddish pink in color, coarse- to medium grained and among its outcrops is the summit of El Bakriyah Mountain, and is traversed by quartz, barite, and fluorite veins (Figure 2c,d). The syenogranite is pale pink to red, medium-grained, and displays sharp contact with the Nubian sandstone. Country rocks for the BRC are mostly monzogranite, which forms moderately elevated hills, pink in color, and coarse- to medium-grained. This monzogranite is hard, massive, and dissected by two different sets of joints, and displays sharp contact with the BRC. In the southern part of the Gebel El Bakriyah area, there is some localized grey granodiorite, which forms low relief topography intruded by monzogranite (Figure 2e). There are occasional xenoliths of monzogranite in some granodiorite outcrops. Fresh gabbro forms sporadic outcrops, which constitute four masses located nearly in the central part and northeast extreme of the area (Figure 1b). The gabbro varieties are hard, coarse-grained, and show no layering. The hornblende gabbro variety shows the highest degree of weathering and alterations compared with troctolite and olivine gabbro. In some instances, troctolite is invaded by NE-SW trending mafic and felsic dykes. The Nubian sandstone, of a Late Cretaceous age, non-conformably caps the Precambrian basement rocks.

Figure 2. Geologic setup and field relationships. (**a**) General view of the granite varieties forming the El Bakriyah ring complex (BRC); namely syenogranite core (SG) and alkali feldspar granite rim (AFG) non-conformably capped by the Nubian sandstone (NS). (**b**) The rugged peak of alkali feldspar granite (AFG) capped by Nubian sandstone (NS) and juxtaposing syenogranite (SG). (**c**) Mineralized fluorite vein (V), with some barite, cross-cutting the alkali feldspar granite (AFG). (**d**) Excavated pit for the extraction of fluorite and barite. (**e**) Monzogranite (MG) intruding weathered granodiorite (GR).

3. Materials and Methods

Integration of field and laboratory works include field trips and a variety of analytical procedures for geochemistry and mineral chemistry, in addition to microscopic investigation. The fieldwork was performed for the collection of fresh samples on a systematic basis and field observations for lithological variation, contacts, and structural elements. For the microscopic study, representative thin- and polished sections were prepared for investigation using transmitted and reflected lights. The petrographic section pays attention to mineralogical composition, including alterations, and textures. For the whole-rock chemical analyses, the selected samples were fresh and pulverized down to less than 40 µm after crushing and coarse grinding. Analyses of major oxides and trace elements were conducted in the Geo Analytical Lab of Washington University, USA. The measurements were done by applying the Thermal Spectrometer of X-ray Fluorescence (XRF) and Agilent 7700 inductively coupled plasma mass spectrometry (ICP-MS). For accuracy and precision, the routine of duplicate samples was followed in which the XRF analytical precision is greater than 1% (2σ) for most major elements and better than 5% (2σ%) for the majority of trace elements (except Ni, Cr, Sc, V, and Cs). ICP-mass spectrometry (ICP-MS) was used to measure the concentrations of REEs in addition to Ba, Sr, Zr, Rb, Sc, Cs, Nb, Y, Hf, Ta, Pb, U, and Th. Most trace elements had detection limits that were better than 5% (2σ). This varied for Th, U, Nb, Ta, Pb, Rb, Cs, and Sc between ±9% to ±17% (2σ). Ten polished thin sections were subjected to mineral chemistry analysis utilizing a JEOL JXA-8500F super probe housed at Washington State University. A ZAF correction program was used in order to rectify the data following the procedure recommended by [27]. Operating conditions were a 15 kV voltage acceleration, a 20 nA current beam, a focused 5-10 μm beam diameter, and on-peak counting times of 20 s. A set of natural and synthetic minerals were used as standards.

4. Petrography

The granitic rocks, including the BRC and its country rocks in the Gebel El Bakriyah area were subjected to petrographic investigation. Different rock varieties can be distinguished among the two main categories: (1) country rocks (granodiorites and monzogranite), and (2) BRC (syenogranite and alkali feldspar granites). Generally, the granodiorites are medium- to coarse-grained and show hypidiomorphic and porphyritic texture. They are essentially composed of quartz, plagioclase, microcline, and biotite with some minor zircon and allanite as accessory minerals. Plagioclase crystals are euhedral to subhedral prisms that show variable degrees of sericitization while the biotite flakes are slightly altered to chlorite with some fine apatite inclusions (Figure 3a). The monzogranite, syenogranite, and alkali feldspar granites are holocrystalline and show hypidiomorphic texture, and they are composed essentially of K-feldspar, plagioclase, quartz, biotite and hornblende while kaolinite and sericite represent secondary minerals. The percentage of mafic minerals in the three granite varieties range from 4% to 8%. They are represented by ferromagnesian minerals such as hornblende and biotite.

Accessory minerals are mostly zircon, apatite, columbite, fluorite, titanite, and Fe-Ti oxides. K-feldspars are represented by medium to coarse, subhedral to anhedral crystals that are characterized by perthite texture. Content of potash feldspars, either homogeneous or in the form of perthite intergrowth, in syenogranite and alkali feldspar granite is greater than in monzogranite. K-feldspars occur as subhedral to anhedral prisms which are represented by microcline perthite and to a lesser extent orthoclase perthite, in addition to their homogeneous counterparts. They appear turbid or cloudy at their cores where partial kaolinitic alteration can be observed. Alteration of feldspars to sericite is also seen (Figure 3b). Sericitized plagioclase and kaolinitized potash feldspars occur either as prismatic phenocrysts or as small flattened crystals with subhedral to anhedral outlines. Perthite is one of the distinctive micro-textures in the investigated syenogranite and alkali feldspar granite where three types of perthite intergrowth can be seen: namely string, vein and patchy. Some perthite crystals contain inclusions of small quartz crystals. Quartz occurs mostly in the form of subhedral phenocrysts or as medium- to coarse-sized aggregates, which fill the interstitial spaces between the feldspars. Several examples show graphic and granophyric intergrowths in quartz and potash feldspar discrete crystals (Figure 3c). Biotite is partially chloritized and occurs as dark brown flakes interstitially between quartz and feldspars. Some biotite flakes show rugged peripheries and are squeezed in between the felsic minerals. Muscovitization of biotite is common (Figure 3d). Hornblende occurs as coarse prismatic crystals that are partially altered to chlorite and encloses flakes of muscovite and perthite (Figure 3e,f). Occasionally, some simply-twinned hornblende crystals can be seen.

Metamict zoning in accessory zircon can be seen especially using high magnification. Fluorite is the most common accessory mineral, found mostly in mineral interstices occurring as subhedral to anhedral isolated crystals or in the form of veinlets. Granites in the Gebel El-Bakriyah area contain 2%–4% opaque mineral content that is represented by Fe-Ti oxides. They are homogeneous magnetite and ilmenite in which the latter shows variable degrees of martitization whereas the latter is partially replaced by secondary titanite (Figure 3g). Figure 3h shows fluorite with Nb-Ta inclusion.

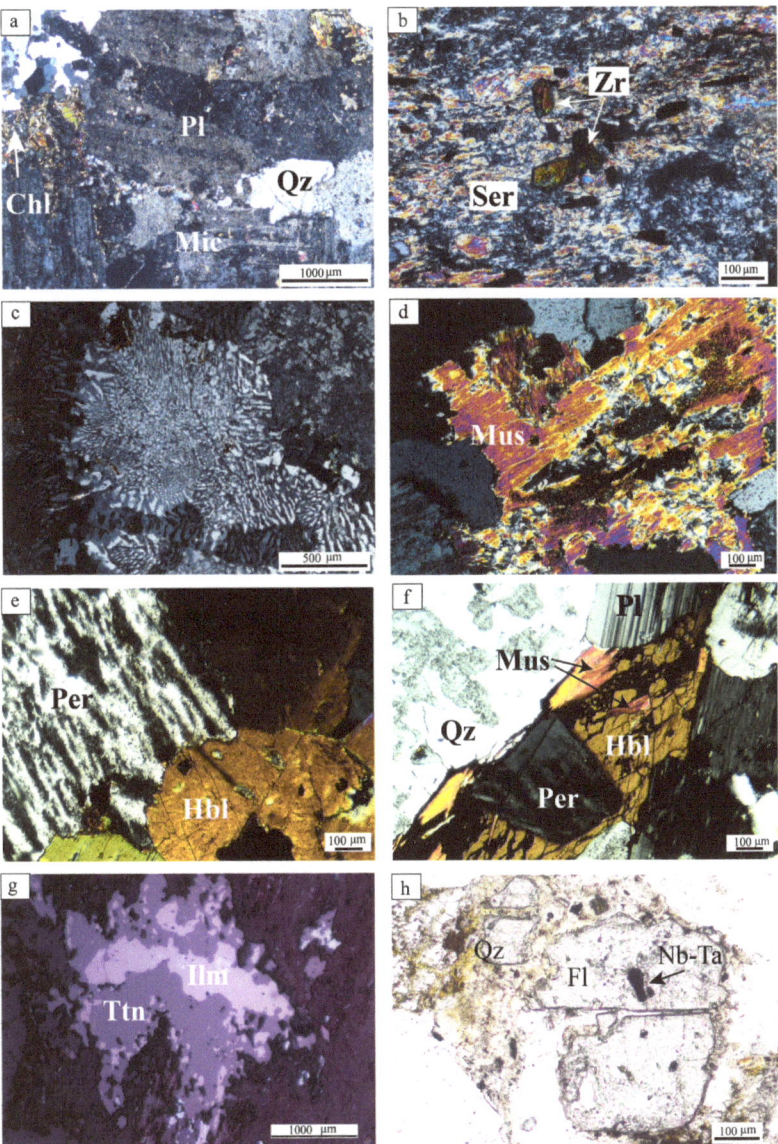

Figure 3. Photomicrographs of Gebel El Bakriyah granitic rocks. (**a**) Large plagioclase crystal (Pl), microcline (Mic), quartz (Qz), and chlorite (Chl) as primary and secondary constituents of granodiorite, Crossed-Nicols (CN). (**b**) Alteration of feldspars to sericite (Ser) and occurrence of euhedral terminated zircon in syenogranite, CN. (**c**) Intergrowths of quartz and K-feldspars to form graphic and granophyric textures in monzogranite, CN. (**d**) Muscovitization (Mus) of primary biotite in syenogranite, C.N. (**e**) Coarse hornblende (Hbl) in contact with perthite (Per) in monzogranite, CN. (**f**) Hornblende crystal (Hbl) surrounded by quartz (Qz) and plagioclase (Pl). Notice that hornblende partially encloses muscovite (Mus) and pethite (Per) in syenogranite, CN. (**g**) Homogeneous ilmenite (Ilm) extensively altered to titanite (Tnt) reaction rim in granodiorite, CN. (**h**) Fluorite (Fl) with typical cubic habit enclosing Nb-Ta minerals.

5. Whole-Rock Geochemistry

Tables 1 and 2 provide the bulk chemical composition (major, trace and REEs) of the selected samples from the country rocks and BRC. Alkali feldspar granites are characterized by SiO_2 = 69.6–72.7 wt.%), CaO (1.08–1.29 wt.%), Al_2O_3 (13.7–15.4 wt.%), TiO_2 (0.07–0.31 wt.%), and MgO (0.19–0.24 wt.%). Syenogranite has (SiO_2 = 74.3 –77.1 wt.%) with lower contents of Al_2O_3 (12.4–13.6 wt.%), CaO (0.7–1.1 wt.%), TiO_2 (0.02–0.09 wt.%), MgO (0.08–0.1 wt.%), P_2O_5 (0.01 wt.%). In the country rocks, granodiorite has a moderate SiO_2 = 63–64 wt.%) with Al_2O_3 of 15.8–16.3 wt.%, high CaO (3.56–3.58 wt.%), TiO_2 (0.63–0.69 wt.%), MgO (0.08–0.1 wt.%), and P_2O_5 (0.14–0.15 wt.%). Monzogranite distinguished by SiO_2 = 73.4–74.5 wt.%), Al_2O_3 (12.6–13.7 wt.%), CaO (0.46–0.56 wt.%), TiO_2 (0.2–0.3 wt.%), MgO (0.27–0.38 wt.%), P_2O_5 (0.04–0.07 wt.%). Also, the concentrations of trace elements in the BRC granite varieties differ from those in the country rocks. In this respect, granites of the BRC (syenogranite and alkali feldspar granite) are rich in Rb, Y, Ga, Nb, Ta, Th, and U compared to the country rocks (granodiorite and monzogranite). However, Zn and Sr are of lower amounts in syenogranite (9–27 ppm and 31–52 ppm, respectively) but high in the alkali feldspar granite (57–96 ppm and 57–70 ppm, respectively) and the country rocks (51–226 ppm and 123–272 ppm, respectively). Ba is much higher in country rocks (378–1012 ppm) compared to the BRC granites (99–326 ppm in syenogranite and 301–468 ppm in alkali feldspar granite). The total concentrations of rare-earth elements ΣREE are high in the BRC granites (ΣREEs = 180–264 in syenogranite and ΣREEs = 296–547 ppm in alkali feldspar granite) compared to those in the country rocks (ΣREEs = 103–265 ppm). To confirm the field and petrographic nomenclature of the different granitic varieties, classification of the Gebel El Bakriyah granites can be ascertained using the total alkalies vs. silica (TAS) diagram (Figure 4a) of [28]. Some of the country rocks plot in the granodiorite field while the others (monzogranite) plot in the granite field. This is supported by the use of the quartz-alkali feldspar-plagioclase (QAP) diagram proposed by [29]. According to this diagram, the country rock samples fall in the granodiorite and monzogranite fields while the BRC samples plot in the syenogranite and alkali feldspar granite fields (Figure 4b). The [30] diagram (Figure 4c) shows that all granitic samples have an alumina saturation index greater than unity indicating that they are peraluminous except granodiorite, which plots at the transition line between the metaluminous and peraluminous fields. This is evidenced by the considerable corundum content of the studied samples up to 5.92% (Supplementary File S1 (T1)). Figure 5a,b present REEs patterns (spider diagrams) normalized using the chondrite values of [31]. The REEs patterns of syenogranite and alkali feldspar granite are comparable with the country rocks in terms of much abundance in the former and magnitude of anomalies. The studied samples are characterized by LREE enrichment with nearly flat HREEs and a strongly negative Eu anomaly, particularly in the case of syenogranite and alkali feldspar granite. Figure 5c,d show spider diagrams of some trace elements normalized to values in the primitive mantle of [32]. The figure shows significant enrichment in large ion lithophile elements (LILE) as well as incompatible elements that act similarly to LILE (such as Th and U). The granites of the BRC and their country rocks are characterized by relative enrichment in Zr, Y, and Ta whereas Ti, Ni, K, and P are low. Also, granodiorite shows slight depletion in Cr.

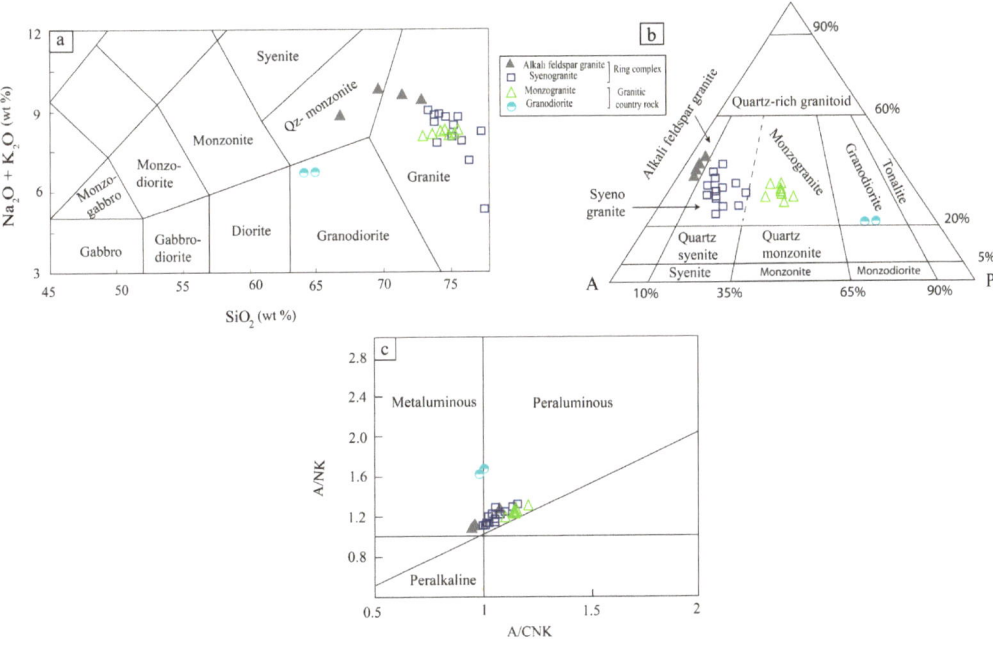

Figure 4. (**a**) Nomenclature of the granitic rocks from the Gebel El Bakriyah area using the total alkali-silica (TAS) classification diagram [28]. (**b**) Plots of the studied granitic rocks, based on their modal compositions, on the QAP diagram [29]. (**c**) Peraluminous nature of the Gebel El Bakriyah younger granites on the A/CNK vs. A/NK diagram [30].

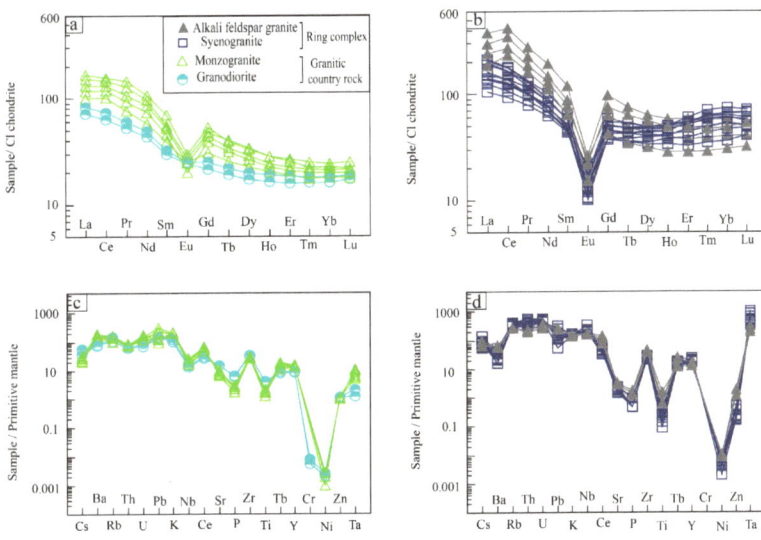

Figure 5. Spider diagrams based on trace elements and rare-earth elements (REEs). (**a**,**b**) Patterns of chondrite-normalized REEs of the Gebel El Bakriyah area [31]. (**c**,**d**) Patterns of primitive mantle-normalized multi-element variation [32].

Table 1. Whole-rock chemical composition of syenogranite of the Gebel El Bakriyah ring complex (BRC).

Sample #	FS1	FS11	FS2	FS5	FS6d	FS7	S26	S28	S29	S30	S31	S33
Major oxides (wt.%)												
SiO_2	74.33	74.74	76.32	73.91	77.88	74.84	77.18	73.56	75.65	73.41	75.28	75.08
TiO_2	0.05	0.05	0.05	0.09	0.05	0.07	0.02	0.06	0.05	0.06	0.06	0.03
Al_2O_3	13.61	13.05	12.42	13.08	12.57	13.37	12.6	13.8	12.96	13.53	13.28	12.96
Fe_2O_3	0.91	1.31	0.34	0.78	0.36	0.57	0.56	1.18	0.76	1.47	0.67	0.79
MnO	0.02	0.02	0.01	0.02	0.02	0.01	0.02	0.01	0.01	0.03	0.01	0.01
MgO	0.14	0.14	0.1	0.15	0.08	0.16	0.07	0.16	0.15	0.17	0.13	0.14
CaO	0.73	0.76	0.75	1.19	0.61	0.83	0.56	0.94	0.64	0.88	0.84	0.77
Na_2O	4.05	2.87	2.66	3.46	1.04	3.92	3.84	4.09	3.44	3.89	3.74	3.96
K_2O	4.69	5.08	4.48	4.24	4.27	4.23	4.31	4.77	4.44	4.72	4.74	4.69
P_2O_5	0.01	0.01	0.01	0.02	0.01	0.01	0.01	0.01	0.01	0.01	0.01	0.01
LOI	1.11	1.42	1.74	2.13	2.33	2.04	1.06	1.87	1.68	1.73	1.43	1.38
Total	99.65	99.45	98.88	99.07	99.22	100.05	100.23	100.45	99.79	99.9	100.19	99.82
Trace elements (ppm)												
Rb	218.94	205.71	247.81	201.62	272.77	212.06	224.59	191.59	212.44	206.48	237.84	215.57
Ba	256.83	326.31	394.28	309.47	278.8	177.01	154.51	371.04	152.34	170.52	156.51	115.23
Sr	47.61	41.79	33.59	51.97	30.75	39.74	30.92	52.09	37.8	50.73	38.58	39.96
Nb	153.67	155.18	171.82	141.83	246.72	164.96	171.76	156.52	170.69	147.39	172.34	168.27
Zr	284.54	285.62	260.56	232.17	320.55	311.42	250.38	285.32	285.19	271.93	314.08	302.31
Y	87.21	87.6	92.42	81.55	106.17	82.68	98.98	79.61	96.54	78.41	90.93	98.17
Zn	9.98	11.83	26.71	17.66	8.35	7.91	9.19	10.73	11.22	24.93	8.89	10.47
Cu	<d.l.	1.55	1.83	2.06	1.38	<d.l.	1.31	<d.l.	<d.l.	<d.l.	<d.l.	<d.l.
Ni	12.32	11.89	7.89	14.67	3.96	10.17	7.98	16.41	8.87	15.23	8.22	10.05
Co	<d.l.	<d.l.	<d.l.	<d.l.	<d.l.	<d.l.	<d.l.	<d.l.	<d.l.	<d.l.	<d.l.	<d.l.
Cr	<d.l.	<d.l.	<d.l.	<d.l.	<d.l.	<d.l.	<d.l.	<d.l.	<d.l.	<d.l.	<d.l.	<d.l.
V	16.07	14.6	12.18	15.76	9.95	12.86	10.68	15.33	13.15	16.03	13.53	12.13
Pb	12.8	11.22	9.1	22.36	7.77	10.74	9.99	17.69	3.93	18.42	13.39	10.25
Ga	40.47	35.31	43.16	33.47	44.84	35.78	40.12	31.82	40.38	34.03	39.29	37.65
Sc	<d.l.	<d.l.	<d.l.	<d.l.	<d.l.	<d.l.	<d.l.	<d.l.	<d.l.	<d.l.	<d.l.	<d.l.
Mo	1.99	1.89	2.27	1.08	4.18	1.58	2.19	1.49	3.24	1.59	1.79	1.77
Cs	0.5	0.64	0.73	0.48	1.09	0.76	0.73	0.27	0.71	0.46	0.78	0.66
Hf	7.11	7.68	9.54	9.54	11.22	9.09	8.31	6.85	7.46	9.48	7.85	11.02
Ta	17.94	18.56	26.13	14.15	42.28	20.37	33.54	17.93	21.57	19.93	20.31	19.67
Th	34.36	33.6	44.12	28.13	48.45	36.93	43.49	28.71	36.45	26.95	33.56	35.82
U	9.33	11.25	11.82	5.75	12.13	11.76	12.02	8.04	11.15	9.65	10.82	9.88
Rare-earth elements (ppm)												
La	47.42	49.33	50.94	24.93	32.51	33.56	29.69	43.86	38.88	45.19	35.73	32.74
Ce	100.95	108.81	110.64	57.33	77.4	78.11	66.13	94	84.15	100.92	85.58	75.65
Pr	12.4	12.59	12.87	7.54	9.7	9.71	8.41	10.6	9.54	11.37	10.94	9.33
Nd	42.8	42.34	47.48	29.12	35.28	36.48	31.47	38.32	32.64	40.61	40.42	35.1
Sm	8.71	8.74	10.09	6.61	7.92	7.43	6.92	8.17	7.14	9.02	8.87	7.93
Eu	0.93	0.91	1.02	0.79	1.19	0.59	0.62	1.03	0.74	0.7	0.78	0.61
Gd	11.15	10.61	12.24	7.67	8.81	7.96	7.95	9.48	9.44	10.31	10.22	9.21
Tb	1.99	1.75	1.95	1.32	1.56	1.42	1.35	1.62	1.59	1.78	1.8	1.56
Dy	12.35	11.9	12.45	8.57	11.35	10.05	8.78	10.95	10.21	12.07	12.53	11.01
Ho	2.76	2.62	2.86	1.91	2.49	2.27	2.16	2.37	2.33	2.66	2.83	2.52
Er	8.94	8.54	9.17	5.36	8.91	7.79	6.43	7.57	7.27	8.59	9.91	9.02
Tm	1.52	1.46	1.55	0.89	1.61	1.25	1.11	1.25	1.22	1.48	1.78	1.55
Yb	10.31	10.4	10.78	6.08	11.16	9.01	7.96	8.81	9.01	10.12	12.52	11.38
Lu	1.42	1.43	1.53	1.04	1.71	1.25	1.23	1.26	1.29	1.47	1.83	1.72
Eu*	10.39	10.49	11.4	7.06	8.76	8.49	7.63	9.31	8.25	10.13	9.85	8.6
Eu/Eu*	4.12	4.04	4.17	4.12	4.03	4.29	4.13	4.12	3.95	4.01	4.1	4.08
$(La/Yb)_n$	3.84	3.93	4.81	5.25	5.44	4.74	6.76	3.79	5.01	3.14	3.39	3.97
$(La/Sm)_n$	2.77	2.67	3.43	3.73	4.99	3.8	5.17	2.71	3.82	2.37	3.07	3.84
$(Gd/Lu)_n$	5.73	5.99	6.51	7.43	4.92	5.94	6.23	6.54	5.85	6.09	4.98	5.12
$(La/Lu)_n$	22.61	23.01	28.46	31.61	30.09	29.54	39.18	22.97	29.88	18.21	18.85	23.11
T1.3	1.04	1.04	0.99	1.02	1.06	1.04	0.99	1.02	1.02	1.04	1.04	1.02
ΣREEs	263.65	271.43	285.57	159.16	211.6	206.88	180.21	239.29	215.45	256.29	235.74	209.33
Y/Ho	31.6	33.44	32.31	42.7	42.64	36.42	45.82	33.59	41.43	29.48	32.13	38.96
T (°C) Zr	842	852	850	824	905	854	834	837	852	835	851	843

Table 2. Whole-rock chemical composition of the other granites forming the Gebel El Bakriyah ring complex (BRC).

	Alkali Feldspar Granite		Monzogranite										Granodiorite	
Sample #	FS13	FS13A	S7b	S35A	S8	S13	S18	S20	FS23	S39	FS12	FS15	S35b	S1
Major oxides (wt.%)														
SiO_2	72.78	69.66	66.64	71.19	73.68	74.37	74.48	74.12	74.53	75.42	74.37	75.52	64.9	63.92
TiO_2	0.07	0.18	0.33	0.13	0.31	0.29	0.36	0.31	0.26	0.21	0.29	0.29	0.66	0.69
Al_2O_3	13.78	14.12	15.48	14.34	13.45	13.09	13.77	12.76	13.17	12.87	13.09	12.61	16.31	15.86
Fe_2O_3	1.82	2.93	3.21	2.28	1.79	1.45	1.28	1.56	1.58	1.3	1.45	1.21	4.72	4.81
MnO	0.03	0.06	0.09	0.03	0.03	0.01	0.02	0.04	0.05	0.03	0.01	0.04	0.08	0.07
MgO	0.19	0.21	0.24	0.2	0.33	0.32	0.38	0.35	0.3	0.27	0.32	0.31	1.72	1.69
CaO	1.02	1.11	1.29	1.08	0.56	0.54	0.55	0.56	0.52	0.46	0.54	0.49	3.56	3.58
Na_2O	4.56	4.64	4.48	4.59	3.49	3.44	3.37	3.43	3.53	3.45	3.44	3.28	4.16	4.41
K_2O	4.88	5.04	4.37	5.01	4.58	4.78	4.54	4.44	4.64	4.75	4.78	4.67	2.74	2.28
P_2O_5	0.02	0.04	0.12	0.03	0.06	0.05	0.07	0.04	0.04	0.04	0.05	0.04	0.15	0.14
LOI	1.22	1.36	2.91	1.23	1.25	1.52	1.12	0.92	0.83	1.02	1.52	1.33	1.14	1.95
Total	100.37	99.35	99.16	100.1	99.53	99.86	99.94	98.53	99.45	99.82	99.86	99.79	100.1	99.4
Trace elements (ppm)														
Rb	179.66	170.87	167.2	172.4	56.26	79.86	96.66	84.48	81.84	90.86	60.48	122.49	80.77	60.48
Ba	300.73	409.67	468.4	402.5	1012.6	765.5	812.5	768.0	551.9	977.3	404.1	741.54	378.2	404.1
Sr	57.14	66.7	69.84	61.42	189.64	175.8	153.1	139.7	126.3	123.5	272.6	133.71	267.9	272.6
Nb	132.41	124.59	112.6	119.6	9.89	9.05	11.71	11.53	13.03	15.21	7.65	19.27	8.38	7.65
Zr	372.31	512.18	507.4	439.5	257.33	256.4	243.6	240.3	236.9	224.8	282.6	217.75	290.2	282.6
Y	88.55	75.33	56.89	76.89	47.68	46.79	48.82	46.28	51.51	54.58	29.97	57.69	30.78	29.97
Zn	57.24	96.41	60.88	92.11	52.79	59.03	203.9	226.3	60.04	51.28	56.46	75.51	57.78	56.46
Cu	<d.l.	<d.l.	<d.l.	<d.l.	5.97	4.79	4.43	3.76	3.68	2.71	10.76	2.33	11.71	10.76
Ni	16.42	18.17	21.05	17.22	5.49	3.91	2.04	<d.l.	<d.l.	5.03	<d.l.	4.15	5.03
Co	<d.l.	2.24	3.1	<d.l.	4.46	3.07	<d.l.	2.14	2.72	1.5	9.52	2.42	10.28	9.52
Cr	<d.l.	<d.l.	<d.l.	<d.l.	<d.l.	<d.l.	<d.l.	<d.l.	<d.l.	<d.l.	24.33	<d.l.	16.47	24.33
V	17.03	18.37	19.62	16.98	34.97	28.74	29.26	21.61	22.3	12.14	47.76	9.18	45.54	47.76
Pb	12.79	16.37	18.66	17.61	27.9	11.66	27.01	37.93	16.03	14.66	18.56	15.02	17.66	18.56
Ga	36.11	31.05	27.29	35.77	20.04	19.23	20.59	21.93	24.04	22.23	21.36	25.13	21.59	21.36
Sc	<d.l.	<d.l.	<d.l.	<d.l.	0.57	0.53	0.46	0.48	0.31	0.24	7.01	0.19	6.86	7.01
Mo	5.05	5.12	4.75	4.03	0.94	1.44	1.88	2.32	2.38	2.04	2.25	3.37	1.57	2.25
Cs	0.83	0.43	0.51	0.79	0.38	0.48	0.41	0.52	0.48	0.59	0.69	0.65	1.05	0.69
Hf	12.68	13.18	12.96	10.53	6.67	6.27	7.09	6.76	6.66	7.11	4.72	7.37	5.57	4.72
Ta	16.55	9.74	8.05	12.23	0.25	0.22	0.28	0.44	0.35	0.47	0.05	0.59	0.09	0.05
Th	26.25	18.71	16.44	20.44	5.27	4.83	5.35	4.74	5.73	5.24	3.96	6.34	4.12	3.96
U	8.75	7.21	5.61	7.66	2.22	2.16	2.48	2.76	2.42	3.21	1.6	3.78	1.25	1.6
Rare-earth elements (ppm)														
La	88.63	57.59	44.24	69.85	27.84	24.03	36.13	31.34	43.27	39.91	16.88	41.65	19.68	16.88
Ce	257.77	166.45	142.7	212.9	73.07	63.12	91.23	81.61	101.8	96.42	38.66	101.95	45.21	38.66
Pr	26.29	17.63	15.35	21.45	8.98	7.27	12.24	10.21	11.6	4.97	13.99	5.7	4.97
Nd	90.14	60.51	52.3	68.63	34.48	27.01	45.41	40.3	41.86	50.34	20.28	52.59	22.97	20.28
Sm	18.25	12.02	9.66	13.49	7.03	5.99	8.44	7.98	9.51	10.8	4.59	11.02	4.95	4.59
Eu	1.52	1.42	0.88	1.28	1.16	1.47	1.58	1.44	1.82	1.74	1.41	1.91	1.45	1.41
Gd	19.9	13.06	8.64	15.46	8.04	6.36	9.78	9.12	11.17	10.94	4.39	13.81	5.17	4.39
Tb	2.81	1.81	1.29	2.37	1.17	0.95	1.53	1.29	1.72	1.48	0.71	2.04	0.82	0.71
Dy	16.16	10.88	7.97	14.01	6.73	5.79	8.72	7.37	9.56	8.46	4.41	11.59	5.04	4.41
Ho	3.31	2.18	1.61	2.77	1.28	1.15	1.6	1.38	1.79	1.6	0.93	2.2	1.08	0.93
Er	9.66	6.16	4.69	7.95	3.42	3.12	4.44	3.67	4.79	4.26	2.63	5.53	3.05	2.63
Tm	1.46	0.99	0.74	1.18	0.51	0.45	0.64	0.55	0.69	0.57	0.41	0.79	0.46	0.41
Yb	9.82	6.76	5.17	8.22	3.38	3.06	4.13	3.59	4.63	3.86	2.74	5.34	3.07	2.74
Lu	1.6	1.03	0.81	1.35	0.51	0.45	0.64	0.55	0.68	0.56	0.44	0.8	0.48	0.44
Eu*	21.9	14.56	12.18	17.01	7.95	6.6	10.16	9.03	10.5	12.23	4.78	12.42	28.54	81.48
Eu/Eu*	4.12	4.16	4.29	4.03	4.34	4.09	4.47	4.46	3.99	4.12	4.25	4.24	0.88	0.27
$(La/Yb)_n$	2.72	3.04	3.51	2.57	1.54	1.55	1.2	1.29	1.2	1.23	1.51	1.15	4.33	5.75
$((La/Sm)_n$	1	1.06	1.07	0.95	0.59	0.66	0.44	0.46	0.49	0.38	0.8	0.45	2.51	3.27
$(Gd/Lu)_n$	12.5	12.14	13.05	11.43	13.78	13.31	13.19	14.51	13.78	18.95	11.2	13.95	1.32	1.4
$(La/Lu)_n$	17.98	18.9	22.22	17.32	10.33	10.73	8.36	8.62	8.3	9.19	9.66	8.03	4.2	5.3
T1.3	1.08	1.08	1.13	1.15	1.05	1.06	1.08	1.04	1.06	1.03	1	1.04	1	1
ΣREEs	547.32	358.49	296.0	441	177.6	150.2	226.5	200.4	244.9	244.7	103.4	265.21	119.1	103.4
Y/Ho	26.75	34.56	35.34	27.76	37.25	40.69	30.51	33.54	28.78	34.11	32.23	26.22	28.5	32.23
T (°C) Zr	851	878	892	858	842	839	842	833	832	827	848	825	821	815

6. Mineral Chemistry

For a better understanding of the mineral composition of the younger granites and its use to deduce some petrogenetic parameters, we analyzed the major silicate minerals and some non-silicates in the four granitic varieties using the electron microprobe. They include feldspars (plagioclase and K-feldspars), biotite, chlorite, amphibole, apatite, zircon, Nb-Ta oxides, and Fe-Ti oxides. Structural formulae were calculated using the Minpet software [33] together with some Excel spreadsheets.

6.1. Feldspars

Electron microprobe analyses (EMPA) of the feldspar-group minerals, their structural formulae, and end-members (mole%) are given in Supplementary File S1 (T2). The analyzed feldspars in the Gabal El Bakriyah younger granites are distinguished into K-feldspars (homogeneous orthoclase/microcline and perthite), in addition to sodic plagioclase. In the monzogranite, homogeneous K-feldspars have a high and narrow K_2O range (15.29–16.22 wt%), and therefore orthoclase component is remarkably high (89–97 mole%) while its albite component never exceeds 8 moles%. On the other hand, the range of K_2O in perthite is lower and wide (3.59–12.36 wt%), which results in a wide range of orthoclase components (20–67 mole%) and high albite components up to 76 moles%. In the syenogranite, its homogeneous K-feldspars have the highest orthoclase (95–98 mole) and lowest albite (up to 5 moles%) owing to a high and narrow K_2O range (15.57–16.32 wt%). The alkali feldspar granites have homogeneous K-feldspars (14.97–16.22 wt% K_2O, 82–97 mole% orthoclase, and up to 17% mole% albite) and few perthites (12.46–12.53 wt% K_2O, 74–75 mole% orthoclase, and up to 25 moles% albite).

The analyzed plagioclase in monzogranite is mainly sodic (9.2–10.3 wt% Na_2O), which is equivalent to 82–88 mole% albite, while anorthite in solid solution is much lower (up to 12.6 mole%) due to the low CaO content (2.2–3.2 wt%). EMPA of homogeneous plagioclase in the three granite varieties shows similarities. In the monzogranite, plagioclase contains 10.13–12.14 wt% Na_2O equivalent to high albite solid solution component amounting to 89–99 mole%. Similarly, the syenogranite (10.39–11.56 wt% Na_2O and 99 moles% albite) and the alkali feldspar granite (10.89–11.65 wt% Na_2O and 96–99 mole % albite). A second generation of homogeneous plagioclase is slightly calcic and has 9.27–10.39 wt% Na_2O and accordingly, the lowest albite component (84–89 mole%), which is typical of an oligoclase composition. On the ternary phase diagram for the classification of feldspars based on their chemistry [34], the analyzed feldspars from BRC granites plot in the orthoclase/microcline and albite field. On the other hand, feldspars from country rock lie in orthoclase/microcline, oligoclase, and anorthoclase fields (Figure 6a).

6.2. Biotite

The major primary ferromagnesian mineral in the investigated granites is biotite. Its EMPA and structural formula are presented in the Supplementary File (T3). Analyses of biotite indicate 9.5–12.8 wt% Al_2O_3, 2.4–3.9 wt% TiO_2, and 0.08–0.5 wt% MnO. Biotite in the alkali feldspar granite has a higher content of FeO^t (30.6–33.7 wt%) compared to the monzogranite (19.5–23.6 wt%) and syenogranite (~20–24.1 wt%). On the other hand, biotite from the alkali feldspar granite and monzogranite has a higher MgO content (8.6–11.9 wt%) than in the alkali feldspar granite (2.9–4.4 wt%). The FeO^t/MgO ratio ranges from 8 to 11 in the alkali-feldspar granite, which is noticeably higher than those in the syenogranite and monzogranite (0.94 to 2). On the TiO_2–FeO^t–MgO ternary diagram of [35], all analyses pertain to primary/magmatic biotite (Figure 6b). According to the classification of [36], biotite from BRC granites plot in Mg and Fe biotite whereas biotite from the country rock granites lie in the Mg-biotite field (Figure 6c).

6.3. Amphibole

For the amphibole EMPA (Supplementary File S1 (T4)), the structural formula is calculated on an anhydrous basis assuming 23 oxygen atoms per half-unit cell. Data in this file indicates that the content of FeO^t (30.41–35.35 wt%) in the monzogranite country rock is much higher than in the syenogranite or the core of the BRC (17.82–17.97 wt%). Similarly, MgO is much higher in the monzogranite (11.56–11.97 wt%) than in the syenogranite (0.44–2.53 wt%). On the classification diagram of [37] (Figure 6d), the analyzed amphibole in BRC granites is Ca-Na amphibole whereas it is Na to Na-Ca amphibole in the granites from the country rocks, i.e., it is an alkali amphibole with some Ca in the double-chain structure, specifically in the Y-sites where divalent cations are allocated. Careful application of some Excel spreadsheets enables the exact nomenclature of amphibole. As given at the Supplementary File S1 (T4), the analyzed amphiboles comprise ferro-richterite, katophorite, ferobarrroisite, and arfvedsonite in the syenogranite. In monzogranite, it has the composition of normal arfvedsonite and Mg-arfvedsonite only.

6.4. Nb-Ta Oxides

It was possible to record and analyze Nb-Ta oxides as primary magmatic phases in the syenogranite variety only (Supplementary File S1 (T5)). These oxides are either columbite or tantalite as end-members of a solid solution series common in rare-metal granites. The analyzed columbite has high Nb_2O_5 content (42.58–69.56 wt%) than those in tantalite (21.35–24.89 wt%). Oppositely, Ta_2O_5 content in tantalite (55.38–59.71 wt%) is much higher than in columbite (8.27–38.96 wt%). MnO content in the columbite is higher than in the tantalite (up to 13.96 wt% and up to 9.77 wt%, respectively). Contents of Si, Al, and Ca oxides in the columbite-tantalite are negligible. The Ta/(Nb + Ta) ratio in tantalite is doubled compared to columbite (0.57–0.63 and up to 0.27, respectively). According to the binary classification diagram of Nb-Ta oxides based on the Mn/(Mn + Fe) vs. Ta/(Ta + Nb) ratios, the analyzed columbite is manganoan (i.e., Mn-columbite) and so is tantalite (Mn-tantalite) as shown in Figure 6e.

6.5. Zircon and Apatite

Zircon is a common accessory mineral in the investigated three granite varieties (Supplementary File S1 (T6)). The analyses show limited variations in the chemical composition in which ZrO_2 content in the BRC granites and its country rocks ranges from about 60.9 to 64.4 wt%. The range of SiO_2 content is also narrow (31 to 32 wt%). This zircon is poor in Al_2O_3 and FeO^t. In general, oxide impurities in the analyzed zircon are negligible and amount < 0.5 wt%. Because Hf_2O_3 was not among the protocol of the electron microprobe session for zircon, the sum of oxides is somehow low (95.62–98 wt %).

Apatite in the monzogranite and alkali feldspar granite varieties was analyzed (Supplementary File S1 (T6)). Apatite in the alkali feldspar granite has lower contents of CaO (44–48 wt%) and P_2O_5 (34–36 wt%) than in the monzogranite (51–56 wt% and 38–41 wt%, respectively). The analyses do not show any significant chemical variations. For example, Mn^{2+} in the apatite structure reaches up to 0.3 atoms per formula unite (apfu) in the apatite from alkali feldspar granite, which is almost the same in the apatite from monzogranite (3.2 pfu). Similarly, apatite in the two granite varieties has almost constant Ca^{2+} (10.7–11.7 apfu). The remaining oxides are present as impurities in trace levels, which do not exceed 0.5 wt%.

6.6. Fe-Ti Oxide

The chemical analyses and structural formulae of the different Fe-Ti oxides are given in the Supplementary File S1 (T7). They include magnetite, ilmenite and goethite. The abundance of magnetite is limited in the monzogranite as well as in the alkali feldspar granite, while ilmenite and goethite are restricted to monzogranite. Magnetite in the BRC granites has FeO^t in the range from 8 wt% to 92.2 wt% whereas the same range in the country rock is 86.5–92.1 wt%.

Negligible amounts of other oxide impurities (<0.5 wt%) are also noticed. In the ilmenite crystals, TiO_2 content is very narrow and ranges from 49 wt% to 50 wt% and has the expected FeO^t content in ilmenite that is noticeably wide in range (29 to 43 wt%). In some instances, the FeO^t content abruptly increases to the amount from 87 to 91 wt% in crystals with minute hematite exsolution (hematite-ilmenite exsolution or primary texture). Appreciable amounts of Mn cations are encountered in the ilmenite structure (3–8.5 wt% MnO) that is a function of a significant amount of manganese in solid solution with end-members (Mn- and Mg-ilmenite); namely geikielite ($MgTiO_3$) or pyrophanite ($MnTiO_3$). The analyzed goethite from the monzogranite shows extremely limited variations in the chemical compositions, in which FeO^t ranges from 59 to 73 wt% and the much lesser SiO_2

content is appreciable (3.8 to 5.5 wt%) and connected to increased silica activity during oxidation of magnetite.

6.7. Chlorite

As a hydrous phyllosilicate mineral, which is a secondary phase replacing biotite, its structural formula is calculated from the EMPA based on 11 oxygen atoms (Supplementary File S1 (T8)). Analysis of chlorite from the monzogranite shows high SiO_2 (27–32 wt%), Al_2O_3 (11–13wt%), FeO^t (24–29 wt%) and MgO (8.8–12.1 wt%). The contents of K_2O (up to 4.3 wt%) are moderate whereas the chlorite is poor in CaO, NaO, TiO_2, and MnO oxides. The ratio of Fe/Fe + Mg in the analyzed chlorite is a bit wide (0.5–0.9 wt%). Fe^{2+} (up to 5.4 apfu) is appreciable and greatly exceeds Fe^{3+} (0.2 apfu). According to the classification by [38], the analyzed chlorite plots in the diabantite field except for a few pycnochlorite compositions (Figure 6f).

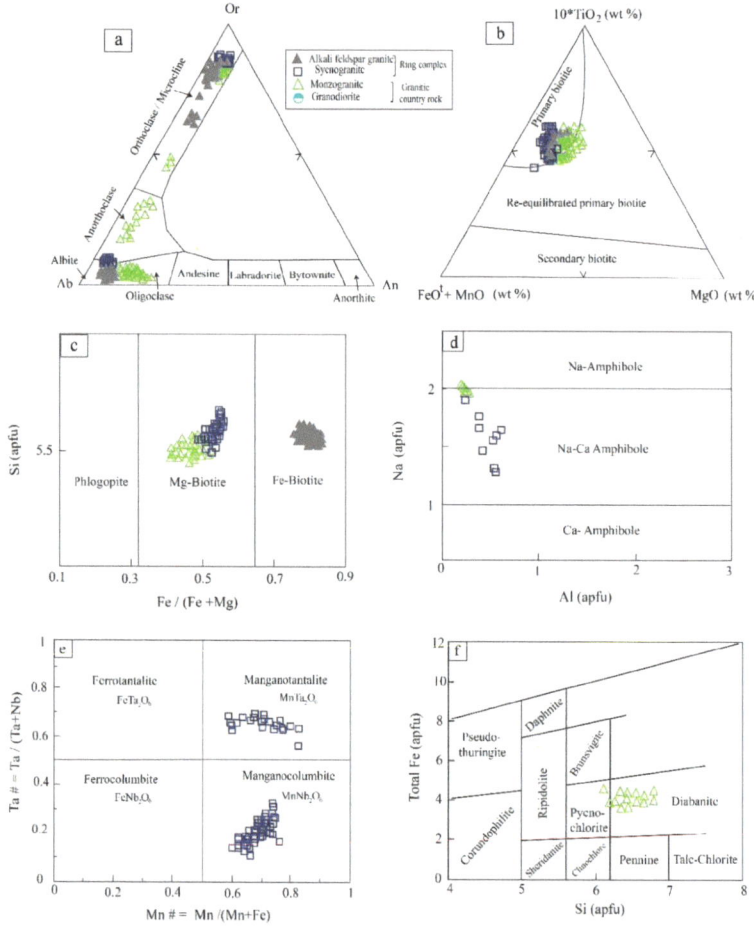

Figure 6. (**a**) Composition of the feldspars in Gebel El Bakriyah granitic rocks using the Ab-Or-An ternary diagram of [34]. (**b**) The primary origin of biotite is based on the (FeO^t + MnO)–10*TiO_2–MgO discrimination diagram of [35]. (**c**) Nomenclature of biotite (Fe- and Mg-bearing) using the classification of [36]. (**d**) Na-Ca amphibole plots after [37]. (**e**) Mn-rich end-members of the Nb-Ta oxides. (**f**) Chlorite composition in the Gebel El Bakriyah granitic rocks [38].

7. Discussion
7.1. Conditions of Magmatic Crystallization

The mineral assemblages and the physico-chemical properties of the magma are closely related during the crystallization process [39]. Temperature (T), pressure (P), and oxygen fugacity (f_{O_2}) all have a significant impact on the mineral compositions and crystallization histories of magmatic systems. Zircon exhibits variable behavior in felsic rocks and is frequently concentrated in residual silicate melts until zircon saturation occurs [40]. For the felsic magmas, [40] used the relationship between the solubility of zircon and crystallization temperature with major element composition in the melt to determine their temperature. According to the calibration of [40], the assessed Zr saturation temperature for the studied granitic rocks from the BRC ranges from 824 to 905 °C while the temperature of the country rocks is between 815–848 °C.

Based on the norm calculations given in the Supplementary File S1 (T1), an Ab-Or-Qz ternary diagram (Figure 7a) was constructed, which shows that the studied granites evolved in extremely low H$_2$O vapor pressures where the samples plot near the lowest point of the diagram between 2–6 kbar. The analyzed biotites in the country rock and the BRC granites have low TiO$_2$, which indicate low-pressure condition of the granite emplacement for all varieties [41]. This gives evidence that the condition of crystallization for the BRC occurred at a shallow depth condition in the crust.

According to [42], oxygen fugacity (f_{O_2}), is an important factor controlling magmatic activity that may be used to determine the redox conditions of melts during petrogenesis. Since magnetite frequently becomes Ti-free during slow cooling and ilmenite generally through one or more oxidation processes, it is difficult to determine the initial oxygen fugacity of the the primary magma [43]. However, amphibole chemistry can be used as a petrogenetic indicator for oxygen fugacity (f_{O_2}). The chemistry of amphiboles in the Gebel El-Bakriyah syenogranite (the core of the BRC) indicates a low f_{O_2} (Figure 7b) that refer to a reducing condition upon magmatic crystallization until the minor metasomatic changes at the late stage, more likely deuteric alterations. Evolution of A-type granite includes fractionation of felsic melts in low f_{O_2} condition [44]. This is supported by low Mg# (0 to 0.04) ratio and low TiO$_2$ along with high Al$_2$O$_3$ content in biotite.

7.2. Magma Type and Tectonic Setting

It is possible to deduce the magma type of the studied granites using either composition of these rocks, i.e., the whole-rock analysis, or with the aid of some rock-forming minerals (e.g., micas) as petrogenetic indicators. The chemical compositions of biotite and saturation index of alumina (Figure 7c) indicated that the investigated country rocks represented by monzogranite have calc alkaline character while the BRC granites have an alkaline nature. Alumina saturation index indicates the peraluminous nature for most of the studied granites, which is confirmed by high normative corundum up to 5% (Figure 4c and Supplementary File S1 (T1). Some geochemical characteristics such as high contents of Nb, Ta, Y, Zr, Th, considerable Ga/Al ratio and remarkable depletion in MgO, CaO, and P$_2$O$_5$ categorize most of the younger granites in the Gebel El-Bakriyah area as A-type granites [10]. Alkali feldspar granite and syenogranite have a ferroan nature representing the alkaline A-type character of the BRC while monzogranite and granodiorite have a magnesian nature (Figure 7d). Some geochemical parameters can be used to examine the Gebel El Bakriyah granitic rocks to determine the tectonic setting for their origins. Based on the whole-rock geochemistry, the Gebel El Bakriyah granites cover the range from volcanic-arc for the granodiorite samples followed by a transitional phase as shown by the monzogranite samples to post-collision (within-plate) granites for the BRC according to the tectonic discrimination diagrams of [45] (Figure 7e).

Generally, the peraluminous granites, which is the case of the El Bakriyah area are often produced as a result of collisional (country rocks) and/or post-collisional (BRC) conditions [46]. However, the A-type rocks of the Gebel El-Bakriyah granites have general geochemical characteristics of calc-alkaline (country rocks) to alkaline (BRC) magma evolved in a collisional to post-collisional stage. This is evidenced by a high content of SiO$_2$, notable depletion in CaO, MgO, and Sr with a high concentration of alkalis content, and considerable enrichment in Nb, Zr, Y, Hf, Th, in addition to REEs patterns with distinct negative Eu anomalies [10,45,47].

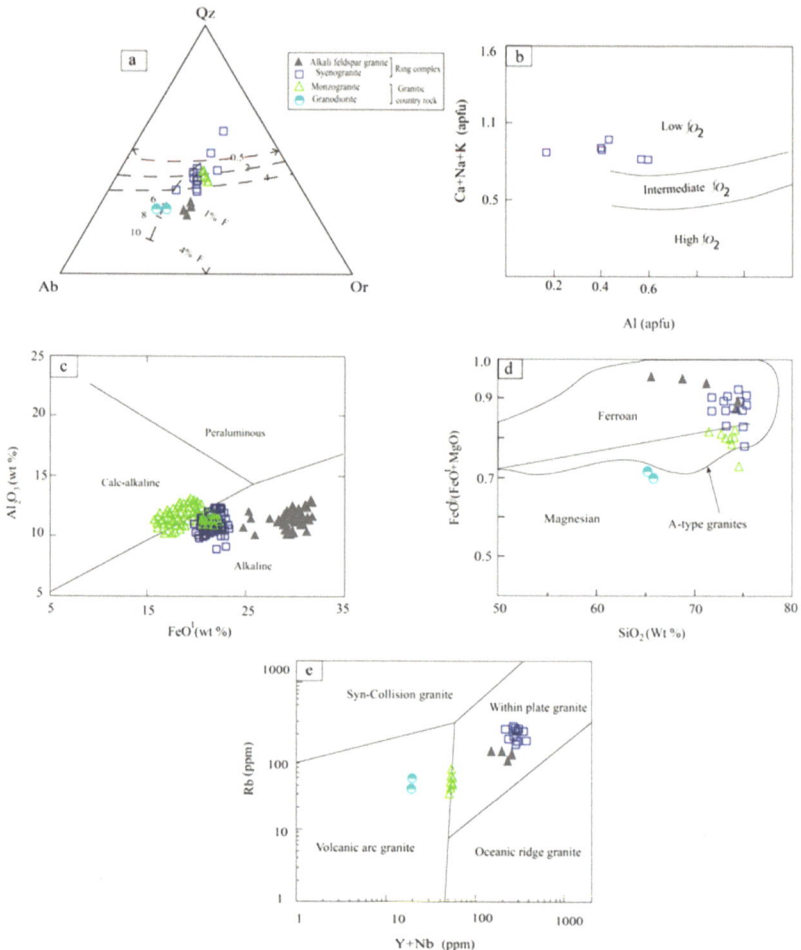

Figure 7. (**a**) The Ab–Or–Qz normative diagram of the studied granitic rocks [48] (over a pressure range from 0.5 kbar to 10 kbar of water-saturated melt) [49]. (**b**) Plots of amphibole composition indicating crystallization in low oxygen fugacity (low f$_{O2}$) magmatic condition [50]. (**c**) Calc-alkaline and alkaline magma composition for the different granite varieties using the contents of FeO^t and Al$_2$O$_3$ in biotite [51]. (**d**) Plots of SiO$_2$ vs. (FeO^t/(FeO^t + MgO) for the Gebel El Bakriyah younger granites show a ferroan nature that confines the A-type granite field [52]. (**e**) Rb vs. Y + Nb tectonic discrimination diagram for the Gebel El Bakriyah granites [45].

7.3. Magma Source and Magmatic vs. Metasomtic Process

The granitic rocks of the Eastern Desert of Egypt show a wide variation in major and trace element concentrations, which strongly suggests that a set of processes and sources were involved in their generation and emplacement [53]. Generally, three major magmatic rock units are present in the Gebel El Bakriyah area including an arc rock association represented by the older granitoids, which are intruded by the fresh post-collisional associations of younger gabbros and the younger granites. The latter are distinguished into (1) monzogranite country rock (side by side with granodiorite of the older granitoids), and (2) syenogranite and alkali feldspar granites forming the BRC.

There are four hypotheses for the petrogenesis of A-type granites as follows: (1) fractional crystallization of lower crustal minerals after partial melting (King 2010 [54]), (2) Mantle-derived mafic magma fractional crystallization [55], (3) lower crustal constituents contaminate mafic mantle-derived melt [47], and (4) melting from the lower crust mixed with some mantle-derived materials [54].

Peraluminous granites can form as a result of fractional crystallization or partial melting of intermediate and mafic rocks or metasedimentary rocks with low total alkali and high Al concentrations, as well as magma mixing and crustal contamination [56,57]. The depletion of Ti, Ba, P, and Eu in the investigated samples of both country rocks and BRC (Figure 5) may support the fractionation of plagioclase and K-feldspar in addition to fractional crystallization of ilmenite and apatite. There is an extensive crystal fractionation in the BRC compared to country rocks, which is evidenced by the enrichment in Rb, Nb, Ta, Zr, Hf, Th, and U in the BRC.

The REEs patterns of the BRC granites show significant degrees of REEs fractionation, with a considerable enrichment in light elements (LREEs) to heavy elements (HREEs) (Figure 5), in addition to a remarkably higher some of the rare-earth elements ($\sum$REEs) compared to the country rock granites. Zircon is spatially associated with thorite, fluorite, and columbite, indicating that these minerals are all formed from a fluid-rich, highly fractionated magma [58]. The Fractional crystallization process in the country rocks and BRC is supported by the plots of the samples on the Th/Nb vs. Zr and Ta vs. Nb diagrams (Figure 8a,b). Normally, the Zr/Hf and Nb/Ta ratios are relatively unchanged in regular magmatic processes, but they significantly change during severe magmatic fractionation when magmatic melt interacted with fluids in particular (e.g., [59]). The BRC granites exhibit a wide range of Nb/Ta ratios (6.11–13.6 and Zr/Hf ratios (16.1–26.7), which indicate fractionation and interaction between fluids and the magmatic melt. The presence of fluorite as a major interstitial accessory mineral in the investigated granites suggests that the magma was F-rich, especially during the late stage of fractional crystallization imposing an impact on the physicochemical characteristics of the felsic magma. Fluorite reduces the solubility of water in magma, as well as its viscosity, and decreases the crystallization temperature in the magma reservoir e.g., [60]. Also, the presence of rare metals, e.g., Nb and Ta, are as used as markers for extensive crystal fractionation in the melt [61]. However, the studied samples exhibit a variable increasing trend of La/Sm and La/Yb ratios versus La abundance (Figure 8c) suggesting that partial melting might have an impact on the evolution of the BRC and its country rocks in the Gebel El Bakriyah area, which is also consistent with the peraluminous nature of the investigated rocks. The role of the fractional crystallization and partial melting in the formation of the country rocks and BRC is supported by the diagram of Ni vs. Rb (Figure 8d). Finally, it can be concluded that fractional crystallization is the most acceptable process for the generation of the Gebel El Bakriyah alkaline rocks. Regardless the limited influence, the role of partial melting cannot be totally discarded, which needs more detailed isotopic evidence in the future hopefully.

The magmatic origin of the studied granitic rocks from El Bakriyah area appears to be mixing of clay-poor, plagioclase-rich, and clay-rich, plagioclase-poor along with amphibolite-rich sources (Figure 9a; [62]), which indicates that the magma originated from a mixture of metasedimentary and igneous sources, including metagraywacke and pelite for the BRC, and slightly mafic-derived melts for the country rocks (Figure 9b). We used some of our mineral chemistry data, particularly those of amphibole and biotite to confirm the source of melt that produced the Gebel El Bakriyah A-type granites. Figure 9c,d suggest post-collisional magma that formed the alkali feldspar granite, and the syenogranite was derived from a crustal source while the monzogranite of the country rocks was derived from crust-mantle mixed source.

Here, we report evidence for a few metasomatic features in the investigated El-Bakriyah granites. They include some hydrothermal alterations by late-stage fluids. In some instances, and based on the petrographic study, graphic texture (Figure 3d) represents the intergrowth of quartz and alkali feldspar formed by a possible metasomatic replacement. Nevertheless, the formation of graphic texture by exclusive eutectic crystallization in proper magmatic conditions cannot be excluded [63]. Isovalence elements have the same radius and charges (CHARC) such as Y/Ho ratios, which are stable in ortho-magmatic melt except might practice metasomatic changes. The Isovalence Y/Ho ratio amounting to 28.1 is very characteristic of chondrites that show a CHARC behavior [32]. The Y/Ho ratio in the BRC granites is higher (average-36.5), which is a non-CHARC character. The non-CHARC character of the investigated El-Bakriyah granites is likely attributed to some metasomatism, which is accepted for highly fractionated magmas exhibiting enrichment of H_2O, Li, F, B, and Cl, which implies a non-CHARC character [64]. Therefore, metasomatism is the best to explain the non-CHARC character induced by an alkaline solution. Primary Mg-Ca-bearing amphiboles (e.g., magneso-arvedsonite) in the Gebel El Bakriyah highly fractionated younger granite are in favor of igneous conditions but some secondary amphiboles enriched in Fe, Na, and K (e.g., ferrobarrosite) indicates a metasomatic origin.

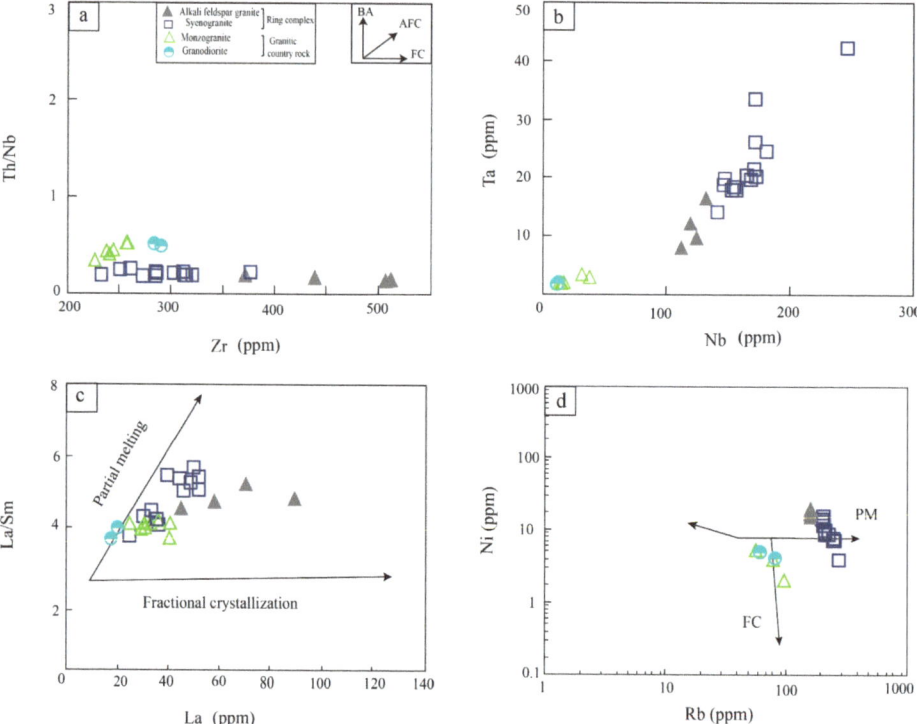

Figure 8. (**a**) Fractional crystallization (FC) and lesser assimilation assignment for the Gebel El-Bakriyah A-type granites based on the Zr vs. Th/Nb diagram [65]. (**b**) Positive correlation of Nb vs. Ta contents as an indicator of fractional crystallization in felsic magma. (**c**) Plots of La vs. La/Sm ratio indicating two processes: namely the partial melting and fractional crystallization for the studied granites (after [66]). (**d**) Ni vs. Rb compositional variation diagram showing fractional crystallization and partial melting trends of the studied granites [67]. (Abbreviations of crystallization trends: FC = Fractional crystallization, AFC = Assimilation-Fractional Crystallization, and BA = Bulk Assimilation (BA).

7.4. Mineralogical Constraints for Petrogenesis

7.4.1. Biotite and Apatite

Fe- and Mg-biotite in the investigated A-type granite are exclusively primary and formed in a proper magmatic condition. None of our biotite analyses plots in the field of re-equilibrated indicating negligible recrystallization during the late magmatic stage [35]. Primary biotite in the Gebel El-Bakriyah A-type granites support the calc-alkaline to alkaline magma sources for the monzogranite country rocks and the two zones of the BRC, respectively (Figure 6c). Content of Ti and the ratio of $Mg/(Mg + Fe^{2+})$ are very sensitive to crystallization temperature in felsic magmas (e.g., [68]. Plots of biotite analyses from the Gebel El-Bakriyah A-type granites on the binary diagram of [68] suggest their crystallization at a temperature between 600 and 700 °C (Figure 9c).

Apatite is essentially a Ca-phosphate mineral that forms in all kinds of rocks in the earth's crust and has a unique structure containing traces of Na, Sr, Fe, Mn, U, Th, Y, and REEs, as well as halogens (F, Cl) and hydroxyl group [69,70]. It is an important accessory mineral in magmatic rocks, which can be used as a petrogenetic indicator in magmatic and hydrothermal systems [71]. On the classification diagram of apatite designed by [72], the majority of apatite analyses from the Gebel El-Bakriyah A-type granite are magmatic (Figure 9d). In addition, apatite is useful to trace the history of magmatic differentiation based on its contents of Mg and Mn [73,74]. According to [74], Mg content in apatite increases proportionally to its concentration in the felsic magma, particularly peraluminous, during magmatic differentiation together with a considerable rise in the Mn content too. Also, the rise of the Mg content in apatite enhances the rate of its crystallization [75].

7.4.2. Nb-Ta Oxides, Zircon, and REEs

In several shield rocks from different ages in the world, some accessory minerals in granitic rocks are very significant petrogenetic indicators, e.g., zircon, Nb-Ta oxides, REE-bearing minerals [76,77]. In nature, particularly the felsic intrusives (e.g., granite and pegmatite), Nb-Ta oxide ore minerals can be found including the columbite-tantalite $(Mn, Fe)Nb_2O_6$ to $(Mn, Fe)Ta_2O_6$ series, the microlite-pyrochlore $(Na, Ca)_2Ta_2O_6(O, OH, F]$ to $(Na, Ca)_2Nb_2O_6(O, OH, F)$ series, and wodginite $[Mn(Sn, Ta, Ti)Ta_2O_8]$ groups [34]. The post-collisional granites in the ANS host some significant concentrations of Nb and Ta [9,78]. Based on our Whole-rock analyses for the Gebel El-Bakriyah highly fractionated granites (Table 1), the concentration of Nb in the BRC ranges from 112.6 to 172.3 ppm and in the country rocks from 7.6–19.2 ppm while the concentration of Ta in the BRC is (8–26.1 ppm) and in the country rocks between (0.05 to 0.4 ppm). The rare-metal granites from the Egyptian Eastern Desert are presumably mica- and Mn-rich [5], which is the case of the investigated BRC that bears Mn-columbite and Mn-tantalite. Zircon is an abundant accessory mineral in granitic rocks which are resistant to weathering processes [76,77]. Neither morphology nor chemical composition of zircon from the Gebel El-Bakriyah younger granites indicates a hydrothermal origin and it is typically magmatic. As we previously mentioned, the whole-rock content of Zr in the BRC and its granitic country rock (averages = 356 ppm and 253 ppm, respectively) are exclusive for proper magmatic condition (T = 816–905 °C, Table 1) based on the calculation of Zr saturation [44,71].

Alkali feldspar granite is characterized by high abundance REE ($\sum$REE = 358–547) than syenogranite ($\sum$REE = 159–285), monzogranite ($\sum$REE = 102–105), and granodiorite ($\sum$REE = 103–119). Syenogranite has enrichment in LREE to HREE (Lan/Lun) = 18.2–39 than alkali feldspar granite (Lan/Lun) = 17.3–22.2 and monzogranite has (Lan/Lun) = 8.02–10.7 and in granodiorite has slightly enrichment in LREE to HREE (Lan/Lun) = 4.2. Eu/Eu* in the investigated granites = 4.03–4.2 except granodiorite Eu/Eu*= 0.27–0.88, which is a function of extensive fractionation of feldspars by magmatic differentiation that stabilizes Eu as Eu^{2+} in reducing condition [79,80]. We present whole-rock analyses (Table 1), which show enrichment of La, Ce, Pr, and Nd in granites constituting the BRC compared with the granitic country rocks. The Tetrad effect (TE) of REEs during the late magmatic stage is meaningful though it is not present in all granites, the ones enriched in fluid-mobile elements such as Li, B, and halogens like F and Cl [81]. The phenomenon of the tetrad effect in the REEs patterns of granite helps to trace either fractional crystallization during differentiation [82] or the interplay of coexisting magma-fluid systems in the late stage of crystallization [82] and this tetrad effect is seen in Figure 9e. We believe that F, evidenced by the presence of interstitial fluorite, plays an important role in the enrichment of REEs in the Gebel El-Bakriyah younger granites. In highly fractionated granites, F influences the enrichment of REEs and HFSE in the form of fluorine-based complexes as the element remains dissolved in the magma until fractionation is ceased and fluorite fills the interstitial spaces in granite [83,84] or forms independent fluorite veins which the case of the Gebel El-Bakriyah area.

7.4.3. Fe-Ti Oxides and Chlorite

Homogeneous Fe-Ti oxides, particularly ilmenite, in the Gebel El Bakriyah A-type granites, alter to secondary titanite in the form of a continuous reaction rim (Figure 3e). From the petrogenetic point of view, this secondary titanite, as well as the sub-graphic rutile-hematite intergrowths after ilmenite, are indicators of late stage deuteric alterations. Mafic and felsic arc-related and post-collisional intrusions in the ANS contain Mn-bearing ilmenite in which Mn increases with the progress of magmatic differentiation [85]. Ilmenite from the Gebel El-Bakriyah A-type granite is Mg-poor (MgO as much as 0.05 wt%) whereas MnO content reaches up to 9.8 wt%. This high Mn content is consistent with the evolution of highly fractionated A-type granites, and the BRC syenogranite and alkali feldspar granite in particular.

Chlorite-group minerals occur as secondary phases, which pseudomorph magmatic biotite. The chemistry of this chlorite is sensitive to temperature and hence the temperature of late stage deuteric alteration can be estimated. In the chlorite structure, occupancy and substitution of Fe, Mg, and Al in the tetrahedral and octahedral sites are common [86,87]. The available chlorite analyses in the Gebel El Bakriyah granitic rocks were possible from the monzogranite country rock and indicate a diabanite composition (Figure 7e). Based on the percentage of Al and its portioning among the octahedral and tetrahedral sites, we used the chlorite geothermometer formulated by [88] to estimate the temperature of alteration numerically. Geothermometric calculations for chlorite in the Gebel El-Bakriyah monzogranite yielded a deuteric alteration temperature of 442 °C.

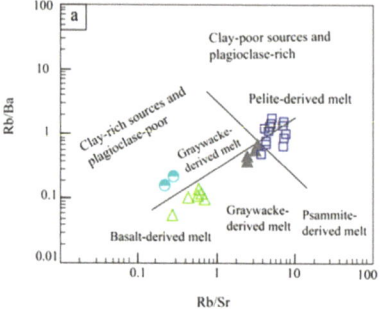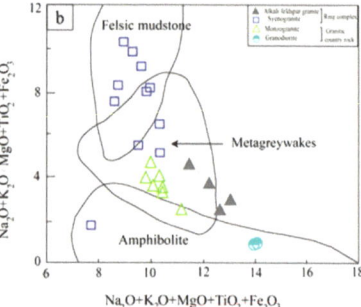

Figure 9. Cont.

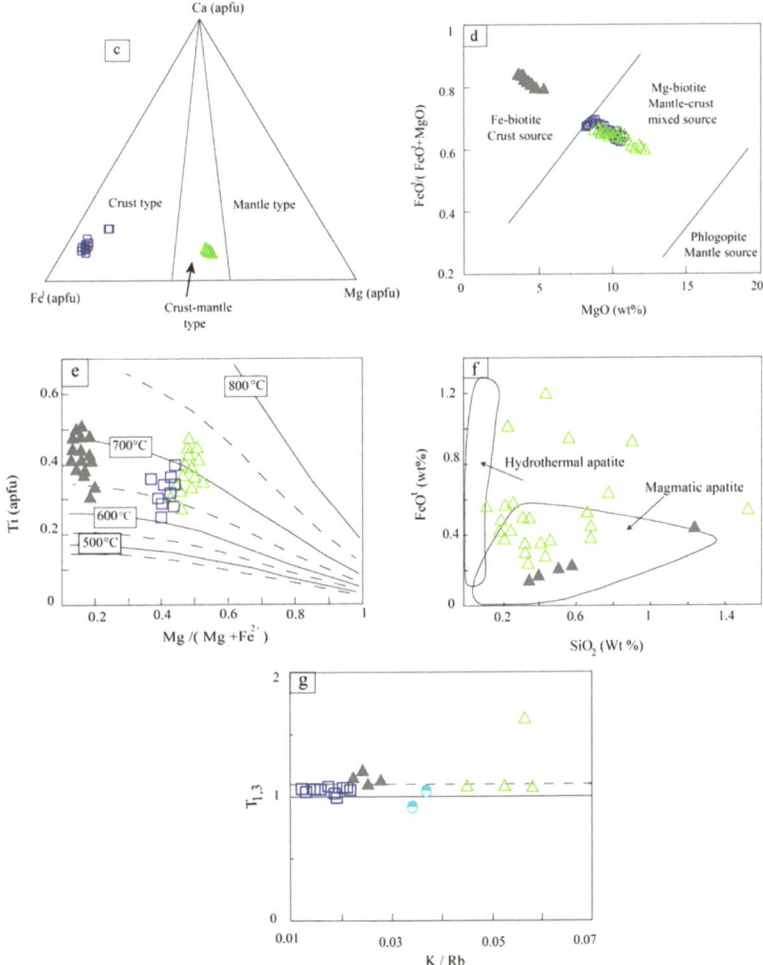

Figure 9. (a) Rb/Ba vs. Rb/Sr for the studied granites [62]. (b) The compositions of the studied granitic rocks, compared to compositional ranges of various experimental metasediment- and amphibolite-derived melts [89]. (c) Crust and mixed crust-mantle source of the Gebel El Bakriyah A-type granite based on amphibole chemistry [90]. (d) Crystallization of Mg-biotite and Fe-biotite from mixed crust-mantle and crust source for the monzogranite + syenogranite and alkali feldspar granite, respectively [91]. (e) The crystallization temperature of biotite in the Gebel El Bakriyah A-type granite [68]. (f) Magmatic origin of apatite from the investigated granites [72]. (g) Lanthanide tetrad effect in some granite varieties [64].

7.5. Geodynamic Model

The ANS represents one of the best examples of Precambrian (Neoproterozoic) juvenile crusts in the world [22,92]. Granitic rocks are major components in the Eastern Desert of Egypt and they represent the emplacement of different magma compositions from the oceanic spread until the post-collisional continental extension. There are several researchers who determined age of different types of granitic rocks in Egypt, particularly in the central part where Gebel El Bakriyah area is located. For example, [93] used Rb/Sr isochron to determine age of emplacement of granodiorite in the adjacent Wadi El-Miyah area, which yielded the age 674 ± 13 Ma contemporaneous with the volcanic arc stage. Also, 614 ± 8 Ma U/Pb zircon was assigned as the age of similar granodiorite and tonalite intrusions, e.g., in Abu Ziran area. This gives an indication of wide time span of arc formation in the central Eastern Desert up to 60 Ma or alternatively there were more than arc. Some other granodiorite intrusions at the Homr Akarem and Homret Mikpid in the southern part of the Eastern Desert were emplaced at 630–620 Ma [94] and at 643 ± 9 Ma for the Um Rus tonalite-granodiorite intrusion [95], and the youngest phase (i.e., 620 Ma) was believed to represent the transition between arc to typical anorogenic setting similar to the case of Gebel El Bakriyah area [96]. On the other hand, typical anorogenic or post-colllisional A-type granite intrusions in the central Eastern Desert of Egypt, e.g., at Um Had area is assigned a U–Pb zircon age of 590–3.1 Ma [97]. Generally, felsic magmatism in the Eastern Desert that produced post-collisional A-type granites, e.g., El-Missikat, Abu Harba, and Gattar yielded single crystal zircon age of ~600 Ma [98].

Bentor [99] divided granites of the Arabian-Nubian Shield into two categories: an older syn- to late-orogenic granites (880–610 Ma), and younger post-orogenic to anorogenic granite (600–475 Ma). Ages determined for the older granites from the Egyptian Eastern Desert are less than 750 Ma [100]. The Egyptian late- to post-tectonic younger granites were formed between 600 and 550 Ma, [93] or 600 and 475 Ma [94]. On the other hand, the 635–580 Ma or 610 and 590 Ma ages are used to distinguish the emplacement of the Egyptian post collisional younger granite [101]. They have been emplaced as two separate suites, although substantially overlapping calc-alkaline and alkaline pulses at 635–590 Ma and 608–580 Ma, respectively [102]. The BRC is located in the central part of the Egyptian Eastern Desert, and it is a part of the northern segment of the Arabian-Nubian Shield that belongs to the third stage (i.e., the post-collisional) in most. And this is characterized by alkaline magmatism including A-type granites. According to [103], a Rb-Sr isochron assigned 520–506 Ma age of Gebel El Bakriyah granites, particularly the BRC.

The younger granites in the Gebel El Bakriyah area formed during the post-collisional stage, i.e., emplacement was controlled by within-plate tectonics in which lithospheric delamination took place to generate mafic and felsic magma batches [24]. Based on our present field and laboratory data, we assume that the formation of the Gondwana assembly during the last stage of subduction (Figure 10a) marked the beginning of the collisional phase, which is marked the first stage of the evolution of El Bakriyah granites represented by granodiorite country rock. This is resulted by the dehydration of the subducted plate comparatively at low pressure, which made it easier for the upper mantle to melt and penetrate the continental crust, forming calc-alkaline volcanic arc magma. The orogenic activity peaked as a result of the water supply being cut off during the transitional period that started when subduction reached its warning stage. This phase denotes the post-orogenic transition from the calc-alkaline to the alkaline stage, despite the anorogenic regime being relaxed and represented by the evolution of monzogranite rocks. We believe that the BRC resulted from decompression melting of crustal source material followed by break-off and delamination of the lithosphere, which was accepted for vast areas in the Eastern Desert of Egypt (e.g., [104]). Mafic melts were produced by melting the residual lithosphere by high heat flux during the transition from the compressional to the extensional phase, and the lithospheric delamination played a significant part in this evolution. The final stage of the evolution of the Gebel El Bakriyah granites is the post-collisional within-plate stage (Figure 10b). When magma from the mantle ascended along the fractured crust during tension and relaxation periods. It causes a relatively greater supply of crustal alkaline magmas to partially melt, which is represented by two successive evolution of syenogranite and alkali feldspar granite.

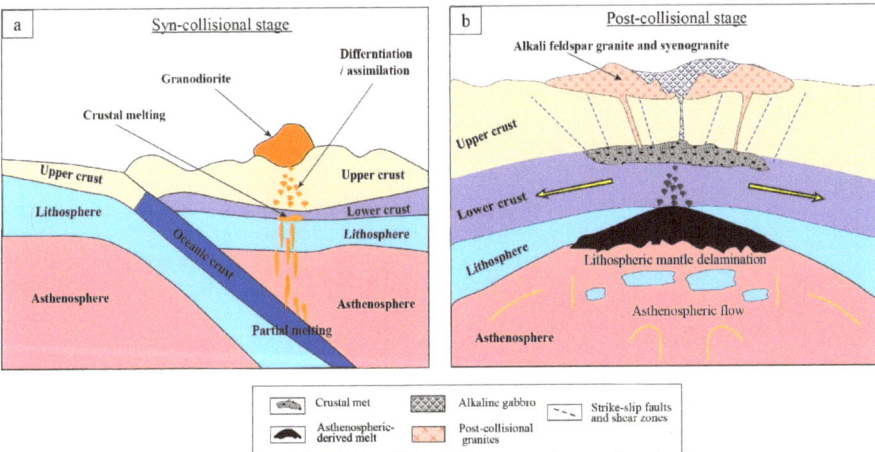

Figure 10. A suggested geodynamic model for the generation of A-type magma by dehydration melting of lithospheric delamination in an extensional tectonic regime during the Late Neoproterozoic in the Arabian-Nubian Shield (ANS). The Gebel El Bakriyah complex witnessed two successive post-collisional magmatic episodes, the first is mafic that formed the younger gabbros [24] and the A-type granites (present study).

Our geochemical data are in favor of progressive magmatic differentiation, fractionation, and little assimilation to form the Gebel El Bakriyah A-type younger granites along the fracture system and this extends to a proper anorogenic/within-plate environment. Other parts in the ANS have A-type younger granites that experienced high fractionation, mostly from a metasedimentary source rock and not tonalite, which is the case of the BRC. Therefore, it is believed that the fractures of the extensional tectonics were shear zones that facilitated the upwelling of two independent suites of calc-alkaline and alkaline felsic magmatism. Similar models have been proposed for the generation of A-type granites in the ANS either in the form of ring complexes or not e.g., [5,13]. A-type younger granites formed as post-collisional derivatives of the delaminated mantle in extensional tectonics form similar ring complexes, e.g., those at Jabal Al-Hassir [93] and Jabal Aja [4] in the Arabian Shield of Saudi Arabia. These ring complexes are parts of Caulderon subsidence structures and calderas above the felsic plutons in the western Arabian Peninsula (i.e., the eastern segment of ANS), and represent the transition from convergent to extensional tectonics [105].

Mineralogy and geochemistry of the Gebel El Bakriyah younger granites provide some evidence for F-and Ba-rich fluids during the late stage of felsic magma fractionation. Interstitial fluorite and fluorite veins are strong evidence for F and Ba enrichment in the felsic injection of dominating alkaline composition in the post-collisional environment at high temperature as low as 830 °C [44,83]. Fluorite in similar A-type granites from the ANS (e.g., [106,107] and others) confirm our observations and conclusion that the F-complexes of REEs, U, and Th, are very characteristic for the highly fractionated granites like the case of Gebel El Bakriyah younger A-type granites. Finally, A-type younger granites in the entire ANS form by delamination of the lithospheric lower crust in different ages that can be as old as Cryogenian (e.g., 686 Ma; [105] or Ediacaran (e.g., 634 Ma to 580 Ma; [92]) or down to 550 Ma [21]. Gebel El Bakriyah A-type granites are peculiar Ediacaran intrusions, which are formed by high-T melting from a crustal source (Cryogenian crust, possibly Tonian too) that fractionated during upwelling along fault planes/shear zone in a typical extensional regime of the Neoproterozoic post-collisional phase of the ANS evolution. The Gebel El Bakriyah A-type granites are equivalent to other calc-alkaline peraluminous granites in the ANS (e.g., [4,21]. In other shield terranes of Neoproterozoic age similar to the ANS, e.g., the Brasiliano, A-type magmatism at ~630 Ma ago is comparable to the Gebel El Bakriyah case. Recycled/reworked ancient oceanic and arc-related crustal materials in the asthenosphere leads to the formation of alkaline magmatism [108–111], and hence it is a potential source of upwelling melt to form A-type granites in a proper post-collisional extension, i.e., within-plate.

8. Conclusions

(1) The Gebel El Bakriyah younger granites comprise a prominent ring complex (BRC) consisting of syenogranite core and alkali feldspar rim. These two granite varieties pertain to typical A-type characteristics. The ring complex was emplaced in a within-plate setting whereas the monzogranite represents a transition between the arc and anorogenic settings.

(2) A calc-alkaline composition is assigned for the three younger granite varieties. All of them have the high-K signature of peraluminous melts that were emplaced in the form of their independent pulses, one produced the monzogranite country rocks and two successive ones formed the syenogranite and alkali feldspar granite, respectively.

(3) The three varieties of younger granites of Ediacaran age in the Gebel El Bakriyah area show enrichment of rare metals (Mn-rich columbite-tantalite), F and Fe, i.e., ferroan granites, which are highly fractionated. This resulted in frequent interstitial fluorite in the granites as well as the formation of excavated fluorite-rich quartz veins.

(4) From the geodynamic point of view, the Gebel El Bakriyah younger granites formed by high-T dehydration melting of a mixed crust-mantle source dominated by metasediments and amphibolite, i.e., delamination of the lithospheric crust. This is followed by high fractionation and upwelling of three independent felsic magma pulses along faults in an extensional tectonic regime.

(5) The BRC and its monzogranite country rocks are peculiar examples of Neoproterozoic (Ediacaran) post-collisional magmatism comparable to those in other Precambrian Shields in the world.

(6) Although most of the fluorite and barite veins are excavated, there are still more exploration efforts that must be carried out to maximize the potentiality of critical materials in the area, including the Nb-Ta resources.

Supplementary Materials: The following supporting information can be downloaded at: https://www.mdpi.com/article/10.3390/min13101273/s1, Supplementary File S1: Whole rock geochemistry and mineral chemistry data [112].

Author Contributions: Conceptualization, A.A.A.E.-F., A.A.S. and A.A.M.; methodology, A.A.A.E.-F.; software, A.A.A.E.-F.; validation, A.A.A.E.-F., A.A.S., M.K.A. and A.A.M.; formal analysis, A.A.A.E.-F. and M.K.A.; investigation, A.A.A.E.-F., A.A.S., M.K.A. and A.A.M.; resources, A.A.A.E.-F.; data curation, A.A.A.E.-F. and A.A.S.; writing—original draft preparation, A.A.A.E.-F.; writing—review and editing, A.A.A.E.-F., A.A.S. and A.A.M.; visualization, A.A.S. and A.A.M.; supervision, A.A.S., M.K.A. and A.A.M.; project administration, A.A.S.; funding acquisition, M.K.A. All authors have read and agreed to the published version of the manuscript.

Funding: Internal Project No. 13010312 of The National Research Centre of Egypt entitled "Evaluation of alkaline ring complexes as a potential source of some strategic mineralization".

Data Availability Statement: Data are available upon request to the corresponding author.

Acknowledgments: The present paper is part of MSc Thesis submitted by the first author to the Faculty of Science, Cairo University. The authors would like to thank Cairo University for the logistics, which enabled us to have field trips and preparation of thin-, polished- and polished-thin sections and pulverization of samples for chemical analyses. Hesham Mokhtar is acknowledged for his patience and sincere revision during the preparation of the manuscript. We also thank Abdallah Atef for his assistance in the field. We would like to thank the editor and three anonymous reviewers for their patience and very fruitful discussion that improved quality of the paper.

Conflicts of Interest: The authors declare no conflict of interest.

References

1. El-Bialy, M.Z.; Omar, M.M. Spatial association of Neoproterozoic continental arc I-type and post-collision A-type granitoids in the Arabian-Nubian Shield: The Wadi Al-Baroud older and younger granites, North Eastern Desert, Egypt. *J. Afr. Earth Sci.* **2015**, *103*, 1–29. [CrossRef]
2. El-Bialy, M.Z. Precambrian Basement Complex of Egypt. In *The Geology of Egypt*; Regional Geology Reviews; Hamimi, Z., El-Barkooky, A., Martínez Frías, J., Fritz, H., Abd El-Rahman, Y., Eds.; Springer: Cham, Switzerland, 2020; pp. 37–79.
3. Abd El-Naby, H.H. The Egyptian granitoids: An up-to-date synopsis. In *Geology of the Egyptian Nubian Shield*; Regional Geology Reviews; Hamimi, Z., Arai, S., Fowler, A.R., El-Bialy, M.Z., Eds.; Springer: Cham, Switzerland, 2021; pp. 239–265.
4. Abuamarah, B.A. Genesis and petrology of post-collisional rare-metal-bearing granites in the Arabian Shield: A case study of Aja ring complex, northern Saudi Arabia. *J. Geol.* **2020**, *128*, 131–156. [CrossRef]
5. Moussa, H.E.; Asimow, P.D.; Azer, M.K.; Maaty, M.A.A.; Adel, I.M.; Yanni, N.N.; Mubarak, H.S.; Wilner, M.J.; Elsagheer, M.A. Magmatic and hydrothermal evolution of highly fractionated rare-metal granites at Gabal Nuweibi, Eastern Desert, Egypt. *Lithos* **2021**, *400–401*, 106405. [CrossRef]
6. Hassaan, M.M.; Desoky, E.H. Granites in the tectonic environs of the Nubian Shield, Egypt: Geochemical characterization and new contributions. *Curr. Res. Earth Sci.* **2016**, *10*, 59–103.
7. Lundmark, A.M.; Andresen, A.; Hassan, A.; Augland, L.E.; Abu El-Rus, M.A.; Boghdady, G.Y. Repeated magmatic pulses in the East African Orogen of Central Eastern Desert, Egypt: An old idea supported by new evidence. *Gondwana Res.* **2012**, *22*, 227–237. [CrossRef]
8. Hargrove, U.S.; Stern, R.J.; Griffin, W.R.; Johnson, P.R.; Abdelsalam, M.G. From island arc to craton: Timescales of crustal formation along the Neoproterozoic Bir Umq Suture zone, Kingdom of Saudi Arabia. In *Saudi Geological Survey*; Technical Report SGS-TR; Saudi Geological Survey: Jeddah, Saudi Arabia, 2006; 69p.
9. Azer, M.K.; Abdelfadil, K.M.; Asimow, P.D.; Khalil, A.E. Tracking the transition from subduction-related to post-collisional magmatism in the north Arabian–Nubian Shield: A case study from the Homrit Waggat area of the Eastern Desert of Egypt. *Geol. J.* **2020**, *55*, 4426–4452. [CrossRef]
10. Eby, G.N. Chemical subdivisions of the A-type granitoids: Petrogenesis and tectonic implications. *Geology* **1992**, *20*, 641–644. [CrossRef]
11. Robinson, F.A.; Bonin, B.; Pease, V.; Anderson, J.L. A discussion on the tectonic implications of Ediacaran late- to post-orogenic A-type granite in the northeastern Arabian Shield, Saudi Arabia. *Tectonics* **2017**, *36*, 582–600. [CrossRef]
12. Heikal, M.T.S.; Khedr, M.Z.; El-Monesf, M.A.; Gomaa, S.R. Petrogenesis and geodynamic evolution of Neoproterozoic Abu Dabbab albite Granite, Central Eastern Desert of Egypt: Petrological and geochemical constraints. *J. Afr. Earth Sci.* **2019**, *158*, 103518. [CrossRef]
13. Abuamarah, B.A.; Azer, M.K.; Seddik, A.M.A.; Asimow, P.D.; Guzman, P.; Fultz, B.T.; Wilner, M.J.; Dalleska, N.; Darwish, M.H. Magmatic and post-magmatic evolution of post-collisional rare-metal bearing granite: The Neoproterozoic Homrit Akarem granitic intrusion, southeastern Desert of Egypt, Arabian-Nubian shield. *Geochemistry* **2022**, *82*, 125840. [CrossRef]
14. Sami, M.; Ntaflos, T.; Farahat, E.S.; Mohamed, H.A.; Hauzenberger, C.; Ahmed, A.F. Petrogenesis and geodynamic implications of Ediacaran highly fractionated A-type granitoids in the north Arabian-Nubian Shield (Egypt): Constraints from whole-rock geochemistry and Sr-Nd isotopes. *Lithos* **2018**, *304–307*, 329–346. [CrossRef]
15. Loiselle, M.; Wones, D. Characteristics and origin of anorogenic granites. In *Abstracts with Programs*; Geological Society of America: Boulder, CO, USA, 1979; Volume 11, p. 468.
16. Moreno, J.A.; Montero, P.; Abu Anbar, M.; Molina, J.F.; Scarrow, J.H.; Talavera, C.; Cambeses, A.; Bea, F. SHRIMP U-Pb zircon dating of the Katerina ring complex: Insights into the temporal sequence of Ediacaran calc-alkaline to per-alkaline magmatism in southern Sinai, Egypt. *Gond. Res.* **2012**, *12*, 887–900. [CrossRef]
17. Mahmoud, M.S. Geological and Geochemical Studies on the Rocks of Gebel El-Hisinat Area, Central Eastern Desert, Egypt. Master's Thesis, Assuit University, Assuit, Egypt, 1984; 109p.
18. El-Sayed, M.M.; Mohamed, F.H.; Furnes, H. Petrological and geochemical constraints on the evolution of late Pan-African Bakriya post-orogenic ring complex, Central Eastern Desert, Egypt. *Neues Jahrb. Mineral. Ab-Handl.* **2004**, *180*, 1–32. [CrossRef]
19. Saleeb-Roufaiel, G.S.; Samuel, M.D.; Hilmy, M.E.; Moussa, H.E. Fluorite mineralization at El-Bakriya, Eastern Desert of Egypt, Egypt. *J. Geol.* **1982**, *26*, 9–18.
20. Eyal, M.; Litvinovsky, B.; Jahn, B.M.; Zanvilevich, A.; Katzir, Y. Origin and evolution of post-collisional magmatism: Coeval Neoproterozoic calc-alkaline and alkaline suites of the Sinai Peninsula. *Chem. Geol.* **2010**, *269*, 153–179. [CrossRef]
21. Be'eri-Shlevin, Y.; Samuel, M.D.; Azer, M.K.; Rämö, O.T.; Whitehouse, M.J.; Moussa, H.E. The late Neoproterozoic Ferani and Rutig volcano-sedimentary successions of the northernmost Arabian–Nubian Shield (ANS): New insights from zircon U-Pb geochronology, geochemistry and O–Nd isotope ratios. *Precambrian Res.* **2011**, *188*, 21–44. [CrossRef]
22. Stern, R.J. Arc assembly and continental collision in the Neoproterozoic East African Orogen: Implications for the consolidation of Gondwanaland. *Annu. Rev. Earth Planet. Sci.* **1994**, *22*, 319–351. [CrossRef]
23. Abdelnasser, A. Genesis of the Gold Mineralization at Atud Area, Central Eastern Desert of Egypt: Geological, Ore Mineralogical and Geochemical Approaches. Ph.D. Thesis, Istanbul Technical University, Istanbul, Turkey, 2016.

24. Azer, M.K.; Surour, A.A.; Madani, A.A.; Ren, M.; Abd El-Fatah, A.A. Mineralogical and geochemical constraints on the post-collisional mafic magmatism in the Arabian-Nubian Shield: An example from the El-Bakriya Area, Central Eastern Desert, Egypt. *J Geol.* **2022**, *130*, 209–230. [CrossRef]
25. Abd El-Fatah, A.A.; Surour, A.A.; Madani, A.A.; Azer, M.K. Integration of Landsat-8 and reflectance spectroscopy data for the mapping of Late Neoproterozoic igneous ring complexes in an arid environment: A case study of the Gebel El-Bakriyah area, Eastern Desert, Egypt. *J. Min. Environ.* **2023**, *14*, 13–31.
26. El-Amin, H. Radiometric and Geological Investigations of El Bakriya Area, Eastern Desert, Egypt. Ph.D. Thesis, Cairo University, Cairo, Egypt, 1975; 224p.
27. Bence, A.E.; Albee, A.L. Empirical correction factors for the electron microanalysis of silicates and oxides. *J. Geol.* **1968**, *76*, 382–403. [CrossRef]
28. Middlemost, E.A.K. Magmas and magmatic rocks: An introduction to igneous petrology. *Geol. Mag.* **1985**, *123*, 87–88.
29. Streckeisen, A. Each plutonic rock has its proper name. *Earth-Sci. Rev.* **1976**, *12*, 1–33. [CrossRef]
30. Maniar, P.D.; Piccoli, P.M. Tectonic discrimination of granitoids. *Geol. Soc. Am. Bull.* **1989**, *101*, 635–643. [CrossRef]
31. Evensen, N.M.; Hamilton, P.J.; Onions, R.K. Rare-earth abundances in chondritic meteorites. *Geochim. Cosmochim. Acta* **1978**, *42*, 1199–1212. [CrossRef]
32. Sun, S.S.; McDonough, W.E. Chemical and isotopic systematics of oceanic basalts: Implications for mantle composition processes. In *Magmatism in the Ocean Basins. Geological Society*; Saunders, A.D., Norry, M.J., Eds.; Special Publications: London, UK, 1989; pp. 313–345.
33. Richard, L.R. *MinPet: Mineralogical and Petrological Data Processing System*, version 2.02; MinPet Geological Software: Québec, QC, Canada, 1995.
34. Deer, W.A.; Howie, R.A.; Zussman, J. *An Introduction to Rock-Forming Minerals*, 2nd ed.; Longman: Harlow, UK, 1992; 696p.
35. Deer, W.A.; Howie, R.A.; Zussman, J. *An Introduction to the Rock-Forming Minerals*, 1st ed.; Longman Scientific and Technical Publishing: Harlow, UK, 1966; 528p.
36. Nachit, H.; Ibhi, A.; Ohoud, M.B. Discrimination between primary magmatic biotites, re-equilibrated biotites, and neoformed biotites. *Comptes Rendus Geosci.* **2005**, *337*, 1415–1420. [CrossRef]
37. Leak, B.E.; Woolley, A.R.; Arps, C.E.S.; Birch, W.D.; Gilbert, M.C.; Grice, J.D.; Hawthorne, F.C.; Kato, A.; Kisch, H.J.; Krivovichev, V.G.; et al. Nomenclature of Amphiboles: Report of the sub-committee on amphiboles of the international mineralogical association, Commission on new minerals and mineral names. *Mineral. Mag.* **1997**, *61*, 295–310. [CrossRef]
38. Hey, M.H. A new review of chlorites. *Mineral. Mag.* **1954**, *30*, 277–292. [CrossRef]
39. Basak, A.; Goswami, B. The physico-chemical conditions of crystallization of the Grenvillian arfvedsonite granite of Dimra Pahar, Hazaribagh, India: Constraints on possible source regions. *Mineral. Petrol.* **2020**, *114*, 329–356. [CrossRef]
40. Watson, E.B.; Harrison, T.M. Zircon saturation revisited: Temperature and composition effects in a variety of crustal magma types. *Earth Planet. Sci. Lett.* **1983**, *64*, 295–304. [CrossRef]
41. Machev, P.; Klain, L.; Hecht, L. Mineralogy and geochemistry of biotites from the Belogradchik pluton—Some petrological implications for granitoid magmatism in north-west Bulgaria: Bulgarian Geological Society, Annual Scientific Conference of the Bulgarian Geological Society. *Geology* **2004**, *16–17*, 48–50.
42. Jayasuriya, K.D.; O'Neill, H.S.C.; Berry, A.; Campbell, S.J. A Mössbauer study of the oxidation state of Fe in silicate melts. *Am. Min.* **2004**, *89*, 1597–1609. [CrossRef]
43. Haggerty, S.E. Opaque mineral oxides in terrestrial igneous rocks. *Mineral. Soc. Am.-Short Course Notes* **1976**, *3*, 101–300.
44. King, P.L.; Chappell, B.W.; Allen, C.M.; White, A.J.R. Are A-type granites the high-temperature felsic granites? Evidence from fractionated granites of the Wangrah Suite. *Aust. J. Earth Sci.* **2010**, *48*, 501–514. [CrossRef]
45. Pearce, J.A.; Harris, N.B.W.; Tindle, A.G. Trace element discrimination diagrams for the tectonic interpretation of granitic rocks. *J. Petrol.* **1984**, *25*, 956–983. [CrossRef]
46. Chappell, B.W.; Bryant, C.J.; Wyborn, D. Peraluminous I-type granites. *Lithos* **2012**, *153*, 142–153. [CrossRef]
47. Whalen, J.B.; Frost, C. The Q-ANOR diagram: A tool for the petrogenetic and tectonomagmatic characterization of granitic suites. In Proceedings of the South-Central Section, 47th Annual Meeting, Austin, TX, USA, 4–5 April 2013; Geological Society of America: Boulder, CO, USA, 2013; Volume 7.
48. Manning, D.A.C. The effect of fluorine on liquidus phase relationships in the system Qz-Ab-Or with excess water at 1 kb. *Contrib. Mineral. Petrol.* **1981**, *76*, 206–215. [CrossRef]
49. Holtz, F.; Johannes, W.; Pichavant, M. Effect of excess aluminium on phase relations in the system Qz-Ab-Or. Experimental investigation at 2 Kbar and reduced H_2O activity. *Eur. J. Mineral.* **1992**, *4*, 137–152. [CrossRef]
50. Anderson, J.L.; Smith, D.R. The effects of temperature and fO2 on the Al-in-hornblende barometer. *Am. Mineral.* **1995**, *80*, 549–559. [CrossRef]
51. Abdel-Rahman, A.M. Nature of biotites from alkaline, calc-alkaline, and peraluminous magmas. *J. Petrol.* **1994**, *35*, 525–541. [CrossRef]
52. Frost, B.R.; Barnes, C.G.; Collins, W.J.; Arculus, R.J.; Ellis, D.J.; Frost, C.D. A geochemical classification for granitic rocks. *J. Petrol.* **2001**, *42*, 2033–2048. [CrossRef]
53. Seddik, A.M.A.; Darwish, M.H.; Azer, M.K.; Asimow, P.D. Assessment of magmatic versus post-magmatic processes in the Mueilha rare-metal granite, Eastern Desert of Egypt, Arabian-Nubian Shield. *Lithos* **2020**, *366–367*, 105542. [CrossRef]

54. Kerr, A.; Fryer, B.J. Nd isotope evidence for crust-mantle interaction in the generation of A-type granitoid suites in Labrador, Canada. *Chem. Geol.* **1993**, *104*, 39–60. [CrossRef]
55. Jarrar, G.H.; Manton, W.I.; Stern, R.J.; Zachmann, D. Late Neoproterozoic A-type granites in the northernmost Arabian-Nubian Shield formed by fractionation of basaltic melts. *Chem. Erde Geochem.* **2008**, *68*, 295–312. [CrossRef]
56. Sylvester, P.J. Post-collisional alkaline granites. *J. Geol.* **1989**, *97*, 261–280. [CrossRef]
57. Sisson, T.W.; Ratajeski, K.; Hankins, W.B.; Glazner, A.F. Voluminous granitic Desert, Egypt. *Sci. J. Fac. Sci. Minufia Univ.* **2005**, *15*, 107–129.
58. Sami, M.; Ntaflos, T.; Farahat, E.S.; Mohamed, H.A.; Ahmed, A.F.; Hauzenberger, C. Mineralogical, geochemical and Sr-Nd isotopes characteristics of fluorite-bearing granites in the Northern Arabian-Nubian Shield, Egypt: Constraints on petrogenesis and evolution of their associated rare metal mineralization. *Ore Geol. Rev.* **2017**, *88*, 1–22. [CrossRef]
59. Ballouard, C.; Poujol, M.; Boulvais, P.; Branquet, Y.; Tartèse, R.; Vigneresse, J.L. Nb-Ta fractionation in per-aluminous granites: A marker of the magmatic-hydrothermal transition. *J. Geol.* **2016**, *44*, 231–234. [CrossRef]
60. Dingwell, D.B. The structures and properties of fluorine-rich magmas: A review of experimental studies. *Can. Inst. Min. Metall. Pet.* **1988**, *39*, 1–12.
61. Wang, R.C.; Wu, F.Y.; Xie, L.; Liu, X.C.; Wang, J.M.; Yang, L.; Lai, W.; Liu, C. A preliminary study of rare-metal mineralization in the Himalayan leucogranite belts, South Tibet. *Sci. China Earth Sci.* **2017**, *60*, 1655–1663. [CrossRef]
62. Chappell, B.W. Aluminium saturation in I-and S-type granites and the characterization of fractionated haplogranites. *Lithos* **1999**, *46*, 535–551. [CrossRef]
63. Khalil, A.E.S.; Obeid, M.A.; Azer, M.K.; Asimow, P.D. Geochemistry and petrogenesis of post-collisional alkaline and peralkaline granites of the Arabian-Nubian Shield: A case study from the southern tip of the Sinai Peninsula, Egypt. *Int. Geol. Rev.* **2018**, *60*, 998–1018. [CrossRef]
64. Irber, W.; Förster, H.J.; Hecht, L.; Möller, P.; Morteani, G. Experimental, geochemical, mineralogical and O-isotope constraints on the late-magmatic history of the Fichtelgebirge granites (Germany). *Int. J. Earth Sci.* **1997**, *86*, 110–124. [CrossRef]
65. Nicolae, I.; Saccani, E. Petrology and geochemistry of the Late Jurassic calc-alkaline series associated to Middle Jurassic ophiolites in the South Apuseni Mountains (Romania). *Swiss J. Geosci.* **2003**, *83*, 81–96.
66. De Souza, Z.S.; Martin, H.; Peucat, J.J.; Jardim de Sá, E.F.; de Freitas Macedo, M.H. Calc Alkaline Magmatism at the Archean-Proterozoic Transition: The Caicoó Complex Basement (NE Brazil). *J. Petrol.* **2007**, *48*, 2149–2185. [CrossRef]
67. Schiano, P.; Monzier, M.; Eissen, J.P. Simple mixing as the major control of the evolution of volcanic suites in the Ecuadorian Andes. *Contrib. Mineral. Petrol.* **2010**, *160*, 297–312. [CrossRef]
68. Henry, D.J.; Guidotti, C.V.; Thomson, J.A. The Ti-saturation surface for low- to medium-pressure metapelitic biotites: Implications for geothermometry and Ti-substitution mechanisms. *Am. Min.* **2005**, *90*, 316–328. [CrossRef]
69. Piccoli, P.M.; Candela, P.A. Apatite in igneous systems. In *Phosphates: Geochemical, Geobiological, and Materials Importance*; Reviews in Mineralogy and Geochemistry; GeoScienceWorld: McLean, VA, USA, 2002; Volume 48, pp. 255–292.
70. Webster, J.D.; Piccoli, P.M. Magmatic apatite: A powerful, yet deceptive, mineral. *Elements* **2015**, *11*, 177–182. [CrossRef]
71. Zeng, L.P.; Li, X.F.; Hu, H.; McFarlane, C. In situ elemental and isotopic analysis of fluorapatite from the Taocun magnetite-apatite deposit, Eastern China: Constraints on fluid metasomatism. *Am. Min.* **2016**, *101*, 2468–2483. [CrossRef]
72. Chen, L.; Yan, Z.; Wang, Z.Q.; Wang, K. Characteristics of Apatite from 160–140 Ma Cu (Mo) and Mo (W) Deposits in East Qinling. *Geol. Acta* **2017**, *91*, 1925–1941.
73. Miles, A.J.; Graham, C.M.; Hawkesworth, C.J.; Gillespie, M.R.; Hinton, R.W.; Bromiley, G.D. Apatite: A new redox proxy for silicic magmas. *Geochim. Cosmochim. Acta* **2014**, *132*, 101–119. [CrossRef]
74. Nathwani, C.L.; Loader, M.A.; Wilkinson, J.J.; Buret, Y.; Sievwright, R.H.; Hollings, P. Multi-stage arc magma evolution recorded by apatite in volcanic rocks. *Geology* **2020**, *48*, 323–327. [CrossRef]
75. Prowatke, S.; Klemme, S. Trace element partitioning between apatite and silicate melts. *Geochim. Cosmochim. Acta* **2006**, *70*, 4513–4527. [CrossRef]
76. Hoskin, P.W.O.; Schaltegger, U. The composition of zircon and igneous and metamorphic petrogenesis. *Rev. Mineral. Geochem.* **2003**, *53*, 27–62. [CrossRef]
77. Erdmann, S.; Wodicka, N.; Jackson, S.E.; Corrigan, D. Zircon textures and composition refractory recorders of magmatic volatile evolution. *Contrib. Mineral. Petrol.* **2013**, *165*, 45–71. [CrossRef]
78. Abuamarah, B.A.; Azer, M.K.; Asimow, P.D.; Shi, Q. Petrogenesis of the post-collisional rare-metal-bearing Ad-Dayheen granite intrusion, Central Arabian Shield. *Lithos* **2019**, *384–385*, 105956. [CrossRef]
79. McKay, G.A. Partitioning of rare earth elements between major silicate minerals and basaltic melts. In *Geochemistry and Mineralogy of Rare Earth Elements*; Lipin, B.R., McKay, G.A., Eds.; De Gruyter: Berlin, Germany; Boston, MA, USA, 1989; pp. 45–78.
80. Lee, S.G.; Asahara, Y.; Tanaka, T.; Lee, S.R.; Lee, T. Geochemical significance of the Rb-Sr, La-Ce, and Sm-Nd isotope systems in A-type rocks with REE tetrad patterns and negative Eu and Ce anomalies: The Cretaceous Muamsa and Weolaksan granites, South Korea. *Geochemistry* **2013**, *73*, 75–88. [CrossRef]
81. London, D. The application of experimental petrology to the genesis and crystallization of granitic pegmatites. *Can. Mineral.* **1992**, *30*, 499–540.

82. Monecke, T.; Kempe, U.; Monecke, J.; Sala, M.; Wolf, D. Tetrad effect in rare earth element distribution patterns: A method of quantification with application to rock and mineral samples from granite-related rare metal deposits. *Geochim. Cosmochim. Acta* **2002**, *66*, 1185–1196. [CrossRef]
83. Abdel-Rahman, A.M.; El-Kibbi, M.M. Anorogenic magmatism: Chemical evolution of the Mount El-Sibai A-type complex (Egypt), and implications for the origin of within-plate felsic magmas. *Geol. Mag.* **2001**, *138*, 67–85. [CrossRef]
84. Agangi, A.; Kamenetsky, V.S.; McPhie, J. The role of fluorine in the concentration and transport of lithophile trace elements in felsic magmas: Insights from the Gawler Range Volcanics, South Australia. *Chem. Geol.* **2010**, *273*, 314–325. [CrossRef]
85. Surour, A.A.; Ahmed, A.H.; Harbi, H.M. Mineral chemistry as a tool for understanding the petrogenesis of Cryo-genian (arc-related) Ediacaran (post-collisional) gabbros in the western Arabian Shield of Saudi Arabia. *Int. J. Earth Sci.* **2017**, *106*, 1597–1617. [CrossRef]
86. Laird, J. Chlorites: Metamorphic petrology. In *Hydrous Phyllosilicates. (Exclusive of Micas)*; Reviews in Mineralogy; Bailey, S.W., Ed.; Walter de Gruyter GmbH & Co KG.: Berlin, Germany, 1988; pp. 405–453.
87. Inoue, A.; Kurokawa, K.; Hatta, T. Application of chlorite geothermometry to hydrothermal alteration in Toyoha geothermal system, Southwestern Hokkaido, Japan. *Resour. Geol.* **2010**, *160*, 52–70. [CrossRef]
88. Jowett, E.C. Fitting iron and magnesium into the hydrothermal chlorite geothermometer. In Proceedings of the GAC/MAC/SEG Joint Annual Meeting, Program with Abstracts, Toronto, ON, Canada, 27–29 May 1991; Volume 16, p. 62.
89. Patiño Douce, A.E. What do experiments tell us about relative contributions of crust and mantle to the origin of granitic magma? In *Understanding Granites: Integrating New and Classical Techniques*; Castro, A., Fernandez, C., Vigneress, J.L., Eds.; Geological Society, London, Special Publications: London, UK, 1999; Volume 168, pp. 55–75.
90. Xie, Y.W.; Zhang, Y.Q. Peculiarities and genetic significance of hornblende from granite in the Hengduansan region. *Acta Mineral. Sin.* **1990**, *10*, 35–45.
91. Zhou, Z.X. The origin of intrusive mass in Fengshandong, Hubei province. *Acta Petrol. Sin.* **1986**, *2*, 59–70.
92. Ali, K.A.; Kröner, A.; Hegner, E.; Wong, J.; Li, S.-Q.; Gahlan, H.A.; Abu El Ela, F.F. U-Pb zircon geochronology and Hf-Nd isotopic systematics of Wadi Beitan granitoid gneisses, Southeastern Desert, Egypt. *Gondwana Res.* **2015**, *27*, 811–824. [CrossRef]
93. Stern, R.J.; Hedge, C.E. Geochronologic constraints on late Precambrian crustal evolution in the Eastern Desert of Egypt. *Am. J. Sci.* **1985**, *285*, 97–127. [CrossRef]
94. Ali, K.A.; Andresen, A.; Stern, R.J.; Manton, W.I.; Omar, S.A.; Maurice, A.E. U-Pb zircon and Sr-Nd-Hf isotopic evidence for a juvenile origin of the El-Shalul Granite, Central Eastern Desert, Egypt. *Geol. Mag.* **2012**, *149*, 783–797. [CrossRef]
95. Zoheir, B.; Goldfarb, R.; Holzheid, A.; Helmy, H.; El Sheikh, A. Geochemical and geochronological characteristics of the Um Rus granite intrusion and associated gold deposit, Eastern Desert, Egypt. *Geosci. Front.* **2019**, *11*, 325–345. [CrossRef]
96. El Bahariya, G.A.; Abu anbar, M.M.; El Galy, M.M. Petrology and geochemistry of Um Rus and Samadi granites, central Eastern Desert, Egypt: Implications for I-type granites of variable magma sources. *Ann. Geol. Surv. Egypt* **2008**, *39*, 1–15.
97. Anderson, P.V.; Kerr, B.J.; Weber, T.E.; Ziemer, C.J.; Shurson, G.C. Determination, and prediction of digestible and metabolizable energy from chemical analysis of corn coproducts fed to finishing pigs. *J. Anim. Sci.* **2012**, *90*, 1242–1254. [CrossRef]
98. Ali, K.A.; Zoheir, B.A.; Stern, R.J.; Andresen, A.; Whitehouse, M.J.; Bishara, W.W. Lu-Hf and O isotopic compositions on single zircons from the Northeastern Desert of Egypt, Arabian- Nubian shield: Implications for crustal evolution. *Gondwana Res.* **2016**, *32*, 181–192. [CrossRef]
99. Bentor, Y.K. The crustal evolution of the Arabo-Nubian massif with special reference to the Sinai Peninsula. *Precambrian Res.* **1985**, *28*, 1–74. [CrossRef]
100. Andresen, A.; El-Rus, M.M.A.; Myhre, P.I.; Boghdady, G.Y.; Corfu, F. U-Pb TIMS age constraints on the evolution of the Neoproterozoic Meatiq Gneiss Dome, Eastern Desert, Egypt. *Int. J. Earth Sci.* **2009**, *98*, 481–497. [CrossRef]
101. Andresen, A.; Abu El-Enen, M.M.; Stern, R.J.; Wilde, S.A.; Ali, K.A. The Wadi Zaghra metaconglomerates of Sinai, Egypt: New constraints on the Ediacaran tectonic evolution of the northernmost Arabian-Nubian Shield. *Int. Geol. Rev.* **2014**, *56*, 1020–1038. [CrossRef]
102. Morag, N.; Avigada, D.; Gerdesb, A.; Belousovac, E.; Harlavand, Y. Crustal evolution and recycling in the northern Arabian-Nubian Shield: New perspectives from zircon Lu–Hf and U-Pb systematics. *Precambrian Res.* **2011**, *186*, 101–116. [CrossRef]
103. El-Manharawy, S.M. Geochronological Investigations of Some Basement Rocks in the Central Eastern Desert, Egypt, between Latitudes 25"-26" N. Ph.D. Thesis, Cairo University, Cairo, Egypt, 1977; 220p.
104. Eliwa, H.A.; Breitkreuz, C.; Murata, M.; Khalaf, I.M.; Bühler, B.; Itaya, T.; Takahashi, T.; Hirahara, Y.; Miyazaki, T.; Kimura, J.I.; et al. SIMS zircon U-Pb and mica K–Ar geochronology, and Sr-Nd isotope geochemistry of Neoproterozoic granitoids and their bearing on the evolution of the northeastern Desert, Egypt. *Gondwana Res.* **2014**, *25*, 1570–1598. [CrossRef]
105. Johnson, P.R.; Andresen, A.; Collins, A.S.; Fowler, A.R.; Fritz, H.; Ghebreab, W.; Kusky, T.; Stern, R.J. Late Cryoge-nian-Ediacaran history of the Arabian-Nubian Shield: A review of depositional, plutonic, structural, and tectonic events in the closing stages of the northern East African Orogen. *J. Afr. Earth Sci.* **2011**, *61*, 167–232. [CrossRef]
106. Moghazi, A.M.; Harbi, H.M.; Ali, K.A. Geochemistry of the Late Neoproterozoic Hadb adh Dayheen ring complex, Central Arabian Shield: Implications for the origin of rare-metal-bearing post-orogenic A-type granites. *J. Asian Earth Sci.* **2011**, *42*, 1324–1340. [CrossRef]
107. Gahlan, H.A.; Azer, M.K.; Al-Hashim, M.H.; Heikal, M.T.S. Highly evolved rare-metal bearing granite overprinted by alkali metasomatism in the Arabian Shield: A case study from the Jabal Tawlah granites. *J. Afr. Earth Sci.* **2022**, *192*, 104556. [CrossRef]

108. Yang, W.-B.; Niu, H.-C.; Hollings, P.; Zurevinski, S.E.; Bo Li, N. The role of recycled oceanic crust in the generation of alkaline A-type granites. *J. Geophys. Res. Solid Earth* **2017**, *122*, 7975–7983. [CrossRef]
109. Mushkin, A.; Navon, O.; Halicz, L.; Hartmann, G.; Stein, M. The petrogenesis of A-type magmas from the Amram Massif, southern Israel. *J. Petrol.* **2003**, *44*, 815–832. [CrossRef]
110. Kessel, R.; Stein, M.; Navon, O. Petrogenesis of late Neoproterozoic dikes in the northern Arabian-Nubian Shield Implication for the origin of A-type granites. *Precambrian Res.* **1998**, *92*, 195–213. [CrossRef]
111. Weissman, A.; Kessel, R.; Oded, N.; Mordechai, S. The petrogenesis of calc-alkaline granites from the Elat massif, Northern Arabian–Nubian shield. *Precambrian Res.* **2013**, *236*, 252–264. [CrossRef]
112. Tindle, A.G.; Webb, P.C. Estimation of lithium contents in trioctahedral micas using microprobe data: Application to micas from granitic rocks. *Eur. J. Mineral.* **1990**, *2*, 595–610. [CrossRef]

Disclaimer/Publisher's Note: The statements, opinions and data contained in all publications are solely those of the individual author(s) and contributor(s) and not of MDPI and/or the editor(s). MDPI and/or the editor(s) disclaim responsibility for any injury to people or property resulting from any ideas, methods, instructions or products referred to in the content.

MDPI AG
Grosspeteranlage 5
4052 Basel
Switzerland
Tel.: +41 61 683 77 34

Minerals Editorial Office
E-mail: minerals@mdpi.com
www.mdpi.com/journal/minerals

Disclaimer/Publisher's Note: The title and front matter of this reprint are at the discretion of the Guest Editors. The publisher is not responsible for their content or any associated concerns. The statements, opinions and data contained in all individual articles are solely those of the individual Editors and contributors and not of MDPI. MDPI disclaims responsibility for any injury to people or property resulting from any ideas, methods, instructions or products referred to in the content.

www.ingramcontent.com/pod-product-compliance
Lightning Source LLC
LaVergne TN
LVHW072332090526
838202LV00019B/2402